Lecture Notes in Computer Scien

Edited by G. Goos, J. Hartmanis and J. van Leeuwen

Springer

Berlin
Heidelberg
New York
Barcelona
Hong Kong
London
Milan
Paris
Singapore
Tokyo

Laurence Pierre   Thomas Kropf (Eds.)

Correct Hardware Design and Verification Methods

10th IFIP WG10.5 Advanced Research
Working Conference, CHARME'99
Bad Herrenalb, Germany, September 27-29, 1999
Proceedings

Springer

Series Editors

Gerhard Goos, Karlsruhe University, Germany
Juris Hartmanis, Cornell University, NY, USA
Jan van Leeuwen, Utrecht University, The Netherlands

Volume Editors

Laurence Pierre
CMI/Université de Provence
39, rue Joliot-Curie, F-13453 Marseille Cedex 13, France
E-mail: laurence@gyptis.univ-mrs.fr

Thomas Kropf
Technische Informatik, Universität Tübingen
Im Sand 3, D-72076 Tübingen, Germany
E-mail: kropf@informatik.uni-tuebingen.de

Cataloging-in-Publication data applied for

Die Deutsche Bibliothek - CIP-Einheitsaufnahme

Correct hardware design and verification methods : 10th IFIP WG
10.5 advanced research working conference ; proceedings / CHARME
'99, Bad Herrenalb, Germany, September 27 - 29, 1999. Pierre
Laurence ; Thomas Kropf (ed.). - Berlin ; Heidelberg ; New York ;
Barcelona ; Hong Kong ; London ; Milan ; Paris ; Singapore ; Tokyo
: Springer, 1999
  (Lecture notes in computer science ; Vol. 1703)
  ISBN 3-540-66559-5

CR Subject Classification (1998): B, F.3.1, D.2.4, F.4.1, I.2.3, J.6

ISSN 0302-9743
ISBN 3-540-66559-5 Springer-Verlag Berlin Heidelberg New York

Typesetting: Camera-ready by author
SPIN: 10704583   06/3142 – 5 4 3 2 1 0   Printed on acid-free paper

Preface

CHARME'99 is the tenth in a series of working conferences devoted to the development and use of leading-edge formal techniques and tools for the design and verification of hardware and systems. Previous conferences have been held in Darmstadt (1984), Edinburgh (1985), Grenoble (1986), Glasgow (1988), Leuven (1989), Torino (1991), Arles (1993), Frankfurt (1995) and Montreal (1997). This workshop and conference series has been organized in cooperation with IFIP WG 10.5. It is now the biannual counterpart of FMCAD, which takes place every even-numbered year in the USA. The 1999 event took place in Bad Herrenalb, a resort village located in the Black Forest close to the city of Karlsruhe.

The validation of functional and timing behavior is a major bottleneck in current VLSI design systems. A predominantly academic area of study until a few years ago, formal design and verification techniques are now migrating into industrial use. The aim of CHARME'99 is to bring together researchers and users from academia and industry working in this active area of research. Two invited talks illustrate major current trends: the presentation by Gérard Berry (Ecole des Mines de Paris, Sophia-Antipolis, France) is concerned with the use of synchronous languages in circuit design, and the talk given by Peter Jansen (BMW, Munich, Germany) demonstrates an application of formal methods in an industrial environment. The program also includes 20 regular presentations and 12 short presentations/poster exhibitions that have been selected from the 48 submitted papers.

The organizers are grateful to IFIP WG 10.5 for its support and to Intel, Siemens, Synopsys, and Verysys for their financial sponsorship, which considerably eased the organization of the conference. We are indebted to Renate Murr-Grobe and Klaus Schneider for their help in organizing this event, to Jörg Berdux for providing the nice layout of the call for papers, and to Eric Gascard for his technical assistance.

September 1999

Laurence Pierre, Program Chair
Thomas Kropf, Conference Chair
CHARME'99

Organization

CHARME'99 was organized in Bad Herrenalb by the University of Karlsruhe and the University of Tübingen (Germany), with the support of IFIP WG10.5.

Program Committee

Conference Chair: Thomas Kropf (Univ. of Tübingen, Germany)
Program Chair: Laurence Pierre (Univ. de Provence, France)
François Anceau (CNAM, France)
Dominique Borrione (Univ. Grenoble, France)
Albert Camilleri (Hewlett-Packard, USA)
Paolo Camurati (Politecnico di Torino, Italy)
Luc Claesen (IMEC, Belgium)
Eduard Cerny (Univ. de Montreal, Canada)
Werner Damm (Univ. Oldenburg, Germany)
Hans Eveking (T.U. Darmstadt, Germany)
Ganesh Gopalakrishnan (Univ. of Utah, USA)
Mike Gordon (Cambridge Univ., UK)
Werner Grass (Univ. Passau, Germany)
Mark Greenstreet (Univ. BC, Canada)
Warren Hunt (IBM, USA)
Steven Johnson (Indiana Univ., USA)
Ramayya Kumar (Verysys, Germany)
Robert Kurshan (Bell Labs, USA)
Tiziana Margaria (Univ. Dortmund, Germany)
Andrew Martin (Motorola, USA)
Ken McMillan (Cadence Berkeley Labs, USA)
Tom Melham (Univ. Glasgow, UK)
Paolo Prinetto (Politecnico di Torino, Italy)
Rajeev Ranjan (Synopsys, USA)
Mary Sheeran (Chalmers Univ., Sweden)
Jørgen Staunstrup (T.U. of Denmark, Denmark)
Sofiene Tahar (Concordia Univ., Canada)

Referees

Magdy Abadir	Per Bjesse	Eduard Cerny
E.M. Aboulhamid	Dominique Borrione	Koen Claessen
Otmane Ait-Mohamed	Olaf Burkart	Abdelkader Dekdouk
Ken Albin	Albert Camilleri	Stephen A. Edwards
F. Anceau	Paolo Camurati	Hans Eveking

Y. Feng	Steve Johnson	Laurence Pierre
Eric Gascard	Michael Jones	K.S. Prasad
Jens Chr. Godskesen	Jens Knoop	Stefano Quer
Ganesh Gopalakrishnan	R. Kurshan	Sriram K. Rajamani
Mike Gordon	V. Levin	Rajeev K. Ranjan
Werner Grass	Panagiotis Manolios	K. Ravi
Mark Greenstreet	Tiziana Margaria	Sophie Renault
Claudia Gsottberger	Andrew Martin	Jun Sawada
Pei-Hsin Ho	Ken McMillan	Klaus Schneider
Stefan Hoereth	Tom Melham	Ken Scott
Peng Hong	Michael Mendler	Mary Sheeran
Ravi Hosabettu	Marcus Müller-Olm	Tom Shiple
Jin Hou	Ratan Nalumasu	Jørgen Staunstrup
Henrik Hulgaard	K. Namjoshi	Terence Stroup
Warren A. Hunt	Félix Nicoli	Sofiene Tahar
M. Jahanpour	Jean-Luc Paillet	Raimund Ubar

Table of Contents

Abstraction and Compositional Techniques

Theorem Proving Related Approaches

Symbolic Simulation/Symbolic Traversal

Specification Languages and Methodologies

Posters

Esterel and Jazz : Two Synchronous Languages for Circuit Design

Gérard Berry

Ecole des Mines de Paris
Centre de Mathématiques Appliquées, INRIA
2004 Route des Lucioles
06560 Sophia-Antipolis, France
Gerard.Berry@sophia.inria.fr

Abstract

We survey two synchronous languages for circuit design. Esterel is dedicated to controllers implemented either in software or in hardware. Esterel programs are imperative, concurrent, and preemption-based. Programs are translated into circuits that are optimized using specific sequential optimization algorithms. A verification system restricted to the pure control part of programs is available. Esterel is currently used by several CAD vendors and circuit design companies.

Jazz is a newer language designed for fancy arithmetic circuits. Jazz resembles ML but has a richer type-system that supports inheritance. The current environment comprises a compiler, simulators, and code generators for the Pamette Xilinx-based board. Both languages are not only formal but based on real mathematics. We discuss why this is essential for good language design.

Design Process of Embedded Automotive Systems – Using Model Checking for Correct Specifications

Peter Jansen

BMW AG, 80788 München, Germany
Peter.Jansen@bmw.de

Abstract. The number of Embedded Control Units (ECUs) in the car is permanently increasing. Also complexity and interconnection is increased. Conventional design processes can not cope with this complexity. In the first part of this paper we show the current development-process at BMW, the second part deals with our results of using model checking to verify STATEMATE-models.

1 Introduction

The increasing demand for dynamically controlled safety features, driving comfort and operational convenience in cars require an intensive use of Embedded Control Units (ECUs). The number of ECUs in the car increases permanently. Also complexity and interconnection increase rapidly. Customary design processes can not cope with this complexity. Systems Engineering at BMW describes the process for transforming a product idea into a real system using (semi-)formal methods for specification, analysis, rapid prototyping and support for test and diagnostics. The use of CASE-Tools for designing ECUs is such a semiformal method. It can also help to reduce the development time and costs. Another advantage of using CASE-Tools is the possibility of detection of errors in an early phase of the development process. However, the use of CASE-Tools alone can not guarantee safety-critical properties of the system. New techniques are necessary. Model checking is such a technique. It is an automatic method for proving that an application satisfies its specification as represented by a temporal-logic formula. It offers a mathematical rigid proof. In contrast to *testing* a system, *model checking* allows to check the system under all possible inputs, a test can only check a limited set of inputs. For large systems, an exhaustive test may not be feasible.

2 The Design-Process for Embedded Control Units Software

There is a large amount of ECUs in today's high-end vehicles. Fig. 1 shows some of them, e.g. an engine control or a park-distance-control. All of these systems

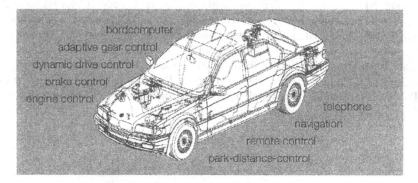

Fig. 1. View of electronic components in a car

are connected via several bus systems. Also the responsibility of these systems and with this the need for reliable systems. As shown in Fig. 2, tomorrow's ECUs could take full command of the car.

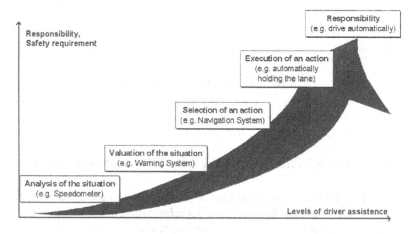

Fig. 2. Increasing responsibility of ECUs in automotive

To develop such systems, traditional design methods are inadequate. New methods, such as the use of CASE-Tools, are needed to create safe systems. Fig. 3 gives an overview on some of the CASE-Tools used at BMW. One such tool, used in an early phase of the development process is STATEMATE[1]. It is used to design state-based systems. With STATEMATE the user can design and simulate the

---

[1] STATEMATE is a registered trademark of i-Logix, Inc.

system. STATEMATE also offers the ability to generate C-code. A tool-set to verify STATEMATE-designs has been developed by OFFIS (Oldenburger Forschungs- und Entwicklungsinstitut für Informatik-Werkzeuge und -Systeme, Oldenburg, Germany) and used for two years at BMW. One problem with STATEMATE is that its generated code is too inefficient to be used with micro-controllers (e.g. a Siemens C167). Therefore, the STATEMATE-Design has to be coded by hand. This is usually done by a supplier like Bosch or Siemens. For the supplier the STATEMATE-Design becomes a specification.

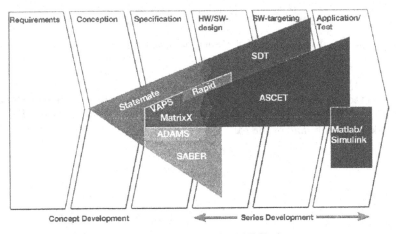

Fig. 3. Overview on CASE-Tools

3   Using Model Checking to Verify Statemate Designs

Model checking is an automatic method for proving that an application satisfies its specification represented by a temporal-logic formula. This offers a mathematical rigid proof. In contrast to testing a system, model checking allows to check the system under *all* possible inputs. No test can do this, because a test can only check a limited set of inputs. A tool-set for automatic verification of STATEMATE-Designs has been developed by OFFIS. With this tool-set it is possible to model check STATEMATE designs. For this a STATEMATE-design has to be translated into a mathematically equivalent model, called FSM (Finite State Machine). Another specification for this design has to be done in temporal-logic formulas. For this, a graphical formalism, called Symbolic Timing Diagrams (STD), has been developed. These diagrams represent an easy to understand way to express properties of a reactive system. The Siemens Model-checker SVE finally verifies whether the STATEMATE design satisfies its specification. Fig. 4 shows the

tool-environment. Model checking assists the engineer in creating software for
controlling systems in an early stage of development, so cost for the development
can be reduced, while making the software "safe".

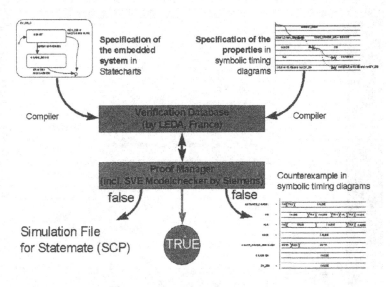

Fig. 4. Overview of CASE-Tools

3.1 Example "Electronic Brake Management"

A recently developed new feature of the brake management is a highly safety-
critical application. Roughly, the main function is the following: If the driver
stops at a red traffic light, he does not have to push the brake-pedal any longer;
the brake is being held automatically. Fig. 5 shows the top-level activity-chart
of the STATEMATE-design. There is a switch in the car to disable this function.
One property for this system is: Whenever the "hold-function" is enabled and
the switch is pressed, the hold-function has to be cancelled immediately. With a
normal test it is not possible to guarantee this property: It can only be checked,
whether the function is cancelled in *some* situations. Model-checking can guarantee that this function is cancelled in *all* situations. Fig. 6 shows the symbolic
timing diagram for this property. The hold-function is represented by the variable *DECELERATION*. The value 10.0 means minimum deceleration, 0 would
mean maximum deceleration.

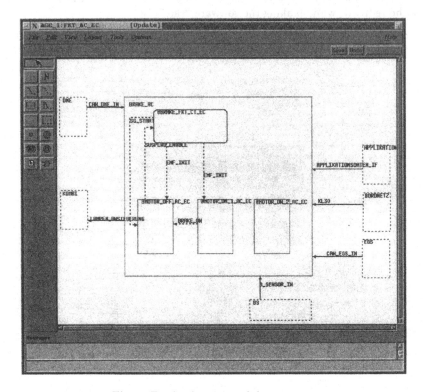

Fig. 5. Top-level activity of the system

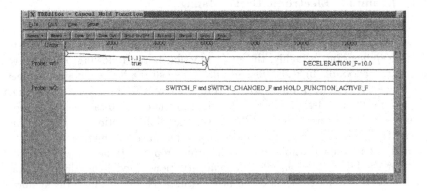

Fig. 6. The property for the brake system as a symbolic timing diagram.

4 Further Information

More information on CASE-Tools and Model-Checking can be found on the following web sites:

- Information on STATEMATE
  http://www.ilogix.com
- Information on the model-checking tool-set
  http://ca.informatik.uni-oldenburg.de/publications/publications.html
- Information on BMW
  http://www.bmw.com

A Proof of Correctness of a Processor Implementing Tomasulo's Algorithm without a Reorder Buffer

Ravi Hosabettu[1], Ganesh Gopalakrishnan[1], and Mandayam Srivas[2]

[1] Department of Computer Science, University of Utah, Salt Lake City, UT 84112,
hosabett,ganesh@cs.utah.edu
[2] Computer Science Laboratory, SRI International, Menlo Park, CA 94025,
srivas@csl.sri.com

Abstract. The *Completion Functions Approach* was proposed in [HSG98] as a systematic way to decompose the proof of correctness of pipelined microprocessors. The central idea is to construct the abstraction function using completion functions, one per unfinished instruction, each of which specifies the effect (on the observables) of completing the instruction. However, its applicability depends on the fact that the implementation "commits" the unfinished instructions in the pipeline in program order. In this paper, we extend the completion functions approach when this is not true and demonstrate it on an implementation of Tomasulo's algorithm without a reorder buffer. The approach leads to an elegant decomposition of the proof of the correctness criterion, does not involve the construction of an explicit intermediate abstraction, makes heavy use of an automatic case-analysis strategy based on decision procedures and rewriting, and addresses both safety and liveness issues.

1 Introduction

For formal verification to be successful in practice, not only is it important to raise the level of automation but is also essential to develop methodologies that scale verification to large state-of-the-art designs. One of the reasons for the relative popularity of model checking in industry is that it is automatic when readily applicable. A technology originating from the theorem proving domain that can potentially provide a similarly high degree of automation in verification is one that makes heavy use of decision procedures for the combined theory of boolean expressions with uninterpreted functions and linear arithmetic [CRSS94,BDL96]. Just as model checking suffers from a state-explosion problem, a verification strategy based on decision procedures suffers from a "case-explosion" problem. That is, when applied naively, the sizes of the terms generated and the number of examined cases during validity checking explodes. Just as compositional model

---

* The first and second authors were supported in part by NSF through Grant no. CCR-9800928. The third author was supported in part by NASA contract NAS1-20334 and ARPA contract NASA-NAG-2-891 (ARPA Order A721).

checking provides a way of decomposing the overall proof and reducing the effort for an individual model checker run, a practical methodology for decision procedure-centered verification must prescribe a systematic way to decompose the correctness assertion into smaller problems that the decision procedures can handle.

In [HSG98], we proposed such a methodology for pipelined processor verification called the *Completion Functions Approach*. The central idea behind this approach is to define the abstraction function[1] as a composition of a sequence of completion functions, one for every unfinished instruction, in their program order. A completion function specifies how a partially executed instruction is to be completed in an atomic fashion, that is, the desired effect on the observables of completing that instruction, assuming those ahead of it in the program order are completed. Given such a definition of the abstraction function in terms of completion functions, the methodology prescribes a way of organizing the verification into proving a hierarchy of *verification conditions*. The methodology has the following attributes:

- The verification proceeds incrementally making debugging and error tracing easier.
- The verification conditions and most of the supporting lemmas (such as the lemma on the correctness of the feedback logic) needed to support the incremental methodology can be generated systematically.
- Every generated verification condition and lemma can be proved, often automatically, using a strategy based on decision procedures and rewriting.
- The verification avoids the construction of an explicit intermediate abstraction as well as the large amount of manual effort required to construct it.

In summary, the completion functions approach strikes a balance between full automation that (if at all possible) can potentially overwhelm the decision procedures, and a potentially tedious manual proof. This methodology is implemented using PVS [ORSvH95] and was applied (in [HSG98]) to three processor examples: DLX [HP90], dual-issue DLX, and a processor that exhibited limited out-of-order execution capability. The proof decomposition that this method achieves and the verification conditions generated in the DLX example is illustrated in Figure 1.

Later, we extended the methodology to verify a truly out-of-order execution processor with a reorder buffer [HSG99]. We observed that regardless of how many instructions are pending in the reorder buffer, the instructions can only be in one of a few (small finite number) distinct states and exploited this fact to provide a single compact parameterized completion function applicable to all the pending instructions in the reorder buffer. The proof was decomposed on the basis of how an instruction makes a transition from its present state to the next state.

However, the applicability of the completion functions approach depends on the fact that the implementation "commits" the unfinished instructions in the

---

[1] Our correctness criteria is based on using an abstraction function, as most others.

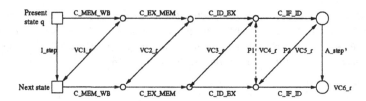

Fig. 1. The proof decomposition in the DLX example using the completion functions approach (C_... are the completion functions for the various unfinished instructions).

pipeline in program order. The abstraction function is defined by composing the completion functions of the unfinished instructions in the program order too. Because of this, it is possible to relate the effect of completing instructions one at a time in the present and the next states and incrementally build the proof of the commutative diagram (See Figure 1). Also, one can provide for every unfinished instruction, an "abstract" state where the instructions ahead of it are completed. This fact is useful in expressing the correctness of the feedback logic. If instructions were to commit out-of-order, it is not possible to use these ideas.

A processor implementing Tomasulo's algorithm without a reorder buffer executes instructions in the data-flow order, possibly committing them to the register file in an out-of-order manner. Hence, the basic premise of the completion functions approach—that instructions commit in the program order—is not true in this case. The implementation maintains the identity of the latest instruction writing a particular register. Those instructions issued earlier and not the latest ones to write their respective destinations, on completing their execution, only forward the results to other waiting instructions but do not update the register file. Observe that it is difficult to support branches or exceptions in such an implementation. (In an implementation supporting branches or exceptions, the latest instruction writing a register can not be easily determined.)

In this paper, we extend the completion functions approach to be applicable in such a scenario. Instead of defining the completion function to directly update the observables, we define it to return the value an instruction computes in the various states. The completion function for a given instruction recursively completes the instructions it is dependent on to obtain its source values. The abstraction function is defined to assign to a register the value computed by the latest instruction writing that register. We show that this modified approach leads to a decomposition of the overall proof of correctness, and we make heavy use of an automatic case-analysis strategy in discharging the different obligations in the decomposition. The proof does not involve the construction of an explicit intermediate abstraction. Finally, we address the proof of liveness properties too.

The rest of the paper is organized as follows: In Section 2, we describe our processor model. Section 3 describes our correctness criteria. This is followed by the proof of correctness in Section 4. We compare our work with others in Section 5 and finally provide the conclusions.

2 Processor Model

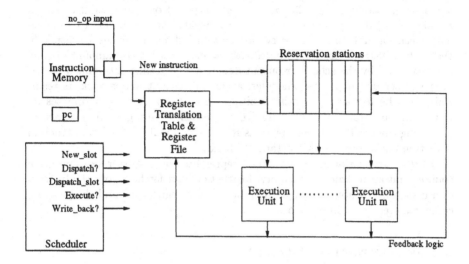

Fig. 2. The block diagram model of our implementation

Figure 2 shows the model of an out-of-order execution processor implementing Tomasulo's algorithm without a reorder buffer used in this paper. The model has z reservation stations where instructions wait before being sent to the execution units. There are m execution units represented by an uninterpreted function. (z and m are parameters to our implementation model.) A register translation table (RTT) maintains the identity of the latest pending instruction writing a particular register (the identity is a "tag"—in this case, the reservation station index). A scheduler controls the movement of the instructions through the execution pipeline (such as being dispatched, executed etc) and its behavior is modeled in the form of axioms (instead of a concrete implementation). Instructions are fetched from the instruction memory (using a program counter which then is incremented); and the implementation also takes a no_op input, which suppresses an instruction fetch when asserted.

An instruction is *issued* by allocating a free reservation station for it (New_slot). No instruction is issued if all the reservation stations are occupied or if no_op is asserted. The RTT entry corresponding to destination of the instruction is updated to reflect the fact that the instruction being issued is the latest one to write that register. If the source operands are not being written by previously issued pending instructions (checked using the RTT) then their values are obtained from the register file, otherwise the tags of the instructions providing the source operands is maintained in the reservation station allocated to the instruction. An issued instruction monitors the execution units to see if they produce the values it is waiting for, by comparing the tags it is waiting on with the tags of the instructions producing the result. An instruction can be *dispatched* when its source operands are ready and the corresponding execution unit is free. `Dispatch?` and `Dispatch_slot` outputs from the scheduler (each a m-wide vector) determine whether or not to dispatch an instruction to a particular execution unit and the reservation station index from where to dispatch. Dispatched instructions get *executed* after a non-deterministic amount of time as determined by the scheduler output `Execute?`. At a time determined by the `Write_back?` output of the scheduler, an execution unit writes back its result which will be forwarded to other waiting instructions. A register updates its value with this result only if its RTT entry matches the tag of the instruction producing the result and then clears its RTT entry. Finally, when an instruction is written back, its reservation station is freed.

At the specification level, the state is represented by a register file, a program counter and an instruction memory. Instructions are fetched from the instruction memory, executed, result written back to the register file and the program counter incremented in one clock cycle.

3 Our Correctness Criteria

Intuitively, a pipelined processor is correct if the behavior of the processor starting in a flushed state (i.e., no partially executed instructions), executing a program and terminating in a flushed state is emulated by an ISA level specification machine whose starting and terminating states are in direct correspondence through projection. This criterion is shown in Figure 3(a) where `I_step` is the implementation transition function, `A_step` is the specification transition function and `projection` extracts those implementation state components visible to the specification (i.e., observables). This criterion can be proved by an easy induction on n once the *commutative diagram* condition (due to Hoare [Hoa72]) shown in Figure 3(b) is proved on a single implementation machine transition (and a certain other condition discussed in the next paragraph holds).

The criterion in Figure 3(b) states that if the implementation machine starts in an arbitrary reachable state `impl_state` and the specification machine starts in a corresponding specification state (given by an abstraction function ABS), then after executing a transition their new states correspond. Further ABS must be chosen so that for all flushed states `fs` the *projection condition* ABS(`fs`) = projection(`fs`) holds. The commutative diagram uses a modified transition function `A_step'`, which denotes zero or more applications of `A_step`, because an implementation transition from an arbitrary state might correspond to executing in the specification machine zero instruction (*e.g.*, if the implementation machine stalls without fetching an instruction) or more than one instruction (*e.g.*, if multiple instructions are fetched in a cycle). The number of instructions executed by the specification machine is provided by a user-defined *synchronization* function on implementation states. One of the crucial proof obligations is to show that this function does not always return zero (*No_indefinite_stutter* obliga-

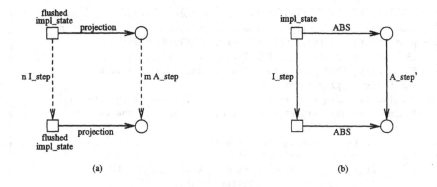

Fig. 3. Pipelined microprocessor correctness criteria

tion). One also needs to prove that the implementation machine will eventually reach a flushed state if no more instructions are inserted into the machine, to make sure that the correctness criterion in Figure 3(a) is not vacuous (*Eventual flush* obligation). In addition, the user may need to discover *invariants* to restrict the set of impl_state considered in the proof of Figure 3(b) and prove that it is closed under I_step.

4 Proof of Correctness

We introduce some notations which will be used throughout this section: q represents the implementation state, s the scheduler output, i the processor input, rf(q) the register file contents in state q and next(q,s,i) the "next state" after an implementation transition. "Primed" variables will be used to refer to the value of a given variable in the next state. Also, we identify an instruction in the processor by its reservation station index (i.e., instruction rsi means instruction at reservation station index rsi). When the instruction in question is clear from the context (say rsi), we use just rs_op to refer to its opcode instead of rs_op(q)(rsi). (rs_op' will refer to rs_op(next(q,s,i))(rsi)). The PVS specifications and the proof scripts can be found at [Hos99].

4.1 Specifying the completion functions

An instruction in the processor can be in one of the three following possible states inside the processor—issued, dispatched or executed. (Once written back, it is no longer present in the processor). We formulate predicates describing an instruction in each of these states and specify the value an instruction computes in each of these states. The definition of the completion function is shown in [1].

```
% state_I : implementation state type; rsindex : reservation station   1
% index type; value : type of the data computed by an instruction.
Complete_instr(q:state_I,rsi:rsindex): RECURSIVE value =
   IF executed_pred(q,rsi) THEN Value_executed(q,rsi)
   ELSIF dispatched_pred(q,rsi) THEN Value_dispatched(q,rsi)
   ELSIF issued_pred(q,rsi) THEN
   % Value_issued(q,rsi) expanded to highlight the recursive call.
   % alu is an uninterpreted function. ''rs_op'' is the opcode.
     alu(rs_op(q)(rsi),
         IF rs_src_ptr1(q)(rsi) = 0 THEN
            rs_src_value1(q)(rsi)
         ELSE Complete_instr(q,rs_src_ptr1(q)(rsi)) ENDIF,
         ''Second operand -- similar definition'')
   ELSE default_value ENDIF
   MEASURE rs_instr_num(q)(rsi)
```

In this implementation, when an instruction is in the executed state, the result value is available in eu_result field of the execution unit, so Value_executed returns this value. We specify Value_dispatched along the same lines. When an instruction is in the issued state, it may be waiting for its source operands to get ready. In determining the value computed by such an instruction, we need the source operands which we specify as follows: When rs_src_ptr1 is zero, the first source operand is ready and its value is available in rs_src_value1, otherwise its value is obtained by completing the instruction it is waiting on (rs_src_ptr1 points to that instruction). Similarly the second source operand is specified.

To specify the completion function, we added three auxiliary variables. The first one maintains the index of the execution unit an instruction is dispatched to. Since the completion function definition is recursive, one needs to provide a measure function to show that the function is well-defined ; the other two auxiliary variables are for this purpose. We should prove that instructions producing the source values for a given instruction rsi have a lower measure than rsi. So we assign a number rs_instr_num to every instruction that records the order in which it is issued and this is used as the measure function. (The counter that is used in assigning this number is the third auxiliary variable).

4.2   Constructing the abstraction function

The register translation table maintains the identity of the latest pending instruction writing a particular register. The abstraction function is constructed by updating every register with the value obtained by completing the appropriate pending instruction, as shown in 2. The synchronization function returns zero if no_op input is asserted or if there is no free reservation station to issue an instruction, otherwise returns one.

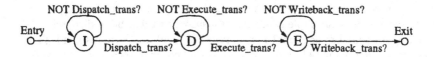

Fig. 4. The various states an instruction can be in and transitions between them, I: issued, D: dispatched, E: executed

```
% If the ''rtt'' field for a given register is zero, then it is      2
% not updated, otherwise complete the instruction pointed to by
% ''rtt'' and update the register with that value.
Complete_all(q:state_I): state_I =
   q WITH [ (rf) := LAMBDA(r:reg):
                        IF rtt(q)(r) = 0 THEN rf(q)(r)
                        ELSE Complete_instr(q,rtt(q)(r)) ENDIF ]

% state_A is the specification state type.
ABS(q:state_I): state_A = projection(Complete_all(q))
```

4.3 Proof decomposition

We first prove a lemma that characterizes the value an instruction computes and then use it in the proof of the commutative diagram. Consider an arbitrary instruction rsi. We claim that the value an instruction computes (as given by Complete_instr) is the same whether in state q or in state next(q,s,i), as long as the instruction is valid in these states. (Intuitively, an instruction is valid as long as it has not computed and written back its result.) This is shown as lemma same_result in ⊡3⊡. We prove this by induction on rsi (induction with a measure function as explained later).

```
% rs_valid means the instruction is valid.                           3
same_result: LEMMA
     FORALL(rsi:rsindex):
     (rs_valid(q)(rsi) AND rs_valid(next(q,s,i))(rsi))
         IMPLIES
     Complete_instr(q,rsi) = Complete_instr(next(q,s,i),rsi)
```

We generate the different cases of the induction argument (as will be detailed shortly) based on how an instruction makes a transition from its present state to its next state. This is shown in Figure 4 where we have identified the conditions under which an instruction changes its state. For example, we identify the predicate Dispatch_trans?(q,s,i,rsi) which takes the instruction rsi from issued state to dispatched state. In this implementation, this predicate is true when there is an execution unit for which Dispatch? output from the scheduler is true and the Dispatch_slot output is equal to rsi. Similarly other "trans" predicates are defined.

Having defined these predicates, we prove that they indeed cause instructions to take the transitions shown. Consider a valid instruction rsi in the issued state i.e., `issued_pred(q,rsi)` holds. We prove that if `Dispatch_trans?(q,s,i,rsi)` is true, then after an implementation transition, `rsi` will be in dispatched state (i.e., `dispatched_pred(next(q,s,i),rsi)` is true) and remains valid. (This is shown as a lemma in [4].) Otherwise (if `Dispatch_trans?(q,s,i,rsi)` is false), we prove that `rsi` remains in the issued state in `next(q,s,i)` and remains valid. There are three other similar lemmas for the other transitions. The sixth lemma is for the case when an instruction `rsi` in the executed state is written back. It states that `rsi` is no longer valid in `next(q,s,i)`.

```
issued_to_dispatched: LEMMA                                          4
    FORALL(rsi:rsindex):
    (rs_valid(q)(rsi) AND issued_pred(q,rsi) AND
                             Dispatch_trans?(q,s,i,rsi))
       IMPLIES
    (dispatched_pred(next(q,s,i),rsi) AND rs_valid(next(q,s,i),rsi))
```

Now we come back to the details of the `same_result` lemma. In proving this lemma for an instruction `rsi`, one needs to assume that the lemma holds for the two instructions producing the source values for `rsi` (Details will be presented later). So we do an induction on `rsi` with `rs_instr_num` as the measure function. As explained earlier in Section 4.1, instructions producing the source values (`rs_src_ptr1` and `rs_src_ptr2` when non-zero) have a lower measure than `rsi`. The induction argument is based on a case analysis on the possible state `rsi` is in, and whether or not it makes a transition to its next state. Assume the instruction `rsi` is in issued state. We prove the induction claim in the two cases—`Dispatch_trans?(q,s,i,rsi)` is true or false—separately. (The proof obligation for the first case is shown in [5].) We have similar proof obligations for `rsi` being in other states. In all, the proof decomposes into six proof obligations.

```
% One of the six cases in the induction argument.                   5
issued_to_dispatched_induction: LEMMA
    FORALL(rsi:rsindex):
    (rs_valid(q)(rsi) AND issued_pred(q,rsi) AND
    Dispatch_trans?(q,s,i,rsi) AND Induction_hypothesis(q,s,i,rsi))
       IMPLIES
    Complete_instr(q,rsi) = Complete_instr(next(q,s,i),rsi)
```

We sketch the proof of `issued_to_dispatched_induction` lemma. We refer to the goal that we are proving—`Complete_instr( ... ) = Complete_instr( ... )`—as the consequent. We expand the definition of the completion function corresponding to `rsi` on both sides of the consequent. In `q`, `rsi` is in the issued state and in `next(q,s,i)`, it is the dispatched state—this follows from the `issued_to_dispatched` lemma. After some rewriting and simplifications in PVS, the left hand side of the consequent simplifies to `Value_issued(q,rsi)` and the right hand side simplifies to `Value_dispatched(next(q,s,i),rsi)`. (The proofs of all the obligations are similar till this point. After this point, it depends on the particular obligation being proved since different invariants are needed for the different obligations.) Proof now proceeds by expanding the definitions of `Value_issued` and `Value_dispatched`, using the necessary invariants and simplifying. We use the PVS strategy `apply (then* (repeat (lift-if))` (bddsimp) (ground) (assert)) to do the simplifications by automatic case-analysis (many times, simply assert will do).

We illustrate the proof of another lemma issued_remains_induction (shown in $\boxed{6}$) in greater detail pointing out how the feedback logic gets verified. As above, the proof obligation reduces to showing that Value_issued(q,rsi) and Value_issued(next(q,s,i),rsi) are the same. (The definition of Value_issued is shown in $\boxed{1}$.) This can be easily proved once we show that the source values of rsi as defined by op_val1 (and a similar op_val2) remain same, whether in q or in next(q,s,i). Proving this lemma op_val1_same (and a similar op_val2_same) establishes the correctness of the feedback logic.

```
% Value of the first operand.                                    6
op_val1(q:state_I,rsi:rsindex): value =
    IF rs_src_ptr1(q)(rsi) = 0 THEN rs_src_value1(q)(rsi)
    ELSE Complete_instr(q,rs_src_ptr1(q)(rsi)) ENDIF

op_val1_same: LEMMA
    FORALL(rsi:rsindex):
    (rs_valid(q)(rsi) AND issued_pred(q,rsi) AND
    NOT Dispatch_trans?(q,s,i,rsi) AND Induction_hypothesis(q,s,i,rsi))
        IMPLIES
    op_val1(q,rsi) = op_val1(next(q,s,i),rsi)

issued_remains_induction: LEMMA
    FORALL(rsi:rsindex):
    (rs_valid(q)(rsi) AND issued_pred(q,rsi) AND
    NOT Dispatch_trans?(q,s,i,rsi) AND Induction_hypothesis(q,s,i,rsi))
        IMPLIES
    Complete_instr(q,rsi) = Complete_instr(next(q,s,i),rsi)
```

In proving op_val1_same lemma, there are three cases. Consider the case when rs_src_ptr1 is zero. We then show that rs_src_ptr1' is zero and rs_src_value1 is the same as rs_src_value1'. Consider the case when rs_src_ptr1 is non-zero. rs_src_ptr1' may or may not be zero. If rs_src_ptr1' is zero, then it implies that in the current cycle, the instruction pointed to by rs_src_ptr1 completes its execution and forwards its result to rsi. So it is easy to prove rs_src_value1' (the value actually written back in the implementation) is the same as the expected value Complete_instr(q,rs_src_ptr1(q)(rsi)). If rs_src_ptr1' is non-zero, then one can conclude from the induction hypothesis that rs_src_ptr1 computes the same value in q and in next(q,s,i).

Proving the commutative diagram Consider the case when no new instruction is issued in the current cycle, that is, the synchronization function returns zero. The commutative diagram obligation in this case is shown in $\boxed{7}$.

```
% sch_rs_slot (i.e., scheduler output New_slot) is valid means no       7
% free reservation stations.
commutes_no_issue: LEMMA
    (no_op?(i) OR rs_valid(q)(sch_rs_slot(s)))
        IMPLIES
    rf(ABS(q)) = rf(ABS(next(q,s,i)))
```

We expand the definition of ABS (shown in $\boxed{2}$) and consider a particular register r. This again leads to three cases as in the correctness of op_val1_same. Consider the case when rtt (i.e., rtt(q)(r)) is zero. We then show that rtt' is zero too and the values of register r match in q and next(q,s,i). Consider the case when rtt is non-zero. rtt' may or may not be zero. If rtt' is zero, then it implies that in the current cycle, the instruction pointed to by rtt completes its execution and writes its result to r. It is easy to show that this value written into r is the same as the expected value Complete_instr(q,rtt(q)(r)). If rtt' is non-zero, then we use same_result lemma to conclude that the same value is written into r in q and next(q,s,i).

The case when a new instruction is issued is similar to the above except when r is the destination register of the instruction being issued. We show that in state next(q,s,i), the new instruction is in issued state, its operands as given by op_val1 and op_val2 equal the ones given by the specification machine and the value written into r by the implementation machine equals the value given by specification machine.

The program counter pc is incremented whenever an instruction is fetched. This is the only way pc is modified. So proving the commutative diagram for pc is simple. The commutative diagram proof for the instruction memory is trivial since it is not modified at all.

The invariants needed We describe in this section *all* the seven invariants needed by our proof. We do not have a uniform strategy for proving all these invariants but we use the automatic case-analysis strategy shown earlier to do the simplifications during the proofs.

- Two of invariants are related to rs_instr_num and instr_counter, the auxiliary variables introduced for defining a measure for every instruction. The first invariant states that the measure of any instruction (rs_instr_num) is less than the running counter (instr_counter). The second one states that for any instruction, if the source operands are not ready, then the measure of the instructions producing the source values is less than the measure of the instruction. The need for these was realized when we decided to introduce the two auxiliary variables mentioned above.

- Two other invariants are related to rs_exec_ptr, the auxiliary variable that maintains the execution unit index an instruction is dispatched to. The first invariant states that, if rs_exec_ptr is non-zero, then that execution unit is busy and its tag (which records the instruction executing in the unit) points to the instruction itself. The second invariant states that, whenever an execution unit is busy, the instruction pointed to by its tag is valid and that instruction's `rs_exec_ptr` points to the execution unit itself. These invariants are very similar to ones we needed in an earlier verification effort [HSG99].

- Two other invariants characterize when an instruction is valid. The first one states that for any register, the instruction pointed to by `rtt` is valid. The second one states that for any given instruction, the instructions pointed to by `rs_src_ptr1` and `rs_src_ptr2` are valid. The final invariant we needed was that `rs_exec_ptr` for any instruction is non-zero if and only if `rs_disp?` (a boolean variable that says whether or not an instruction is dispatched) is true. The need for these three invariants was realized during the proofs of other lemmas/invariants.

PVS proof timings: The proofs of all the lemmas and the invariants discussed so far takes about 500 seconds on a 167 MHz Ultra Sparc machine. [2]

4.4 Other obligations - liveness properties

We provide a sketch of the proof that the processor eventually gets flushed if no more instructions are inserted into it. The proof that the synchronization function eventually returns a nonzero value is similar. The proofs involve a set of obligations on the implementation machine, a set of fairness assumptions on the inputs to the implementation and a high level argument using these to prove the two liveness properties. All the obligations on the implementation machine are proved in PVS. In fact, most of them are related to the "instruction state" transitions shown in Figure 4 and the additional obligations needed (not proved earlier) takes only about 15 seconds on a 167 MHz Ultra Sparc machine. We now provide a sketch of the high level argument which is being formalized in PVS.
Proof sketch: The processor is flushed if for all registers r, $rtt(q)(r) = 0$.

- First, we show that "any valid instruction in the dispatched state eventually goes to the executed state and be valid" and "any valid instruction in the executed state eventually gets written back and its reservation station will be freed". Consider a valid instruction `rsi` in the dispatched state. If in state `q`, `Execute_trans?(q,s,i,rsi)` is true, then `rsi` goes to the executed state in `next(q,s,i)` and remains valid (refer to Figure 4). Otherwise it continues to be in the dispatched state and remains valid. We observe that when `rsi` is in the dispatched state, the scheduler inputs that determine when an instruction should be executed are enabled and these remain enabled as long as `rsi` is in the dispatched state. By a fairness assumption on the scheduler, it eventually decides to execute the instruction (i.e., `Execute_trans?(q,s,i,rsi)` will be true) and in `next(q,s,i)`, the instruction will be in the executed state and be valid. By a similar argument, it eventually gets written back and the reservation station gets freed.

---

[2] The manual effort involved in doing the proofs was one person week. The authors had verified a processor with a reorder buffer earlier [HSG99] and most of the ideas/proofs carried over to this example.

- Second, we show that "every busy execution unit eventually becomes free and stays free until an instruction is dispatched on it". This follows from the observation that whenever an execution unit is busy, the instruction occupying it is in the dispatched/executed state and that such an instruction eventually gets written back (first observation above).

- Third, we show that "a valid instruction in the issued state will eventually go to the dispatched state and be valid". Here, the proof is by induction (with rs_instr_num as the measure) since an arbitrary instruction rsi could be waiting for two previously issued instructions to produce its source values. Consider a valid instruction rsi in the issued state. If the source operands of rsi are ready, then we observe that the scheduler inputs that determine dispatching remain asserted as long as rsi is not dispatched. Busy execution units eventually get free and remain free until an instruction is dispatched on it (second observation above). So by a fairness assumption on the scheduler, rsi eventually gets dispatched. If a source operand is not ready, then the instruction producing it has a lower measure. By the induction hypothesis, it eventually goes to the dispatched state, eventually gets written back (first observation) forwarding the result to rsi. By a similar argument as above, rsi eventually gets dispatched.

- Finally, we show that "the processor eventually gets flushed". We observe that every valid instruction in the processor eventually gets written back freeing its reservation stations (third and first observations). Since no new instructions are being inserted, free reservation stations remain free. Whenever rtt(q)(r) is non-zero, it points to an occupied reservation station. Since, eventually all reservation stations get free, all rtt entries become zero and the processor is flushed.

5   Related Work

The problem of verifying the control logic of out-of-order execution processors has received considerable attention in the last couple of years using both theorem proving and model checking approaches. In particular, prior to our work, one theorem prover based and three model checking based verifications of a similar example—processor implementing Tomasulo's algorithm without a reorder buffer—have been carried out.

The theorem prover based verification reported in [AP98] is based on *refinement* and the use of "predicted value". They introduce this "predicted value" as an auxiliary variable to help in comparing the implementation against its specification without constructing an intermediate abstraction. However there is no systematic way to generate the invariants and the obligations needed in their approach. And they do not address liveness issues needed to complete the proof.

A model checking based verification of Tomasulo's algorithm is carried out in [McM98]. He uses compositional model checking and aggressive symmetry reductions to manually decompose the proof into smaller correctness obligations via refinement maps. Setting up the refinement maps requires information similar to that provided by the completion functions in addition to some details of the design. However the proof is dependent on the configuration of the processor (number of reservation stations etc) and also on the actual arithmetic operators.

Another verification of Tomasulo's algorithm is reported in [BBCZ98] where they combine symbolic model checking with uninterpreted functions. They introduce a data structure called reference file for representing the contents of the register file. While they abstract away from the data path, the verification is for a fixed configuration of the processor and they is no decomposition of the proof.

Yet another verification based on assume-guarantee reasoning and refinement checking is presented in [HQR98]. The proof is decomposed by providing the definitions of suitable "abstract" modules and "witness" modules. However the proof can be carried out for a fixed small configuration of the processor only.

Finally, verification of a processor model implementing Tomasulo's algorithm with a reorder buffer, exceptions and speculative execution is carried out in [SH98]. Their approach relies on constructing an explicit intermediate abstraction (called MAETT) and expressing invariant properties over this. Our approach avoids the construction of an intermediate abstraction and hence requires significantly less manual effort.

6 Conclusion

We have showed in this paper how to extend the completion functions approach to be applicable in a scenario where the instructions are committed out-of-order and illustrated it on a processor implementation of Tomasulo's algorithm without a reorder buffer. Our approach lead to an elegant decomposition of the proof based on the "instruction state" transitions and did not involve the construction of an intermediate abstraction. The proofs made heavy use of an automatic case-analysis strategy and addressed both safety and liveness issues.

We are currently developing a PVS theory of the "eventually" temporal operator to mechanize the liveness proofs presented here. We are also working on extending the completion functions approach further to verify a detailed out-of-order execution processor (with a reorder buffer) involving branches, exceptions and speculative execution.

References

[AP98]   T. Arons and A. Pnueli. Verifying Tomasulo's algorithm by refinement. Technical report, Weizmann Institute, 1998.

[BBCZ98]   Sergey Berezin, Armin Biere, Edmund Clarke, and Yunshan Zu. Combining symbolic model checking with uninterpreted functions for out-of-order processor verification. In Ganesh Gopalakrishnan and Phillip Windley, editors, *Formal Methods in Computer-Aided Design, FMCAD '98*, volume 1522 of *Lecture Notes in Computer Science*, pages 369–386, Palo Alto, CA, USA, November 1998. Springer-Verlag.

[BDL96]   Clark Barrett, David Dill, and Jeremy Levitt. Validity checking for combinations of theories with equality. In Mandayam Srivas and Albert Camilleri, editors, *Formal Methods in Computer-Aided Design, FMCAD '96*, volume 1166 of *Lecture Notes in Computer Science*, pages 187–201, Palo Alto, CA, November 1996. Springer-Verlag.

[CRSS94]   D. Cyrluk, S. Rajan, N. Shankar, and M. K. Srivas. Effective theorem proving for hardware verification. In Ramayya Kumar and Thomas Kropf, editors, *Theorem Provers in Circuit Design, TPCD '94*, volume 910 of *Lecture Notes in Computer Science*, pages 203–222, Bad Herrenalb, Germany, September 1994. Springer-Verlag.

[Hoa72]   C.A.R. Hoare. Proof of correctness of data representations. In *Acta Informatica*, volume 1, pages 271–281, 1972.

[Hos99]   Ravi Hosabettu. The Completion Functions Approach homepage, 1999. At address http://www.cs.utah.edu/~hosabett/cfa.html.

[HP90]   John L. Hennessy and David A. Patterson. *Computer Architecture: A Quantitative Approach*. Morgan Kaufmann, San Mateo, CA, 1990.

[HQR98]   Thomas Henzinger, Shaz Qadeer, and Sriram Rajamani. You assume, we guarantee: Methodology and case studies. In Hu and Vardi [HV98], pages 440–451.

[HSG98]   Ravi Hosabettu, Mandayam Srivas, and Ganesh Gopalakrishnan. Decomposing the proof of correctness of pipelined microprocessors. In Hu and Vardi [HV98], pages 122–134.

[HSG99]   Ravi Hosabettu, Mandayam Srivas, and Ganesh Gopalakrishnan. Proof of correctness of a processor with reorder buffer using the completion functions approach. 1999. Accepted for publication in the Conference on Computer Aided Verification, Trento, Italy.

[HV98]   Alan J. Hu and Moshe Y. Vardi, editors. *Computer-Aided Verification, CAV '98*, volume 1427 of *Lecture Notes in Computer Science*, Vancouver, BC, Canada, June/July 1998. Springer-Verlag.

[McM98]   Ken McMillan. Verification of an implementation of Tomasulo's algorithm by compositional model checking. In Hu and Vardi [HV98], pages 110–121.

[ORSvH95]   Sam Owre, John Rushby, Natarajan Shankar, and Friedrich von Henke. Formal verification for fault-tolerant architectures: Prolegomena to the design of PVS. *IEEE Transactions on Software Engineering*, 21(2):107–125, February 1995.

[SH98]   J. Sawada and W. A. Hunt, Jr. Processor verification with precise exceptions and speculative execution. In Hu and Vardi [HV98], pages 135–146.

Formal Verification of Explicitly Parallel Microprocessors

Byron Cook, John Launchbury, John Matthews, and Dick Kieburtz
{byron,jl,johnm,dick}@cse.ogi.edu

Oregon Graduate Institute

Abstract. The trend in microprocessor design is to extend instruction-set architectures with features—such as parallelism annotations, predication, speculative memory access, or multimedia instructions—that allow the compiler or programmer to express more instruction-level parallelism than the microarchitecture is willing to derive. In this paper we show how these instruction-set extensions can be put to use when formally verifying the correctness of a microarchitectural model. Inspired by Intel's IA-64, we develop an explicitly parallel instruction-set architecture and a clustered microarchitectural model. We then describe how to formally verify that the model implements the instruction set. The contribution of this paper is a specification and verification method that facilitates the decomposition of microarchitectural correctness proofs using instruction-set extensions.

1 Introduction

Simple instruction sets were once the fashion in microprocessor design because they gave a microarchitect more freedom to exploit instruction-level parallelism [29]. However, as microarchitectural optimizations have become more complex, their effect on microprocessor performance has begun to diminish [5,7,9,31]. The trouble is that sophisticated implementations of impoverished instruction sets come at a cost. Superscalar out-of-order microarchitectures [22], for example, lead to larger and hotter microprocessors. They are difficult to design and debug, and typically have long critical paths, which inhibit fast clock speeds.

This explains why microprocessor designers are now adding constructs to instruction sets that allow the explicit declaration of instruction level parallelism. Here are a few examples:

Parallelism annotations declare which instructions within a program can be executed out of order. This feature appears in Intel's IA-64 [5, 8, 12, 15, 24, 34], and indications are that Compaq [37] and others are exploring similar approaches.

Predication [2] expresses conditional execution using data dependence rather than branch instructions. IA-64 and ARM [20] are examples of predicated instruction-sets.

Speculative loads [8] behave like traditional loads—however exceptions are raised only when and only if the data they fetch is used. IA-64 and PA-RISC [21] both support speculative loading mechanisms.

Multimedia instructions denote where optimizations specific to multimedia computation can be applied. MMX [3], AltiVec [13], and 3DNow [14, 32] are examples of multimedia specific instruction-set extensions.

What do these new instruction-sets look like? How will we verify microarchitectural designs against them? These are the questions that we hope to address. In this paper we develop a modeling and verification method for extended instruction-sets and microarchitectures that facilitates the decomposition of microarchitectural correctness proofs. Using this method we construct a formal specification for an instruction-set architecture like Intel's IA-64; and a microarchitectural design that draws influence from Compaq's 21264 [11] and Intel's Merced [34]. We demonstrate how to decompose the microarchitectural correctness proof using the extensions from the specification. We then survey related work, and propose several directions for future research.

2 OA-64: An explicitly parallel instruction set

In this section we develop a specification of an explicitly parallel instruction-set, called the *Oregon Architecture* (OA-64), which is based on Intel's IA-64. OA-64 extends a traditional reduced instruction set with parallelism annotations and predication.

To see how these extensions fit into OA-64, look at Fig. 1 which contains an OA-64 implementation of the factorial function:

- An OA-64 program is a finite sequence of *packets*, where each packet consists of three instructions. Programs are addressed at the packet-level. That is, instructions are fetched in packets, and branches can jump only to the beginning of a packet.
- Instructions are annotated with *thread identifiers*. For example, the 0 in the load instruction declares that instructions with thread identifiers that are not equal to 0 can be executed in any order with respect to the load instruction.
- Packets can be annotated with the directive FENCE, which directs the machine to calculate all in-flight instructions before executing the following packet.
- Instructions are predicated on boolean-valued predicate registers. For example, the load instruction will only be executed if the value of p5 is true in the current predicate register-file state.

One way to view the parallelism annotations (thread identifiers and fences) is with directed graphs whose nodes are sets of threads that occur between fence directives. We call the sets of threads *regions*. The idea is that an OA-64 machine will execute one region at a time. In this manner, all values computed in previously executed regions are available to all threads in the current region.

```
101:  r2 ← load 100 if p5 in 0
      r3 ← r2 if p5 in 0
      r1 ← 1 if p5 in 1
FENCE

102:  p2,p3 ← r2 != 0 if p5 in 0
      r3 ← r2 if p5 in 1
      nop
FENCE

103:  r1 ← r1 * r3 if p2 in 0
      r2 ← r2 - 1 if p2 in 1
      pc ← 102 if p2 in 2


104:  store 401 r1 if p3 if 3
      pc ← 33 if p3 in 4
      nop
FENCE
```

Fig. 1. OA-64 factorial function.

For example, the first packet loads data into r2 and initializes the values of registers r1 and r3. Because r3 depends on the value of r2, the load instruction must be executed before writing to r3 — this is expressed by placing the same thread-identifier (0) on the two instructions. The calculation of r1, however, can be executed in any order with respect to the 0 thread.

The fence directive in packet 101 instructs the machine to retire the active threads before executing the following packet. Assuming that packet 100 also issues a fence directive, packet 101 forms its own region:

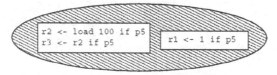

where boxes represent threads. Instructions within a thread must be executed in order. Threads, however, can be executed in any interleaving-order with other threads. Because packet 101 is a region the machine is required to synchronize the state before executing the next packet.

Because packet 102 also issues a fence, it forms its own region:

The comparison instruction sets the predicate register p2 to true if r2 is not equal to 0. The value of p3 is set to the negation of p2.

Because packet 103 is not fenced, but packet 104 is, the next region is formed from packets 103 and 104:

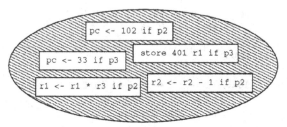

This region contains 5 singleton threads. Note that, if both p2 and p3 were true, two threads would write to the program counter (pc) in an arbitrary order. However, because p2 and p3 are the negation of one another, for a given run of the region only one thread will write to pc.

Assignments to the program counter within a region are visible to the machine's fetch mechanism only after a fence directive has been issued. That is, assignments to pc tell the machine where to fetch from after executing the next fence. Therefore, a trace of an OA-64 program can be viewed as an infinite path through the finite directed graph formed by regions and their successors:

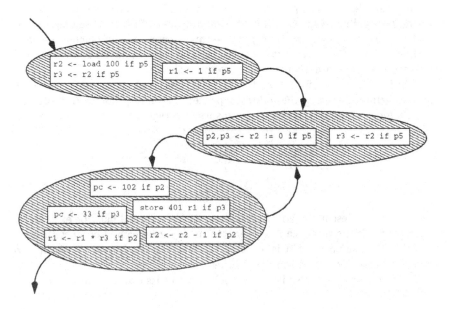

2.1 An OA-64 specification

As is standard practice, we specify OA-64 as a state machine. However, rather than a single monolithic machine, OA-64's specification is the composition of functions from machines to machines:

$$\texttt{oa64}_p = \texttt{fnt}_p \circ \texttt{cls} \circ \texttt{prd} \circ \texttt{risc}$$

We call these functions *state machine transformers*.

Each transformer specifies an extension such as instruction caching, parallelism or predication. The transformer `risc`, for example, ignores the incoming state machine and constructs a classic reduced instruction-set. The instructions defined by `risc` are then predicated by `prd`. The parallelism transformer `cls` then adds a notion of threads, thread identifiers, and synchronization. The final transformer `fnt` builds a state machine with program memory, a program counter, and instruction fetching.

Transformers accept and return state machines with the type constructor `SM` in their type. $\texttt{SM } \Sigma \; \Gamma \; \Delta$ represents the set of state machines with state Σ, input-type Γ and observation-type Δ (Σ, Γ, and Δ are polymorphic type variables). Given a state machine $m :: \texttt{SM } \Sigma \; \Gamma \; \Delta$, three projections are defined:

- $\texttt{initial}_m :: \Sigma$ is the initial state.
- $\texttt{observe}_m :: \Sigma \to \Delta$ is an observation function.
- $\texttt{next}_m :: \Gamma \to \Sigma \to \{\Sigma\}$ is a next-state relation indexed by Γ.

For example, the transformer `prd`, which has type:

$$\texttt{prd} :: \texttt{BUBBLE} \in \Gamma \wedge \texttt{BIND } \Psi \; \Phi \in \Gamma \Rightarrow \texttt{SM } \Sigma \; \Gamma \; (\texttt{Maybe } (\Psi \to \Phi)) \to$$
$$\texttt{SM } (\texttt{Prd\_St } \Psi \; \Sigma) \; (\texttt{Prd\_Instr } \Psi \; \Gamma) \; (\texttt{Maybe } (\Psi \to \Phi))$$

expects a state machine with an environment observation-type and returns a new state machine with the same observation-type but slightly richer input- and state-types. The type constructor `Maybe` adds a "bottom" element to a type:

$$\texttt{Maybe } (\Psi \to \Phi) = \texttt{Nothing} \mid \texttt{Just } (\Psi \to \Psi)$$

That is, an environment of type $\texttt{Maybe } (\Psi \to \Phi)$ is either available (`Just` f where the value f has type $\Psi \to \Phi$) or unavailable (`Nothing`). The type constructor `Prd_Instr`, when given an instruction type Γ and register-type Ψ, returns a type that represents tagged expressions that can be either register to predicate-register moves (`TO` Ψ Ψ), predicate-register to register moves (`FROM` Ψ Ψ), or instructions predicated by a predicate register (`IF` Γ Ψ):

$$\texttt{Prd\_Instr } \Psi \; \Gamma = \texttt{TO } \Psi \; \Psi \mid \texttt{FROM } \Psi \; \Psi \mid \texttt{IF } \Gamma \; \Psi \mid \ldots$$

The type constructor `Prd_St`, when given a state-type and register-type, returns the state-type paired with the type that represents a predicate register file:

$$\texttt{Prd\_St } \Psi \; \Sigma = (\Sigma, \texttt{Env } \Psi \; \texttt{Bool})$$

The function `prd`, found in Fig. 2, takes a state machine and uses its components to build a new predicated state machine.

```
prd m =
  { observe = λ(s,p). observe_m s
  , initial = (initial_m, emptyEnv True)
  , next = λi. λ(s,e). if stall_m s then (next_m BUBBLE s,e)
      else case i of
        BIND r w → (next_m (BIND r w) s,e)
      | BUBBLE r w → (next_m BUBBLE s,e)
      | TO r r' → (s, updateEnv e (r, read_m s r' ≠ 0))
      | FROM r r' → if readEnv e r' then (next_m (BIND r 1) s,e)
                    else (next_m (BIND r 0) s,e)
      | IF i r → if readEnv e r then (next_m i s,e) else (s,e)
  }
```

Fig. 2. Predication state machine transformer

As is the case for prd, we have found that most transformers expect that state machines make visible an environment and a stalling-bit. Let $m :: \text{SM } \Sigma \, \Gamma$ (Maybe $(\Psi \to \Phi)$), we can define the following functions on the states of m:

$read_m \ :: \Sigma \to \Psi \to \Phi$
$read_m \ = \lambda s.$ **case** $observe_m \ s$ **of** Just $e \to e$ | Nothing → **undefined**

$stall_m \ :: \Sigma \to$ Bool
$stall_m = \lambda s.$ **case** $observe_m \ s$ **of** Just $e \to$ **False** | Nothing → **True**

The partial function $read_m$ when given a state of type Σ and a register reference of type Ψ returns a value of type Φ.

Transformers usually expect that BIND $\Psi \, \Phi$ is an instruction in the instruction-type that allows for the update of the environment. That is,

$$\forall s,r,w. \ \neg stall_m \ s \Rightarrow read_m(next_m \ s \ (\text{BIND } r \ w)) \ r = w$$

We also expect that a BIND cannot bring a machine out of a stalling state:

$$\forall s,r,w. \ stall_m \ s \Rightarrow stall_m(next_m \ s \ (\text{BIND } r \ w))$$

3 Columbia: An OA-64 microarchitecture

This section describes the OA-64 microarchitecture—named Columbia—pictured in Fig. 3. Columbia employs three independent execution clusters (buffered execution pipelines); though, in principle, it could use many more clusters. Columbia fetches packets from the instruction cache (**ICache**) and feeds them to the clusters. In the case that a packet contains a fence directive, the machine stops fetching instructions until all of the clusters have been flushed. The **Route** unit schedules fetched instructions into execution clusters based on their thread-identifier (modulus 3). The fetch logic uses the register-file's program counter

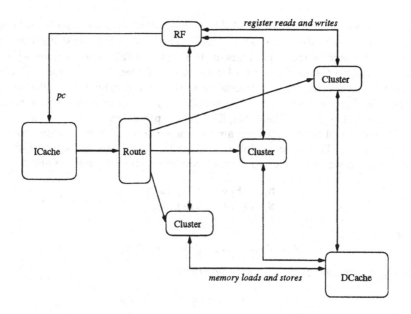

Fig. 3. Columbia data paths— pictured with three clusters

value. It determines, based on whether or not the machine is still servicing a fence directive, if the program counter should be used (i.e. the machine has finished processing a region).

Like OA-64, we can construct Columbia's formal model as the composition of transformers:

$$\texttt{columbia}_p = \texttt{fnt}_p \circ \texttt{cls}' \circ \texttt{prd}' \circ \texttt{pipeline}$$

The transformer `pipeline` constructs a pipeline like that found in Hennessy & Patterson [16]. Then `prd'` adds a predicate register-file and appropriately locks the pipeline in the case of an instruction stream like:

$$\texttt{r2} \leftarrow \cdots; \quad \texttt{p2} \leftarrow \texttt{r2}$$

The transformer `cls'` maintains three parallel buffered and pipelines. That is, rather than finding one instruction to issue at each cycle, `cls'` maintains an instruction buffer for each pipeline from which it issues up to three instructions in parallel every cycle.

4 Decomposing correctness with transformers

In this section we introduce *transformer decomposition*—which allows us to verify machines by point-wise verification of their transformers. For example, does pipeline implement risc?, does cls' implement cls?, etc. We assume that the correctness criteria, abstractly denoted as $\sqsubseteq$, is a reflexive and transitive relation over SM (this could be trace containment [1], simulation [27], correspondence [38], etc). We assume the existence of a family of transformers F that represent the microarchitectural model, and transformers G that represent the architectural model. We also assume that $F_1 \ldots F_K$ and $G_1 \ldots G_K$ are all defined up to some K. In the case of OA-64 and Columbia, $K = 4$, $F_1 = \text{pipeline}$, $G_1 = \text{risc}$, etc.

If a model and its specification are the composition of state machine transformers (i.e. $K > 1$), we can exploit this structure to decompose the proof of $\sqsubseteq$. From F and G, we construct two indexed families of state machines M and N:

$$M_j = (F_j \circ \ldots \circ F_1) \text{ unit}$$
$$N_j = (G_j \circ \ldots \circ G_1) \text{ unit}$$

Proposition 1.
$$F_j \sqsubseteq G_j \wedge \ldots \wedge F_1 \sqsubseteq G_1 \Rightarrow M_j \sqsubseteq N_j$$

where $\sqsubseteq$ is extended point-wise over functions:

$$f \sqsubseteq g \equiv \forall x, y.\ x \sqsubseteq y \Rightarrow f\ x \sqsubseteq g\ y$$

Proof. By induction on j. If $j = 0$ then, by the reflexivity of $\sqsubseteq$, unit $\sqsubseteq$ unit. If $j > 0$ then, by the inductive hypothesis, $M_{j-1} \sqsubseteq N_{j-1}$. By assumption, $F_j \sqsubseteq G_j$, meaning that:

$$M_{j-1} \sqsubseteq N_{j-1} \Rightarrow M_j \sqsubseteq N_j$$

□

4.1 Using uninterpreted next-state relations

When proving that $\sqsubseteq$ holds for two state machines at level $j \in \{1 \ldots K\}$, we can use uninterpreted functions to represent the composition of levels 1 to $j - 1$. One caveat: the implementation transformer, F_j, must be monotonic.

Proposition 2. $\forall j \in \{1 \ldots K\}$ *where F_j is monotonic with respect to $\sqsubseteq$,*

$$(M_{j-1} \sqsubseteq N_{j-1}) \wedge (F_j\ N_{j-1} \sqsubseteq G_j\ N_{j-1}) \Rightarrow (F_j\ M_{j-1} \sqsubseteq G_j\ N_{j-1})$$

Proof. By the monotonicity of F_j, and by the assumption that $M_{j-1} \sqsubseteq N_{j-1}$,

$$F_j\ M_{j-1} \sqsubseteq F_j\ N_{j-1}$$

By assumption,

$$F_j\ N_{j-1} \sqsubseteq G_j\ N_{j-1}$$

Therefore, by the transitivity of $\sqsubseteq$,

$$F_j\ M_{j-1} \sqsubseteq G_j\ N_{j-1}$$

□

Notice the subtle difference (of one letter) between $F_j\ N_{j-1} \sqsubseteq G_j\ N_{j-1}$ and $F_j\ M_{j-1} \sqsubseteq G_j\ N_{j-1}$. By Proposition 1 we need to show that $F_j\ M_{j-1} \sqsubseteq G_j\ N_{j-1}$ for each j in the set $\{1\ldots K\}$. During the proof for a particular j, if you can use the same state machine N_{j-1} as arguments to both F_j and G_j, then the proof is easier to construct. Why? Because the underlying next-state relation is shared by the two machines $F_j\ N_{j-1}$ and $G_j\ N_{j-1}$. In pictures, Proposition 2 states that for all machines x, x', y, y', and z:

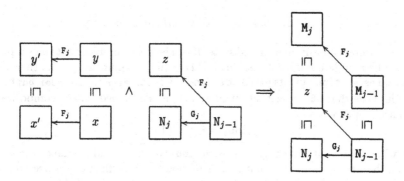

5 Related work

5.1 Completion functions

The approach to correspondence checking proposed by Hosabettu et al. [19] replaces Burch & Dill's flushing abstraction [4] with a composition of *completion functions*—each of which contains knowledge about how a pipeline stage behaves. For example, the abstraction function for a DLX pipeline [16] might be

$$\text{observe}_m \circ \text{IF\_ID} \circ \text{ID\_EX} \circ \text{EX\_WB}$$

which results in the commuting diagram:

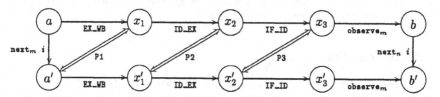

P1, P2 and P3 assert properties about the effect that pipeline stages have on the implementation state. For example:

$$\text{P1 } x\ y \equiv \text{observe}_m(\text{next}_m\ i\ x) = \text{observe}_m(\text{EX\_WB}\ y)$$

Therefore, the proof is decomposed into the obligations P1, P1 $\Rightarrow$ P2, and P2 $\Rightarrow$ P3. Transformers are quite similar to completion functions: whereas a completion function modifies the behavior of a next-state relation, a transformer modifies behavior *and* changes the structure of the next-state relation's state.

5.2 Compositional reasoning

Transformer decomposition involves reasoning about the composition of "higher-order" state machines—which, to our knowledge, is new to the literature on processor verification. However, reasoning about the *parallel composition* of state machines is not new [26]. For example, the Pentium III's microarchitecture (called P6) [10] could hypothetically be modeled as three concurrent processes (for some assembly-language program p):

$$p6_p = \texttt{inst\_fetch}_p \parallel \texttt{reorder\_buffer} \parallel \texttt{res\_station}$$

that communicate through shared state. If run in isolation, $\texttt{res\_station}$'s inputs would be non-deterministicly assigned. When run in conjunction with other processes, $\texttt{res\_station}$'s inputs would be assigned by $\texttt{inst\_fetch}$ and $\texttt{reorder\_buffer}$.

Suppose that $p6$'s specification could also be developed as: the composition of parallel processes:

$$x86_p = \texttt{F} \parallel \texttt{R} \parallel \texttt{E}$$

Showing that $\texttt{inst\_fetch} \sqsubseteq \texttt{F}$, $\texttt{reorder\_buffer} \sqsubseteq \texttt{R}$, and $\texttt{res\_station} \sqsubseteq \texttt{E}$ would be sufficient to show that $p6 \sqsubseteq x86$, however its difficult to prove properties about processes in isolation. The assume-guarantee principle [35] provides a more useful form of decomposition:

$$(q_1 \parallel p_2) \sqsubseteq p_1 \ \wedge \ (p_1 \parallel q_2) \sqsubseteq p_2 \implies (q_1 \parallel q_2) \sqsubseteq (p_1 \parallel p_2)$$

Assuming that the surrounding context meets its specification, if each process in the model is proved to meet the process' specification then the model meets the specification. When the assume-guarantee principle is applied to $p6 \sqsubseteq x86$, the proof is decomposed into cases such as:

$$(\texttt{inst\_fetch}_p \parallel \texttt{R} \parallel \texttt{E}) \sqsubseteq \texttt{F}$$

Like transformer decomposition, the assume-guarantee principle is valid if $\sqsubseteq$ denotes trace containment or simulation [17].

5.3 Intermediate models

In a proof that $\texttt{cls}$ contains the traces of $\texttt{cls}'$ we constructed an an intermediate model which simplified the construction of a simulation relation. This is a common approach in microarchitecture verification. For example, when relating microarchitectural models to their specifications, it is often helpful to carry around more state than necessary using *auxiliary models*. Sawada & Hunt [30] built a microarchitectural model that includes a *trace table* to record the behavior of speculative execution, internal exceptions and external interrupts. The model does not use the trace table to make decisions—only to record the past. Sawada & Hunt then established microarchitectural correctness by demonstrating properties about the trace table.

Abstract models are sometimes used to reduce the complexity of microarchitectural models. That is, an abstract model is constructed from the microarchitecture with a simpler and more general next-state relation. The concrete next-state relation is then proved to meet the specification of the abstract one, and it is then shown that the intermediate abstraction implements the specification. Damm & Pnueli [6] and Skakkebaek et al. [33] both use this technique.

5.4 Other extended instruction-sets

The instruction set of the Java virtual machine [25] includes facilities for multi-threaded execution. However, to date, the formalizations of the Java virtual machine have concentrated on type-safety ([28], for example) or have assumed a single-threaded semantics (such as [36]).

Jones et al. [23] and Ho et al. [18] have both based case studies on a dual-issue VLIW microprocessor—called *the protocol processor*—which is found in the Stanford FLASH multiprocessor. The published papers, which focus on validity checking and test-case generation, do not provide many details on the formal specification of the extended instruction-set.

6 Discussion

In this paper we have demonstrated that instruction-set extensions, when modeled as state-machine transformers, can be used to decompose the proof of a correctness relation on extended state machines. In the future we hope to pursuit research in the following directions:

Architectural relevance: Beyond parallel and predicated instructions, what can transformers specify? We should explore the boundaries by developing transformers that specify other architectural phenomena such as multimedia extensions, operating system support instructions, speculative loads, rotating register-files, etc.

Microarchitectural relevance: The microarchitectural transformers that we have thus far developed are not very exciting. We should demonstrate that transformers, and the theory that facilitates their decomposition, can model microarchitecturally interesting optimizations—such as branch predication, or trace caches. In cases where optimizations break the "transformer barrier", we should develop techniques for constructing and verifying intermediate-level models with large K-values against microarchitectural models with lower K.

Correctness relations: It is tempting to believe that properties (such as reflexivity or transitivity) are maintained when lifted point-wise on functions. Unfortunately this is not so. We should develop techniques to characterize sets of transformers over which the lifted relations have useful properties.
For example, it can probably be proved that any transformer that is completely polymorphic in its state-type is monotonic with respect to simulation. Simulation lifted to the set of transformers with polymorphic state-types is then a reflexive order.

Connections to existing techniques: We should demonstrate that other popular techniques—such as symbolic model checking, symmetry reduction, abstraction, or compositional reasoning—can be used on decomposition obligations. For example, can we use symbolic model checking and uninterpreted functions to build refinement mappings between prd′ and prd? As in McMillan's correctness proof of a Tomasulo-based model [26], can we use a symmetry argument to show that any number of clusters in cls′ correctly implement cls? Rather than composition with fnt, is parallel composition the right way to specify the front-end of the machine? For example:

$$\text{oa64}_p \; m = \text{inst\_fetch}_p \; \| \; (\text{cls} \circ \text{prd} \circ \text{risc}) \; m$$

Automation and structuring techniques: We should explore the existence of new automation and structuring techniques for transformer verification. For example, for certain fixed correctness relations, such as correspondence, we might develop special techniques that help automate or structure verification.

Liveness properties: We should investigate how the addition of liveness properties would effect transformer decomposition.

7 Acknowledgments

Mark Aagaard, Todd Austin, Nancy Day, Sava Krstić, Bee Lavender, Tim Leonard, Abdelillah Mokkedem, John O'Leary, Mark Shields, and the anonymous reviewers : thank you for your comments, suggestions and support.

Funding for this research was provided by Intel Corporation, the U.S. National Security Agency, and the U.S. Air Force Material Command (F19628-93-C-0069). John Matthews is supported by a fellowship from the U.S. National Science Foundation.

References

1. ABADI, M., AND LAMPORT, L. The existence of refinement mappings. *Theoretical Computer Science 2*, 82 (1991), 253–284.

2. ALLEN, J., KENNEDY, K., PORTERFIELD, C., AND WARREN, J. Conversion of control dependence to data dependence. In *The 10th ACM Symposium on Principles of Programming Languages* (Jan. 1983).

3. BISTRY, D., DELONG, C., GUTMAN, M., JULIER, M., KEITH, M., MENNEMEIR, L. M., MITTAL, M., PELEG, A. D., AND WEISER, U. *The Complete Guide to MMX Technology*. McGraw-Hill, 1997.

4. BURCH, J., AND DILL, D. Automatic verification of pipelined microprocessor control. In *6th International Conference of Computer Aided Verification* (Stanford, California, June 1994).

5. CASE, B. IA-64's static approach is controversial. *Microprocessor Report 11*, 16 (1997).

6. DAMM, W., AND PNUELI, A. Verifying out-of-order executions. In *Conference on Correct Hardware Design and Verification Methods* (Montreal, Canada, 1997).

7. DEIFENDORFF, K. WinChip 4 thumbs nose at ILP. *Microprocessor Report 12*, 16 (1998).

8. DELONG, C. The IA-64 architecture at work. *IEEE Computer 31*, 7 (1998).

9. DIEFENDORFF, K. Microarchitecture in the ditch. *Microprocessor Report 12*, 17 (1998).

10. GWENNAP, L. Intel's P6 uses decoupled superscalar design. *Microprocessor Report 9*, 2 (1995).

11. GWENNAP, L. Digital 21264 sets new standard. *Microprocessor Report 14*, 10 (1996).

12. GWENNAP, L. Intel, HP make EPIC disclosure. *Microprocessor Report 11*, 14 (1997).

13. GWENNAP, L. AltiVec vectorizes PowerPC. *Microprocessor Report 12*, 6 (1998).

14. GWENNAP, L. AMD deploys K6-2 with 3DNow. *Microprocessor Report 12*, 7 (1998).

15. GWENNAP, L. Intel outlines high-end roadmap. *Microprocessor Report 12*, 14 (1998).

16. HENNESSY, J. L., AND PATTERSON, D. A. *Computer Architecture: A Quantitative Approach*. Morgan Kaufmann, 1995.

17. HENZINGER, T. A., QADEER, S., RAJAMANI, S. K., AND TASIRAN, S. An assume-guarantee rule for checking simulation. In *Formal Methods in Computer-Aided Design* (Palo Alto, California, 1998).

18. HO, R. C., YANG, C. H., HOROWITZ, M. A., AND DILL, D. Architecture validation for processors. In *Proceedings of the 22nd Annual International Symposium on Computer Architecture* (Santa Margherita Ligure, Italy, 1995).

19. HOSABETTU, R., SRIVAS, M., AND GOPALAKRISHNAN, G. Decomposing the proof of correctness of pipelined microprocessors. In *International Conference on Computer-Aided Verification* (Vancouver, Canada, July 1998).

20. JAGGAR, D. *Advanced RISC Machines Architectural Reference Manual*. Prentice Hall, 1997.

21. JOHNSON, D. Techniques for mitigating memory latency in the the the PA-8500 processor. In *Hot Chips 10* (Palo Alto, Aug. 1998).

22. JOHNSON, M. *Superscalar Microprocessor Design*. Prentice Hall, 1991.

23. JONES, R. B., DILL, D. L., AND BURCH, J. R. Efficient validity checking for processor verification. In *Proceedings of the 1995 International Conference on Computer-Aided Design* (San Jose, 1995).

24. KATHAIL, V., SCHLANSKER, M., AND RAU, B. R. HPL PlayDoh architecture specification: Version 1.0. Tech. Rep. HPL-93-80, Hewlett Packard Laboratories, 1993.

25. LINDHOLM, T., AND YELLIN, F. *The Java Virtual Machine Specification*. Addison Wesley, 1997.

26. MCMILLAN, K. Verification of an implementation of Tomasulo's algorithm by compositional model checking. In *International Conference on Computer-Aided Verification* (Vancouver, Canada, July 1998).

27. MILNER, R. An algebraic definition of simulation between programs. In *Proceedings of 2nd International Joint Conference on Artificial Intelligence* (The British Computer Society, 1971).

28. QIAN, Z. A formal specification of a large subset of Java virtual machine instructions for objects, methods and subroutines. In *Formal Syntax and Semantics of Java*. Springer-Verlag, 1998.

29. RAU, B. R., AND FISHER, J. A. Instruction-level parallel processing: History, overview and perspective. *The Journal of Supercomputing 7*, 1 (1993).

30. SAWADA, J., AND HUNT, W. Processor verification with precise exceptions and speculative execution. In *International Conference on Computer-Aided Verification* (Vancouver, Canada, July 1998).

31. SCHLANSKER, M., RAU, B. R., , MAHLKE, S., KATHAIL, V., JOHNSON, R., ANIK, S., AND ABRAHAM, S. G. Achieving high levels of instruction-level parallelism with reduced hardware complexity. Tech. Rep. HPL-96-120, Hewlett Packard Laboratories, 1996.

32. SHRIVER, B., AND SMITH, B. *The Anatomy of a High-Performance Microprocessor: A Systems Perspective.* IEEE Computer Society Press, 1998.

33. SKAKKEBAEK, J., JONES, R., AND DILL, D. Formal verification of out-of-order execution using incremental flushing. In *International Conference on Computer-Aided Verification* (Vancouver, Canada, July 1998).

34. SONG, P. Demystifying EPIC and IA-64. *Microprocessor Report 12*, 1 (1998).

35. STARK, E. A proof technique for rely/guarantee properties. In *Proceedings of the 5th Conference on Foundations of Software Technology and Theoretical Computer Science* (Aug. 1985).

36. STEPHENSON, K. Towards an algebraic specification of the Java virtual machine. In *Prospects for Hardware Foundations.* Springer-Verlog, 1998.

37. TULLSEN, D. M., EGGERS, S. J., EMER, J. S., LEVY, H. M., LO, J. L., AND STAMM, R. L. Exploiting choice: Instruction fetch and issue on an implementable simultaneous multithreading processor. In *23rd Annual International Symposium on Computer Architecture* (Philadelphia, PA, May 1996).

38. VELEV, M. N., AND BRYANT, R. E. Bit-level abstraction in the verification of pipelined microprocessors by correspondence checking. In *Formal Methods in Computer-Aided Design* (Palo Alto, California, 1998).

Superscalar Processor Verification Using Efficient Reductions of the Logic of Equality with Uninterpreted Functions to Propositional Logic[1]

Miroslav N. Velev[*]

mvelev@ece.cmu.edu
http://www.ece.cmu.edu/~mvelev

Randal E. Bryant[‡, *]

randy.bryant@cs.cmu.edu
http://www.cs.cmu.edu/~bryant

[*]Department of Electrical and Computer Engineering
[‡]School of Computer Science
Carnegie Mellon University, Pittsburgh, PA 15213, U.S.A.

Abstract. We present a collection of ideas that allows the pipeline verification method pioneered by Burch and Dill [5] to scale very efficiently to dual-issue superscalar processors. We achieve a significant speedup in the verification of such processors, compared to the result by Burch [6], while using an entirely automatic tool. Instrumental to our success are exploiting the properties of positive equality [3][4] and the simplification capabilities of BDDs.

1 Introduction

The properties of positive equality [3][4] were proposed as a way to increase the computational efficiency of a decision procedure for the logic of Equality with Uninterpreted Functions and Memories (EUFM). EUFM was introduced by Burch and Dill [5] for verifying of pipelined processors. In collaboration with German [3][4], we recently showed that by extending the syntax of EUFM and by applying certain abstractions, it is possible to use distinct values for all the instruction addresses and data operands. The result is a significantly increased computational efficiency of EUFM.

The main contribution of this paper is in presenting an entirely automatic tool that works on term-level models and is able to handle complex processors, including a dual-issue superscalar DLX [10] with two complete pipelines. We employ a variety of techniques that enhance the performance at each level, including an automatic detection of positive equality comparisons, the encoding method of Goel *et al.* [8] modified to account for positive equality, and an automatic BDD variable ordering. By comparison, in our previous work on positive equality [3][4][20], we only demonstrated the potential of the logic by verifying efficiently single-issue DLX processors implemented at the bit-level. Furthermore, the user was required to define the initial pipeline state and to give hints for the BDD variable ordering.

Our earlier work [18] showed the overhead for verifying bit-level processors with functional units (FUs) implemented at the gate level to be prohibitive for a BDD-based tool, due to the complexity of the generated formulas. The major sources of complexity were the symbolic modeling of all the bits of data in the data path and the feedback loops, created by the forwarding logic. We then employed abstraction and an efficient

---

1. This research was supported in part by the SRC under contract 99-DC-068.

encoding technique for representing word-level values [19]. While that allowed us to verify more complex designs, we ran into BDD blow-up, due to contradictory BDD variable ordering requirements. Using positive equality and exploiting techniques that make the FUs different for each executed instruction, we succeeded in verifying pipelined processors with very large instructions set architectures [20]. Yet, later we were not successful in scaling these techniques for verifying dual-issue superscalar processors. In this paper we examine abstract term-level models of processors, as has most of the work in this field [5][6][8][12][13]. An area for future research will be to prove that the correctness of an abstract term-level model implies the correctness of the original bit-level design.

The correctness criterion of Burch and Dill's method is presented in Fig. 1. The implementation with transition function F_{Impl} is verified by comparison against a specification with transition function F_{Spec}. On each clock cycle the implementation initiates the execution of between 0 and m instructions, where m is bounded by the issue rate of the processor. In each transition, the specification executes 1 instruction. We use F_{Spec}^{m} to denote m applications of function F_{Spec}. It is assumed that the implementation and the specification start from a pair of matching initial states - Q_{Impl} and Q_{Spec}, respectively - where the match is determined according to an abstraction function Abs. The correctness criterion is that the two transition functions should yield a pair of matching final states - Q'_{Impl} and Q'_{Spec}, respectively - where the match is determined by the same abstraction function. In other words, the abstraction function should make the diagram commute. This correctness criterion is due to Hoare [8] who used it (in a version where $m = 1$) for verifying computations on abstract data types in software.

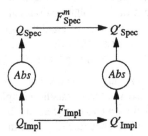

Fig. 1. Commutative diagram for the correctness criterion

The correctness criterion, as formulated by Burch [6], is expressed by:

$$\forall Q_{Impl} \ \exists m \ [Abs(F_{Impl}(Q_{Impl})) = F_{Spec}^{m}(Abs(Q_{Impl}))]. \tag{1}$$

Burch and Dill's contribution [5] is a conceptually elegant way to automatically compute the abstraction function Abs that maps the pipeline state of a processor to its matching state in the specification by symbolic simulation of the hardware design. Namely, starting from a general symbolic initial state Q_{Impl} they simulate a *flush* of the pipeline by feeding it with bubbles for a sufficient number of cycles to allow all partially executed instructions to complete. Then, they consider the resulting state of the user-visible memories (e.g., the register file and the program counter) to be the matching state Q_{Spec}. Experiments by Isles *et al.* [13] to verify a single-issue pipelined DLX, without using flushing as an abstraction function, ran out of memory, given 1 GB was available.

Burch [6] has extended the method to superscalar processor verification by proposing a flushing mechanism suitable for multi-issue processors and by decomposing the commutative diagram into three commutative diagrams which are easier to verify. A correctness proof of this decomposition is presented in [21]. The weakness of his work is that it requires extensive manual intervention in both decomposing the commutative diagram and in identifying case-splitting expressions, used to speed up the validity checking of the correctness criterion formulas.

Pnueli *et al.* [15] also exploit the positive and negative equality structure in order to reduce the complexity of the decision procedure. Their and our method are not directly comparable, since they do the analysis after eliminating function applications by Ackermann's method [1]. That typically introduces both positive and negative equalities for domain variables, which would only appear as positive equalities in our scheme. Hence, they will not exploit the benefits of positive equality as efficiently as we do by using distinct values for a large number of domain variables.

In the remainder of the paper, Sect. 2 reviews the logic of EUFM. Sect. 3 summarizes the benefits of exploiting positive equality. Sect. 4 presents our algorithm for transforming an EUFM formula into a propositional formula, whose validity implies the validity of the original EUFM formula. Sect. 5 explains our manipulation of the EUFM DAG during the transformation. Experimental results are presented in Sect. 6.

2 Logic of Equality with Uninterpreted Functions and Memories

The logic of Equality with Uninterpreted Functions and Memories (EUFM) presented by Burch and Dill [5] can be expressed by the following syntax:

$$
\begin{aligned}
term \quad ::= \quad & ITE(formula, term, term) \\
& | \ function\text{-}symbol(term, \ldots, term) \\
& | \ read(memory, term) \\
memory \quad ::= \quad & memory\text{-}symbol \\
& | \ write(memory, term, term) \\
& | \ ITE(formula, memory, memory) \\
formula \quad ::= \quad & \textbf{true} \ | \ \textbf{false} \ | \ (term = term) \ | \ (memory = memory) \\
& | \ (formula \wedge formula) \ | \ (formula \vee formula) \ | \ \neg formula \\
& | \ read(memory, term) \ | \ predicate\text{-}symbol(term, \ldots, term)
\end{aligned}
$$

In this logic, *formulas* have truth values while *terms* have values from some arbitrary domain. *Memories* can be viewed as mappings from domain values, representing addresses, to domain or Boolean values (as determined by the type of the memory), representing data. Terms are formed by applications of uninterpreted function symbols, and by applications of *ITE* (for "if-then-else") and *read* operators. The *ITE* operator chooses between two terms based on a Boolean control value, i.e., *ITE*(**true**, x_1, x_2) yields x_1 while *ITE*(**false**, x_1, x_2) yields x_2. The *read* operator takes two arguments, the first of which is a memory, and the second one a term that serves as an address. This operator returns a term for the value of the given memory at the location specified by the address term. A *memory* can be a memory symbol, representing an initial memory state. It can also be the result of an *ITE* operator that selects between two memories. This can be used to express conditional writes to a memory. Finally, a memory can be a *write* operator that takes three arguments, the first of which is a memory, the second is a term that represents a memory address to be updated with the third - a term, representing the new data value for that address. Semantically, *read*(*write*(*memory*, *waddr*, *wdata*), *raddr*) is equivalent to *ITE*(*raddr* = *waddr*, *wdata*, *read*(*memory*, *raddr*)), i.e., a *read* that follows a *write* to a memory returns the value of the *write* when the read and write addresses are equal, and the value of the memory at the read address otherwise. The base case for the *read* operator is to read from the initial state of a memory, represented by a memory symbol m, in which case there are no writes to account for, so that the *read* operator can be represented as an uninterpreted function f_m that is specific for memory symbol m.

Formulas are formed by comparing two terms for equality, by comparing two memories for equality, by applying the *read* operator to return the contents of the argument memory at the address specified by the argument term, by applying an uninterpreted predicate symbol to a list of terms, and by combining formulas using Boolean connectives. A formula expressing equality between two terms or two memories is called an *equation*. Equations with memory arguments are allowed to occur only in the top-level verification condition to express the equivalence of memory states in the implementation and the specification. The rules for eliminationg reads from memories of type formula are analogous to those for reads from memories of type term, as defined in the previous paragraph, except that reads from initial memory state are represented as uninterpreted predicates.

Every function symbol f has an associated *order*, denoted *ord*(f), indicating the number of terms it takes as arguments. Function symbols of order zero are referred to as *domain variables*. We use the shortened form v, rather than $v()$ to denote an instance of a domain variable. Similarly, every predicate p has an associated order *ord*(p). Predicates of order zero are referred to as *propositional variables*.

The truth of a formula is defined relative to a domain D of values and an interpretation I of the function, predicate, and memory symbols. An interpretation I assigns to each function symbol of order k a function from D^k to D, to each predicate symbol of order k a function from D^k to {**true**, **false**}, and to each memory symbol a function from D to D or from D to {**true**, **false**}, depending on the type of the memory. Given an interpretation I of the function and predicate symbols and an expression E, we can define the *valuation* of E under I, denoted $I[E]$, according to its syntactic structure. $I[E]$ will be an element of the domain when E is a term, and a truth value when E is a formula.

A formula F is said to be *true under interpretation* I when $I[F]$ equals **true**. It is said to be *valid over domain D* when it is true for all interpretations over domain D. F is said to be *universally valid* when it is valid over all domains. It can be shown that if a formula is valid over some suitably large domain, then it is universally valid [1]. In particular, it suffices to have a domain as large as the number of syntactically distinct function application terms occurring in F.

3 Positive Equality

In collaboration with German, we have recently shown [3][4] that major improvements can be obtained by exploiting the polarity of the equations in the original formula F before replacing any function applications with domain variables. Let us introduce some notation regarding the polarity of equations and their dependent function symbols. For a formula F of the form $T_1 = T_2$, we say that this equation is a positive equation of F. For formula F of the form $\neg F_1$, any positive equation of F_1 is a negative equation of F, and any negative equation of F_1 is a positive equation of F. For formula F of the form $F_1 \wedge F_2$ or $F_1 \vee F_2$, any positive (respectively, negative) equation of either F_1 or F_2 is a positive (respectively, negative) equation of F as well. Note that all equations of a formula that controls an *ITE* operator will be both positive and negative equations of a formula containing the *ITE*, since such equations are implicitly negated when selecting the "else-expression" of an *ITE*. We call equations which are both positive and negative in a formula F, *general equations* of F. Equations of the form $m_1 = m_2$, where m_1 and m_2 are memories, are allowed to occur only as positive equations.

For term T of the form $f(T_1, \ldots, T_k)$, function symbol f is said to be a data symbol of T. For term T of the form $ITE(F, T_1, T_2)$, any function symbol that is a data symbol of either T_1 or T_2 is also a data symbol of T.

A function symbol f is said to be a "p-function" (positive function) symbol of a formula F if there are no negative or general equations in F for which f is a data symbol of one of the equation arguments. All other function symbols are said to be "g-function" (general function) symbols of F. Using appropriate abstractions, we can represent all processor operations involving instruction addresses and data operands with p-function symbols, leaving only register identifiers as g-function symbols.

We can exploit the presence of p-function symbols to greatly reduce the number of interpretations that must be considered to determine the universal validity of the original formula. Let Σ denote a subset of the function symbols occurring in F. We say that interpretation I is diverse with respect to Σ for F when for any function application term $f(S_1, \ldots, S_k)$ where $f \in \Sigma$ and any other function application term $g(U_1, \ldots, U_l)$ we have $I[f(S_1, \ldots, S_k)] = I[g(U_1, \ldots, U_l)]$ iff $f = g$ and $I[S_i] = I[U_i]$ for $1 \leq i \leq k$. Interpretation I is said to be "maximally diverse" if it is diverse with respect to the set of all p-function symbols in F. The following result is from [3][4]:

Theorem 1. *A formula F is universally valid iff it is true in all interpretations that are maximally diverse for F.*

The essential idea behind this theorem is that a maximally diverse interpretation forms a worst case as far as determining the validity of a formula is concerned. For any less diverse interpretation I, we can systematically derive a maximally diverse interpretation I' such that among the equations, only the positive ones can change their valuations under I', and these can only change from **true** to **false**. Therefore, the valuation of F under the two interpretations must either be equal or satisfy $I[F] =$ **true** and $I'[F] =$ **false**. The proof of the above theorem is presented in [3][4].

4 Transforming an EUFM Formula to a Propositional Formula

We proceed through a series of transformations, starting from the initial EUFM formula, expressing the correctness criterion, and ending with a propositional formula whose validity implies the validity of the original one. At each step we apply various optimizations and simplifications, with the major steps being ordered as:

1. Replace equations of the form $m_1 = m_2$, where m_1 and m_2 are memories, by the equation $read(m_1, a) = read(m_2, a)$, where a is a new domain variable. As defined earlier, such equations can appear only as positive equations in the top-level formula when checking that the two sides of the commutative diagram updated the initial state of a memory in exactly the same way. Since the new domain variable represents an arbitrary address, it is easy to see that if the two sides of the commutative diagram modified that address identically, then they would have modified all addresses identically.

2. Eliminate all *read* operators from updated memory state, as explained in Sect. 2. In our tool we perform this step dynamically as we parse the expressions of the EUFM formula. The result will be that the original read will be replaced by a nested *ITE* expression with a read from the initial state of the memory as a leaf of the expression.

3. Identify the p-function symbols and general function symbols (see Sect. 3).

4. Eliminate UFs and reads from initial memory state (see Sect. 4.1).

5. Translate the reduced EUFM formula to a propositional formula (see Sect. 4.2).

6. Check that the resulting propositional formula is a tautology.

4.1 Elimination of Reads from Initial Memory State and of UFs

Reads from initial memory state and applications of UFs are eliminated in a depth-first way, after all their argument expressions have their reads from initial memory state and UFs eliminated. Specifically, UFs are eliminated by our method of using nested *ITEs* for imposing consistency of the function outputs [19]. Given an UF symbol, say *ALU*, which takes two arguments, with the first eliminated application of this UF being $ALU(T_{11}, T_{12})$, where T_{11} and T_{12} are terms, that UF application is replaced by a new domain variable v_1. Then, the second eliminated application of the same UF, $ALU(T_{21}, T_{22})$, is replaced by $ITE((T_{21} = T_{11}) \wedge (T_{22} = T_{12}), v_1, v_2)$, where v_2 is a new domain variable introduced for the case where the new pair of arguments does not equal the previous pair of arguments. Similarly, the third eliminated application of the same UF, $ALU(T_{31}, T_{32})$, is replaced by $ITE((T_{31} = T_{11}) \wedge (T_{32} = T_{12}), v_1, ITE((T_{31} = T_{21}) \wedge$

$(T_{32} = T_{22})$, v_2, v_3)), where v_3 is a new domain variable introduced for the case where the new pair of arguments does not equal any of the previous pairs of arguments. One can see that the above scheme achieves consistency of the UF's outputs: when $T_{21} = T_{11}$ and $T_{22} = T_{12}$, the second application of the UF *ALU* will evaluate to the value of the first application of *ALU* - the domain variable v_1. The same technique can be used to eliminate applications of an uninterpreted predicate, using new propositional variables instead of domain variables. This transformation is defined formally in [4].

Although a read from initial memory state is semantically equivalent to an uninterpreted function, we handle the translation differently. If the memory is addressed by p-function symbols only, the reads from its initial state are eliminated as applications of an uninterpreted function. However, if a memory is addressed by a g-function symbol, then the reads from its initial state are eliminated by pushing every such read to the leaves of the nested *ITE* address term, i.e., until reaching a domain variable, and introducing a new domain variable for the initial state of the memory at that address. For example, *read(RegFile, ITE(F, reg1, reg2))*, where *reg1* and *reg2* are two domain variables, is transformed to *ITE(F, read(RegFile, reg1), read(RegFile, reg2))* after pushing the read to the leaves of the address term, and *read(RegFile, reg1)* is replaced by the new domain variable *data1*, while *read(RegFile, reg2)* is replaced by the new domain variable *data2*, so that the resulting expression is *ITE(F, data1, data2)*. This can be viewed as initializing the memory for every distinct domain variable that can be selected to be an address term. Note that this technique does not result in equations between two domain variables used as addresses; it is also a conservative approximation since it does not enforce the constraint that the equality of two addresses implies the equality of their initial states. This is one of the keys to the efficiency of our tool. The same scheme can be applied to eliminating uninterpreted function applications as well.

4.2 Translation of the Reduced EUFM Formula to a Propositional Formula

Let F^* be the translation of the original EUFM formula F, resulting after the elimination of *read* and *write* operators, as well as function and predicate applications. Then F^* contains only logic connectives, equations, and *ITE*s, as well as domain and propositional variables.

Our method [3][4] can exploit positive equality by considering only distinct interpretations of the domain variables that are generated when eliminating the p-function symbols. Let Vp be the union of the set of domain variables occurring in F that are p-function symbols, and the set of all new domain variables generated when eliminating the applications of each p-function symbol f. Similarly, let Vg be the union of the set of domain variables occurring in F that are g-function symbols, and the set of all new domain variables generated when eliminating the applications of each g-function symbol h. Let V denote the set of all domain variables in F^*. The following theorem was developed in [3][4]:

Theorem 2. *EUFM formula F is universally valid iff its translation F^* is **true** under all interpretation I^* that are diverse over Vp.*

The algorithm that we present next is a modification of the one proposed by Goel *et al.* [8], extended to account for positive equality by considering a variable in Vp to be equal only to itself.

Let $Dep(T)$, the *dependency set of term T*, be the set of domain variables that T may evaluate to. For example, if $T = ITE(b_1, v_1, ITE(b_2, v_2, v_3))$, where v_1, v_2, and v_3 are domain variables, then $Dep(T)$ is $\{v_1, v_2, v_3\}$. For each term T and each variable $v \in Dep(T)$, we generate the formula $E(T, v)$ that represents the conditions under which T would evaluate to v.

For each formula G, we generate a formula $\hat{G}$ which is a propositional translation of G. In the base case for $E(T, v)$, when T is the domain variable v, $E(T, v)$ is **true**. For a term T of the form $ITE(G, T_1, T_2)$, the formula $E(T, v)$ is defined as $\hat{G} \wedge E(T_1, v) \vee \neg\hat{G} \wedge E(T_2, v)$. The method of translating G into $\hat{G}$ is as follows:

1. if G is $\neg G_1$ then $\hat{G} \doteq \neg\hat{G}_1$;

2. if G is $G_1 \bullet G_2$ then $\hat{G} \doteq \hat{G}_1 \bullet \hat{G}_2$, where $\bullet$ is either $\wedge$ or $\vee$;

3. if G is $T_1 = T_2$ then

$$\hat{G} \doteq \bigvee_{v \in Dep(T_1) \cap Dep(T_2)} E(T_1, v) \wedge E(T_2, v) \quad \vee \quad \bigvee_{\substack{v_i \in Dep(T_1) \cap Vg, \\ v_j \in Dep(T_2) \cap Vg, \\ i \neq j}} E(T_1, v_i) \wedge E(T_2, v_j) \wedge e_{min(i,j), max(i,j)}$$

where e_{ij} is a propositional variable introduced to express the equality relation between the g-function domain variables v_i and v_j. Note that we introduce an e_{ij} variable only when v_i and v_j are syntactically distinct variables in Vg. Also, we exploit positive equality by considering variables in Vp to be equal only to themselves -- they are used only in the left disjunct of the above formula.

Our propositional formulas do not enforce the transitivity constraints $e_{ik} \wedge e_{kj} \Rightarrow e_{ij}$, and none of our correct models needed such constraints in verifying them. Note that if a formula F evaluates to **true** without transitivity constraints, it will also evaluate to **true** when such constraints are imposed, e.g., by implication: $(e_{ik} \wedge e_{kj} \Rightarrow e_{ij}) \Rightarrow F$ where F is already **true**. However, when using BDDs for evaluation of the final propositional formula, we employ the strategy by Goel *et al.* [8] in order to check that a counterexample is not due to a violation of the transitivity constraints. Namely, when the final BDD is not **true**, it is negated in order to express all counterexamples. Given an implicant in the resulting BDD, our tool automatically checks that for each negated variable e_{ij}, there is no sequence $e_{ik_1}, e_{k_1 k_2}, ..., e_{k_n j}$ of positive variables that would imply that the negated variable e_{ij} should evaluate to **true**, thus canceling the implicant. The first implicant that is not canceled is printed as a counterexample.

Note that our way of eliminating reads from initial memory state by pushing the reads to the leaves of the address term expressions does not create equations between register identifier domain variables used as address terms in reads from the register file of a processor. This would not be the case if the consistency of the initial memory state was imposed by Ackermann constraints [1], $read\_addr1 = read\_addr2 \Rightarrow init\_state1 = init\_state2$, as done in [8][15], or by our scheme of using nested *ITE*s [19], where $ITE(read\_addr1 = read\_addr2, init\_state1, init\_state2)$ is returned as the initial state of address $read\_addr2$ given that $init\_state1$ was already introduced as the initial state of address $read\_addr1$. The result is a reduced number of e_{ij} variables encoding equality relations between domain variables used as register identifiers, which translates into an increased efficiency when evaluating the final propositional formula.

Observe that we are using a conservative approximation by not enforcing consistency of the initial memory state. This makes the verification results sound, but not complete, i.e., false positives would not occur, although false negatives are possible. However, by employing a conservative approximation in our verification, we simplify considerably the propositional formula that has to be checked for being a tautology and, hence, we gain efficiency. We can informally argue that this optimization is complete when verifying our processor models (see Sect. 6) since they do not have direct comparisons of source registers in their control logic. Then, the only way for two source registers to be equal is for them to be simultaneously equal to the same destination register. However, the forwarding logic will then select the result associated with that destination register and would prevent the initial state of the register file from being used. Hence, the consistency of the register file initial state will not matter.

As an implementation note, we can view the set of formulas $E(T, v)$ for all $v \in V$, as a very sparse set, i.e., it will simply be **false** for many entries. The usual way to represent such sets is as a list maintained in some canonical order with respect to the domain variables. Then the various operations described above can be implemented by processing these lists to generate either a new list or a single formula.

5 Manipulating the EUFM DAG

When building and transforming the EUFM DAG, we impose several simple structural restrictions in order to achieve maximal sharing of identical expressions. Similar to BDDs [2], we create only one node equal to the constant **true** value and only one node equal to the constant **false** value. We allow only the logic connectives $\wedge$ and $\vee$, from the possible multi-input connectives. Their inputs are sorted in some canonical order, with duplicates and non-controlling values (**true** for $\wedge$, and **false** for $\vee$) removed. Expressions of the form $c = a \wedge b$, where $b = d \wedge e$, are rewritten as $c = a \wedge d \wedge e$, in order to increase the sharing of logically identical expressions. Similar rewritings are done for expressions with the logic connective $\vee$. The presence of a controlling value (**false** for $\wedge$, and **true** for $\vee$), or the presence of both a and $\neg a$ as inputs, results in returning the controlling value. Otherwise, the list of sorted inputs, together with the type of the connective, forms a key, which is used to search an Operations Hash table for the same expression created previously. If such an expression is not found, it is created and inserted into the Operations Hash table with the formed key.

Other types of expressions -- *ITE*s, equations, uninterpreted function applications, and uninterpreted predicate applications, as well as the *read* and *write* operators -- also have a key formed in some canonical way in order to access the Operations Hash table. When creating an expression that is the negation of another expression, e.g., $b = \neg a$, where a is not a constant Boolean value, such that the key $\neg a$ is not in the Operations Hash table, we insert two keys in that table: $\neg a$ pointing to expression b, and $\neg b$ pointing to expression a. In this way we ensure that if an expression $c = \neg b$ is created later, it will be identified as expression a. Standard simplifications of *ITE* expressions are also employed, which we omit due to lack of space.

6 Experimental Results

We started with a 5-stage single-issue pipelined DLX [10] model, 1×DLX-C, capable of fetching up to 1 new instruction every clock cycle and implementing the 6 instruction types considered by Burch and Dill [5][6]: register-register, register-immediate, load, store, branch, and jump. The 5 pipeline stages are Fetch, Decode, Execute, Memory, and Write-Back. The pipelined model and its non-pipelined specification were described in our own HDL that uses the primitives of EUFM. Namely, it has support for basic logic gates, multiplexors (*ITEs*), equality comparators, memories, latches, uninterpreted functions, and uninterpreted predicates. The implementation and the specification were simulated with our term-level simulator in order to form an EUFM formula for the correctness criterion. This formula was generated in the SVC script format [16].

The instruction memory of both the implementation and the specification was modeled to produce abstract instructions, consisting of 2 source register identifier terms, 1 destination register identifier term, an immediate datum term, an operation-code term, and 3 Boolean variables used to determine the instruction type. The 3 Boolean variables were decoded by a gate-level PLA to produce the pipeline control signals for the different stages of the pipeline, such that each instruction type gets encoded with a unique binary pattern of the 3 variables (e.g., the register-register instructions are encoded with the pattern 000, the register-immediate with 001, and so on). Therefore, the fetched instructions were restricted to be of only one instruction type, although no assumptions were made about the sequences of executed instructions.

We did not impose any restrictions on the initial state of the pipeline latches, as we did in our previous work with bit-level models [18][20]. Hence, we allow the instruction that is initially in a given pipeline latch to be of all the instruction types simultaneously. Furthermore, we consider initial pipeline states that can never arise in actual operation assuming the pipeline interlocks are correct. By not placing any constraints on the initial state, we cover a larger set of states than is required, but also avoid the need to prove any invariants about the state. Note that if a processor is verified without imposing any restrictions on the instructions in flight, it will also be correct when such restrictions are enforced, e.g., by using the restrictive condition to imply the formula for the correctness criterion, where the formula is already valid. The reason why the processors were verified to be correct without imposing invariants for their initial state is twofold. First, the control logic of our models was not designed to depend on any invariant property of the pipeline state. Second, the pipeline latches that are affected by the interlocks, namely the latches before the Execute, Memory, and Write-Back pipeline stages, get their state reflected on the user-visible memory elements identically along the two sides of the commutative diagram. Note that these latches cannot be stalled and only transfer their data forward. Hence, the identical initial state of the user-visible memory elements, that the two sides of the commutative diagram start from, is modified in the same way by the state of these three pipeline latches, resulting in new identical state of the user-visible memory elements. Therefore, imposing the invariant properties that hold for a correct pipelined processor was not necessary for the verification of our models.

The operation-code term, produced by the instruction memory for each instruction, was used to identify the instruction sub-type to functional units by being used as an input to functional units, just as some control bits are in actual pipelined processors. Specifically, it was carried though the pipeline stages and used as an input to the ALU in the execution stage (e.g., to discriminate an add from a subtract instruction) and to the uninterpreted predicate determining the condition for a branch to be taken based on the comparison of two data operands (e.g., to discriminate a branch on less than from a branch on greater than). Since the operation-code is not used as an argument to interpreted equality comparators, it gets identified as a p-function symbol by our translation algorithm. Hence, functional units taking the operation-code as an argument get transformed into distinct functional units for each executed instruction after the UF elimination by means of nested *ITEs*. This was observed in our previous work [20], where the same effect was achieved by using the sequential PC (equal to PC + 4) which is also a p-function symbol that uniquely identifies each executed instruction. The result is an increased efficiency of the computation, since the functional consistency of ALUs can be imposed with nested *ITEs* of a few levels of nesting for each executed instruction. Therefore, the overall DAG for the correctness criterion ends up being much simpler, compared to the one where the consistency is maintained by nested *ITEs* for the entire executed instruction sequence.

The data memory was modeled as a Finite State Machine with a latch for storing the present state, as explained in [3]. The result fetched by load instructions was produced by an uninterpreted function, *DMem_Read*, that takes as arguments the present data memory state, the load address, and the operation-code term of the instruction (in this way we modeled byte-level memory accesses). The next data memory state was produced by an uninterpreted function, *DMem_Update*, taking the same three arguments in addition to the data operand which is to be written to the data memory by a store instruction. The next data memory state gets written to the FSM latch under the condition that the instruction is a store instruction that was not squashed by taken branches or jumps. The reason to model the data memory in this way is to prevent the outputs of the ALU from being classified as g-terms, due to their role as addresses of the data memory.

Later, we designed a set of dual-issue superscalar DLX models with in-order execution, having 2 pipelines of 5 stages each:

2×DLX-AA has two arithmetic pipelines (implementing register-register and register-immediate instructions), such that either 1 or 2 new instructions are fetched every clock cycle, conditional on the second instruction in the Decode stage having (or not) a data dependency on the first instruction in that stage;

2×DLX-SA can execute arithmetic and store instructions by the first pipeline and arithmetic instructions by the second pipeline, so that in addition to the case of the above data dependency, 1 instruction will be fetched also when the second instruction in the Decode stage is a store (i.e., there is a structural hazard);

2×DLX-LA can execute arithmetic, store, and load instructions by the first pipeline and arithmetic instructions by the second pipeline, so that 2 load interlocks come into play now (between the instruction in Execute in the first pipeline and the two instructions in Decode) and 0, 1, or 2 new instructions can be fetched each cycle;

2×DLX-CA has a complete first pipeline, capable of executing the 6 instruction types, and an arithmetic second pipeline, such that 0, 1, or 2 new instructions can be fetched each cycle - equivalent to Burch's processor [6];

2×DLX-CS has a complete first pipeline, and a second pipeline that can execute arithmetic and store instructions, such that 0, 1, or 2 new instructions can be fetched each cycle;

2×DLX-CL has a complete first pipeline, and a second pipeline that can execute arithmetic, store, and load instructions, such that 0, 1, or 2 new instructions can be fetched each cycle, conditional on 4 possible load interlocks (between a load in Execute in either pipeline and an instruction in Decode in either pipeline) and the resolution of the structural hazard of branches and jumps in Decode of pipeline two, which need to wait for pipeline one;

2×DLX-CC has two complete pipelines, 4 possible load interlocks, but no structural hazards, such that 0, 1, or 2 new instructions can be fetched each cycle.

Our results are presented in Tables 1, 2, and 3. The experiments were performed on a Sun4 with 10 UltraSPARC-II processors of 336 MHz, having 6 GB of physical memory, and running Solaris 2.6, although we used the computer in a single processor mode. The tautology checking of the final propositional logic formula was done with the Colorado University BDD package [7]. We applied a very simple BDD-variable ordering heuristic. The nodes in the final propositional logic DAG are sorted in decreasing order of their fanout counts, such that if a node is the complement of a Boolean variable, then the fanout count of that node is added to the fanout count of the variable. Note that this merging of fanout counts is done only for Boolean variables. The nodes get their BDDs built according to the sorted order in a depth-first way until either a node with a computed BDD is encountered, or a Boolean variable is reached, which gets declared last in the BDD variable order. Furthermore, the recursive BDD computations for the inputs of an AND (OR) node was discontinued as soon as an input's BDD was evaluated to be 0 (1). In the case of an ITE node, the BDD of the controlling input was computed first, such that if it evaluated to a constant 0 or 1 BDD, only the BDD for the corresponding selected input was computed. Also, we freed BDDs for internal nodes as soon as the BDDs were no longer needed, i.e., as soon as a usage count became equal to the fanout count of the node.

Generating the EUFM formula for the correctness criterion by using our term-level simulator, required less than 1.1 MB of memory and 0.1 seconds of CPU time for all the processors. We used Burch's controlled flushing [6], where auxiliary inputs are introduced and used only during the flushing of the implementation in order to prevent the pipeline interlocks from introducing uncertainty in the instruction flow during flushing. The controlled flushing significantly reduces the complexity of the expressions for the state of the user-visible memory elements. We found the reduction to be as much as 10 times, in terms of both memory and CPU time, while the effort to add the auxiliary control inputs was negligible, given a familiarity with the designs.

Processor	DAG Node Counts			Topological Levels in Final Propositional Logic DAG
	Initial EUFM DAG	After eliminating reads and UFs	Final propositional logic DAG	
1×DLX-C	299	1,198	334	35
2×DLX-AA	517	2,168	490	21
2×DLX-SA	560	2,534	601	27
2×DLX-LA	623	3,641	1,029	32
2×DLX-CA	759	4,856	1,383	60
2×DLX-CS	779	5,077	1,469	60
2×DLX-CL	796	5,657	1,608	60
2×DLX-CC	850	5,998	1,732	67

Table 1. Statistics from different stages of the translation to a propositional formula. The topological levels in the final propositional logic DAG are computed by assigning a level of 1 to the Boolean variables (the leaves of the DAG). The nodes in the final propositional logic DAG are of types $\neg$, $\wedge$, $\vee$, and ITE.

Processor	Final Domain Variables		Final Vg		Propositional Variables	
	$|Vp|$	$|Vg|$	Source Registers	Destination Registers	e_{ij}	Other
1×DLX-C	52	13	7	6	27	36
2×DLX-AA	41	19	9	10	66	16
2×DLX-SA	53	19	9	10	66	20
2×DLX-LA	65	19	9	10	72	25
2×DLX-CA	87	25	13	12	116	46
2×DLX-CS	92	25	13	12	116	48
2×DLX-CL	96	25	13	12	116	51
2×DLX-CC	102	25	13	12	120	57

Table 2. Variable statistics during the translation of the EUFM DAG to a propositional formula. The final p-function domain variable set Vp and the final g-function domain variable set Vg were obtained after eliminating the reads and the UFs from the EUFM DAG.

Processor	BDD variables	Max. BDD Nodes	Memory [MB]	CPU Time [s]
1×DLX-C	63	2,121	5.8	0.25
2×DLX-AA	82	8,979	6.9	0.46
2×DLX-SA	86	8,319	7.2	0.49
2×DLX-LA	97	11,393	8.3	1
2×DLX-CA	162	163,782	15.4	9
2×DLX-CS	164	188,557	16.3	9
2×DLX-CL	167	236,770	15.9	18
2×DLX-CC	177	433,658	18.2	35

Table 3. Checking the final propositional logic DAG for being a tautology by using BDDs. The BDD variables count is the sum of the counts of e_{ij} and other propositional variables from Table 2.

The best results were obtained after applying an optimization for eliminating common subexpressions in the top-level equations. Given an equation $T_1 = T_2$, where T_1 is a term of the form $ITE(f_{11}, T_{11}, ITE(f_{12}, T_{12}, ... ITE(f_{1k}, T_{1k}, T_3)))$, T_2 is a term of the form $ITE(f_{21}, T_{21}, ITE(f_{22}, T_{22}, ... ITE(f_{2l}, T_{2l}, T_3)))$, and T_3 is a term of nested ITEs -- $ITE(f_{31}, T_{31}, ITE(f_{32}, T_{32}, ...))$ -- that is shared by both T_1 and T_2, then T_3 is replaced by a new domain variable if $Dep(T_3) \notin Dep(T_{1i})$ for $i = 1, ..., k$ and $Dep(T_3) \notin Dep(T_{2j})$ for $j = 1, ..., l$. It can be proved that this optimization is both sound and complete.

As Table 3 shows, our verification times range from less than a second for the single-issue case, up to 35 seconds for the dual-issue superscalar cases. The memory requirement (often the limiting factor for BDD-based applications) ranges from 5.8 to 18.2 MB. The number of propositional variables ranges from 63 to 177, with between 27 and 120 comprising the e_{ij} variables encoding the equality relations between register identifiers. The number of domain variables, identified as p-function domain variables, is between 2 and 4 times greater than that of the g-function domain variables, as illustrated in Table 2.

It should be pointed out that our design 2×DLX-CA is comparable to that used by Burch [6], who could verify his model only after devising 3 different commutative diagrams, providing 28 manual case splits, and using around 30 minutes of CPU time on a SUN4. Therefore, we achieved a speedup of two orders of magnitude. And what is most important of all, we achieve this speedup by an entirely automatic tool.

In order to compare our results to those by Goel *et al.* [8], who proposed the e_{ij} encoding, we ran experiments for verifying the CMU-Pipe, also used by Burch and Dill [5]. CMU-Pipe is a 3-stage pipelined data path, which implements only register-register instructions with 2 source registers and 1 destination register. It has 3 pipeline stages, and 1 level of multiplexors in the forwarding logic. Our tool required less than 1.1 MB of memory and 0.02 seconds of CPU time to verify this benchmark, including the time to simulate it and generate the EUFM formula for the correctness criterion.

We did not use Burch's controlled flushing, which is not applicable since the design does not have interlocks. Furthermore, the correctness criterion formula was evaluated to be valid as soon as our tool was done parsing it, due to our strategy of automatically eliminating reads from updated memory state and using a maximally shared EUFM DAG. The two terms which represent the state of the register file after exercising the implementation and the specification, respectively, simply happen to have exactly the same structure. Using the Operations Hash table helps identify them as exactly the same term, so that when the final equation expression is parsed and its two argument terms are found to be exactly the same expression, the equation is automatically evaluated to be **true**. Hence, BDDs were not used at all. Also, we did not have to exploit positive equality. Goel *et al.* [8] reported CPU time of 0.5 seconds and needed over 130,00 BDD nodes. They used extensive manual intervention in order to impose the constraints for: 1) consistency of the ALU outputs, 2) consistency of the register file initial state, and 3) reflecting a sequence of writes on the initial state of memories. They do not present results from other benchmarks.

Modeling the data memory as a Finite State Machine, as explained earlier in this section, was crucial to the efficiency of our methodology. An alternative way for representing the data memory is to use an uninterpreted function that will serve as a "translation box," accepting the output of the ALU as an input and producing an output that is used to address a regular memory, representing the state of the data memory. In this way, the output of the ALU will still be automatically classified as a p-term, while it will be mapped to a g-term, via the translation box, that will address the memory. Byte-level memory accesses can be modeled by a read-modify-write strategy, by using an uninterpreted function to change the present state of the address. However, when verifying such versions of 2×DLX-CA and 2×DLX-CC, the BDD package ran out of memory after 8 and 5 hours, respectively.

In order to assess the performance of BDDs when verifying incorrect designs, we created 100 versions of 2×DLX-CC, each with a different error. They were all detected by usually using up to twice the CPU time, memory, and BDD nodes required for the verification of the correct processor. However, in the worst case, one of these models did need 1,600 seconds of CPU time, 168 MB of memory, and 8,100,000 BDD nodes.

We also ran experiments using the Stanford Validity Checker (SVC) [17] to evaluate the validity of the EUFM formula for 1×DLX-C. SVC did not finish within 24 hours. Computing the automatically generated correctness criterion (1) using a non-BDD-based validity checker for the logic of EUFM results in a considerable increase in complexity, due to the prohibitive number of case splits that are required even for a simple 5-stage DLX processor. In our BDD-based tool, evaluating the Boolean expression for (1) is made trivial by the simplification capabilities of the BDD package.

Using the SAT-checker GRASP [9][14] as a tautology checker instead of BDDs, resulted in 2 seconds of CPU time for verifying 1×DLX-C, 70 seconds for verifying 2×DLX-AA, and 224 seconds for verifying 2×DLX-SA, and 1:50 hours for verifying 2×DLX-LA. Prover, a commercial SAT/tautology-checker based on Stålmarck's method [16], required 10 seconds of CPU time for verifying 1×DLX-C, 60 seconds for verifying 2×DLX-AA, 5.5 hours for verifying 2×DLX-SA, and more than 24 hours (the run time limit) for verifying 2×DLX-LA. None of the SAT-checkers was able to verify 2×DLX-CC within 24 hours. Then, we applied these tools to verifying an incorrect version of our last model, 2×DLX-CC. SVC, Prover, and GRASP could not produce a counterexample within 24 hours, while using BDDs for checking the formula of the same incorrect design resulted in generating a counterexample in 37 seconds, consuming 18 MB of memory. Experiments with another SAT-checkers -- SATO [22] -- showed that it was not more successful than GRASP and Prover. Therefore, BDDs were the most efficient means to verify both correct and erroneous processors.

7 Conclusions

We have achieved considerable speedup in the verification of dual-issue superscalar DLX processors, compared to the result by Burch [6]. Furthermore, our tool is entirely automatic and does not require manual intervention, compared to previous work based on the logic of Equality with Uninterpreted Functions and Memories (EUFM) [5][6][8]. The keys to our success were: 1) exploiting the properties of positive equality [3][4], which allow domain variables used in non-negated equality comparisons to be treated as distinct from any other domain variable; 2) using e_{ij} Boolean variables [8] to represent the outcome of those domain variable equality comparisons, which are used both negated and non-negated in the formula, when translating the EUFM formula to a propositional formula; 3) eliminating the reads from the initial state of memories in a way that does not create equality comparisons between two read addresses; 4) defining the ALUs in the abstract models in a way that will turn them into distinct functional units for each executed instruction, based on the properties of positive equality; 5) manipulating the EUFM DAG in a way that results in a maximal sharing of nodes; and, 6) using BDDs to evaluate the resulting Boolean formula, by applying an efficient BDD variable ordering heuristic.

We also showed BDDs to be unmatched by SVC [17], applied to the original EUFM formula, and by Prover [16], SATO [22], and GRASP [9][14], used as alternative tautology checkers of the propositional logic formulas generated by our tool. In contrast to these four methods based on combinatorial search, BDDs capture the full structure of a problem as a single data structure, rather than repeatedly enumerating and disproving possible counterexamples.

Acknowledgements

We would like to thank Amit Goel for implementing a translation procedure from propositional logic to CNF, the input format of SATO and GRASP. We express our gratitude to Steven German of IBM for his detailed comments on this paper. We also extend our thanks to João Marques-Silva of the Technical University in Lisbon, Portugal, for his help with GRASP, to Clark Barrett of Stanford University for his help with SVC, to G. Stålmarck of Prover Technology AB, Sweden (URL: http://www.prover.com), for licensing a copy of Prover to Carnegie Mellon University, and to Arne Borälv of the same company for helping us use Prover efficiently.

References

[1] W. Ackermann, *Solvable Cases of the Decision Problem*, North-Holland, Amsterdam, 1954.

[2] R.E. Bryant, "Symbolic Boolean Manipulation with Ordered Binary-Decision Diagrams," ACM Computing Serveys, Vol. 24, No. 3 (September 1992), pp. 293-318.

[3] R.E. Bryant, S. German, and M.N. Velev, "Exploiting Positive Equality in a Logic of Equality with Uninterpreted Functions,"[2] *Computer-Aided Verification (CAV'99)*, LNCS, Springer-Verlag, June 1999.

[4] R.E. Bryant, S. German, and M.N. Velev, "Processor Verification Using Efficient Reductions of the Logic of Uninterpreted Functions to Propositional Logic,"[2] Technical Report CMU-CS-99-115, Carnegie Mellon University, 1999.

[5] J.R. Burch, and D.L. Dill, "Automated Verification of Pipelined Microprocessor Control," *Computer-Aided Verification (CAV'94)*, D.L. Dill, *ed.*, LNCS 818, Springer-Verlag, June 1994, pp. 68-80. Available from: http://sprout.stanford.edu/papers.html.

[6] J.R. Burch, "Techniques for Verifying Superscalar Microprocessors," *33rd Design Automation Conference (DAC'96)*, June 1996, pp. 552-557.

[7] CUDD-2.3.0, URL: http://vlsi.colorado.edu/~fabio.

[8] A. Goel, K. Sajid, H. Zhou, A. Aziz, and V. Singhal, "BDD Based Procedures for a Theory of Equality with Uninterpreted Functions," *Computer-Aided Verification (CAV'98)*, A.J. Hu and M.Y. Vardi, *eds.*, LNCS 1427, Springer-Verlag, June 1998, pp. 244-255.

[9] GRASP, URL: http://andante.eecs.umich.edu.

[10] J.L. Hennessy, and D.A. Patterson, *Computer Architecture: A Quantitative Approach*, 2nd edition, Morgan Kaufmann Publishers, San Francisco, CA, 1996.

[11] C.A.R. Hoare, "Proof of Correctness of Data Representations," *Acta Informatica*, 1972, Vol.1, pp. 271-281.

[12] R. Hojati, A. Kuehlmann, S. German, and R.K. Brayton, "Validity Checking in the Theory of Equality with Uninterpreted Functions Using Finite Instantiations," *International Workshop on Logic Synthesis*, May 1997.

[13] A.J. Isles, R. Hojati, and R.K. Brayton, "Computing Reachable Control States of Systems Modeled with Uninterpreted Functions and Infinite Memory," *Computer-Aided Verification (CAV'98)*, A.J. Hu and M.Y. Vardi, *eds.*, LNCS 1427, Springer-Verlag, June 1998, pp. 256-267.

[14] J.P. Marques-Silva, and K.A. Sakallah, "GRASP: A Search Algorithm for Propositional Satisfiability," *IEEE Transactions on Computers*, Vol. 48, No. 5, May 1999, pp. 506-521.

[15] A. Pnueli, Y. Rodeh, O. Shtrichman, and M. Siegel, "Deciding Equality Formulas by Small-Domain Instantiations," *Computer-Aided Verification (CAV'99)*, LNCS, Springer-Verlag, June 1999.

[16] G. Stålmarck, "A System for Determining Propositional Logic Theorems by Applying Values and Rules to Triplets that are Generated from a Formula", Swedish Patent No. 467 076 (approved 1992), U.S. Patent No. 5 276 897 (1994), European Patent No. 0403 454 (1995), 1989.

[17] Stanford Validity Checker (SVC), URL: http://sprout.stanford.EDU/SVC.

[18] M.N. Velev, and R.E. Bryant, "Verification of Pipelined Microprocessors by Correspondence Checking in Symbolic Ternary Simulation,"[2] *International Conference on Application of Concurrency to System Design (CSD'98)*, IEEE Computer Society, March 1998, pp. 200-212.

[19] M.N. Velev, and R.E. Bryant, "Bit-Level Abstraction in the Verification of Pipelined Microprocessors by Correspondence Checking,"[2] *Formal Methods in Computer-Aided Design (FMCAD'98)*, G. Gopalakrishnan and P. Windley, *eds.*, LNCS 1522, Springer-Verlag, November 1998, pp. 18-35.

[20] M.N. Velev, and R.E. Bryant, "Exploiting Positive Equality and Partial Non-Consistency in the Formal Verification of Pipelined Microprocessors,"[2] *36th Design Automation Conference (DAC'99)*, June 1999, pp. 397-401.

[21] P.J. Windley, and J.R. Burch, "Mechanically Checking a Lemma Used in an Automatic Verification Tool," *Formal Methods in Computer-Aided Design (FMCAD'96)*, M. Srivas and A. Camilleri, *eds.*, LNCS 1166, Springer-Verlag, November 1996, pp. 362-376.

[22] H. Zhang, "SATO: An Efficient Propositional Prover," *International Conference on Automated Deduction (CADE'97)*, LNAI 1249, Springer-Verlag, 1997, pp. 272-275. Available from: http://www.cs.uiowa.edu/~hzhang/sato.html.

---

2. Available from: http://www.ece.cmu.edu/~mvelev

Model Checking TLA$^+$ Specifications

Yuan Yu[1], Panagiotis Manolios[2], and Leslie Lamport[1]

[1] Compaq Systems Research Center
yuanyu@pa.dec.com, lamport@pa.dec.com
[2] Department of Computer Sciences, University of Texas at Austin
pete@cs.utexas.edu

Abstract. TLA$^+$ is a specification language for concurrent and reactive systems that combines the temporal logic TLA with full first-order logic and ZF set theory. TLC is a new model checker for debugging a TLA$^+$ specification by checking invariance properties of a finite-state model of the specification. It accepts a subclass of TLA$^+$ specifications that should include most descriptions of real system designs. It has been used by engineers to find errors in the cache coherence protocol for a new Compaq multiprocessor. We describe TLA$^+$ specifications and their TLC models, how TLC works, and our experience using it.

1 Introduction

Model checkers are usually judged by the size of system they can handle and the class of properties they can check [3, 16, 4]. The system is generally described in either a hardware-description language or a language tailored to the needs of the model checker. The criteria that inspired the model checker TLC are completely different. TLC checks specifications written in TLA$^+$, a rich language with a well-defined semantics that was designed for expressiveness and ease of formal reasoning, not model checking. Two main goals led us to this approach:

- The systems that interest us are too large and complicated to be completely verified by model checking; they may contain errors that can be found only by formal reasoning. We want to apply a model checker to finite-state models of the high-level design, both to catch simple design errors and to help us write a proof. Our experience suggests that using a model checker to debug proof assertions can speed the proof process. The specification language must therefore be well suited to formal reasoning.
- We want to check the actual specification of a system, written by its designers. Getting engineers to specify and check their design while they are

developing it should improve the entire design process. It will also eliminate the effort of debugging a translation from the design to the model checker's language. Engineers will write these specifications only in a language powerful enough to express the wide range of abstractions that they normally use.

We want to verify designs of concurrent and reactive systems such as communication networks and cache coherence protocols. These designs are typically one or two levels above an RTL implementation. Their specifications are usually not finite-state, often containing arbitrary numbers of processors and unbounded queues of messages. *Ad hoc* techniques for reducing such specifications to finite-state ones for model checking are sensitive to the precise details of the system and are not robust enough for industrial use. With TLC it is easy to choose a finite model of such a specification and to check it exhaustively.

The language TLA$^+$ is based on TLA, the Temporal Logic of Actions. TLA was developed to permit the simplest, most direct formalization of assertional correctness proofs of concurrent systems [11, 12]. More than twenty years of experience has shown that this style of proof, based on the concept of invariance [2, 15], is a practical method of reasoning about concurrent algorithms. Good programmers (an unfortunately rare breed) routinely think in terms of invariants when designing multithreaded programs.

TLA assumes an underlying formalism for "ordinary mathematics". TLA$^+$ embodies TLA in a formal language that includes first-order logic and Zermelo-Fränkel set theory along with support for writing large, modular specifications. It can be used to describe both low-level designs and high-level correctness properties. TLA$^+$ is described in [13]; recent developments in TLA and TLA$^+$ are posted in [9].

Any language that satisfies our needs will be too expressive to allow all specifications to be model checked. TLC can handle a subclass of TLA$^+$ specifications that we believe includes most specifications of actual system designs.[1] This subclass also seems to include many of the high-level specifications that characterize the correctness of a design. However, it might not be able to handle a large enough model of such a specification to detect any but the simplest errors.

In TLA, specifications are formulas. Correctness of a design means that its specification implies the high-level specification of what the system is supposed to do. The key step in proving correctness is finding a suitable invariant—that is, a state predicate true of all reachable states. Experience indicates that verifying invariance is the most effective proof technique for discovering errors. We believe that it is also the most effective way to find errors with a model checker. TLC can also be used to check step-simulation under a refinement mapping [12], the second most important part of a TLA proof. This requires checking that every

---

[1] A simple explanation of how TLC evaluates expressions and computes successor states makes it clear what specifications TLC can handle; TLC generates an explanatory error message when it encounters something that it can't cope with.

step of the implementation is an allowed step of the specification, after appropriate state functions are substituted for the specification's internal variables. This substitution is expressed in TLA$^+$ by an INSTANCE construct, which will be supported only in the next version of TLC. With the current version, the substitution must be done by hand to check step-simulation.

TLC does not yet check liveness properties. We hope to add liveness checking in the future, but we do not yet know if it will be practical for real industrial examples.

The design and implementation of TLC were partly motivated by the experience of the first and third authors, Mark Tuttle, and Paul Harter in trying to prove the correctness of a complicated cache coherence protocol. A two man-year effort generated a 1900-line TLA$^+$ specification of the protocol and about 6,000 lines of proof—mostly, proofs of invariance. We saw that a model checker could be used to check proposed invariants. Our first thought was to translate from TLA$^+$ to the input language of an existing model checker. A translator from TLA specifications to S/R, the input language of COSPAN [6], already existed [7]. However, it required specifications to be written in a primitive language that was far from TLA$^+$, so it was no help. We considered using SMV [14] and Murφ [5], but neither of them were up to the task. Among their problems is that their input languages are too primitive, supporting only finite-state programs that use very simple data types. It was tedious, and in some cases seemed impossible, to translate the mathematical formulas in the TLA$^+$ specification into these languages. For nontechnical reasons, we did not try to use COSPAN, but we expect that S/R would have presented the same problems as the input languages for the other model checkers. Translating from TLA$^+$ into a hardware-description language seemed like an unpromising approach.

Because the goal of TLC is finding errors in the specification of an actual design rather than completely verifying a specially-written model, we are willing to sacrifice some speed in order to handle a reasonably large class of TLA$^+$ specifications. TLC therefore simulates the specification rather than compiling it. It is also coded in Java, rather than in language like C that would be more efficient but harder to program. Despite these inefficiencies, TLC is still acceptably fast. Preliminary tests have found that, depending on the example, TLC runs between two and ten times slower than Murφ, which is coded in C and compiles the specification.[2] TLC is also multithreaded and can take advantage of multiprocessors.

While willing to compromise on speed, we do not want to limit the size of specifications that TLC can handle. Model checkers that keep everything in main memory are usually limited by space rather than time. TLC therefore keeps all data on disk, using main memory as a cache. It makes efficient use of disk with sophisticated disk access algorithms. As far as we know, TLC is the first model checker that explicitly manages all disk access. Murφ [20] uses disk, but relies on virtual memory to handle the state queue, a data structure which contains

---

[2] The comparison with Murφ is for a single-threaded version of TLC.

the unexamined states. As noted by Stern [19] and by us in Section 7, this queue can get very large.

This paper is organized as follows. Sections 2 and 3 describe TLA$^+$ specifications and their TLC models. Section 4 sketches how TLC works. Section 5 describes a compact method of representing states, and Section 6 describes how TLC accesses the disk. Section 7 discusses our initial experience using TLC and presents some preliminary performance data. Finally, Section 8 describes TLC's current status and our plans for it.

2 TLA$^+$

Although TLA is expressive enough to permit a wide variety of specification styles, most TLA system specifications have the form $Init \land \Box[Next]_v \land L$, where $Init$ specifies the initial state, $Next$ specifies the next-state relation, v is the tuple of all specification variables, and L is a liveness property written as the conjunction of fairness conditions on actions.[3] TLC does not yet handle liveness properties, so $Init$ and $Next$ are all that concern us.

We do not attempt to describe TLA$^+$ here, but give an idea of what it is like by presenting small parts of an imaginary system specification. The specification includes a set $Proc$ of processors and variables st and inq, where $st[p]$ is the internal state of processor p and $inq[p]$ is a sequence of messages waiting to be received by p. Mathematically, st and inq are functions with domain $Proc$. Since TLA$^+$ is untyped, this means that the values of st and inq in any reachable state are functions. The specification also uses a constant N that is an integer parameter and a variable x whose purpose we ignore. The specification consists of a module, which we name $ImSystem$, that begins:

EXTENDS $Naturals$, $Sequences$

This statement imports the standard modules $Naturals$, which defines the set of natural numbers and operations like $+$, and $Sequences$, which defines operations on sequences. A large specification might be broken into several modules. Next come two declaration statements:

CONSTANT $Proc$, N VARIABLE st, inq, x

The rest of the specification is a series of definitions. After defining the constant expressions $stInit$ and $xInitSet$, the module defines $Init$ by

$$Init \quad \stackrel{\Delta}{=} \quad \land\ st = stInit$$
$$\land\ inq = [p \in Proc \mapsto \langle\,\rangle]$$
$$\land\ x \in xInitSet$$

This predicate, which specifies the initial state, asserts that st equals the value $stInit$; that $inq[p]$ equals the empty sequence $\langle\,\rangle$, for all $p \in Proc$; and that x

---

[3] The specification may also use temporal existential quantification $\exists$ to hide some variables; such hiding is irrelevant here and will be ignored.

may equal any element of the set *xInitSet*. The $\wedge$-list denotes the conjunction of the three subformulas; indentation is used to eliminate parentheses when conjunctions and disjunctions are nested. (We show "pretty-printed" specifications. The actual TLA$^+$ input is an ascii approximation—for example, one types $/\backslash$ for $\wedge$ and \in for $\in$.)

The bulk of the specification is a sequence of definitions of parts of the next-state relation *Next*. One part is the subaction *RcvUrgent(p)* that represents the receipt by processor p of the first message of type *Urgent* in *inq*[p]. The action removes the message from *inq*[p], makes some change to *st*[p], and leaves x unchanged. A sequence of length n is a function whose domain is $1 .. n$, the set of integers 1 through n. A message is represented as a record with a *type* field that is a string. (Mathematically, a record is a function whose domain is a set of strings, and *r.type* stands for $r[$"type"$]$.) The action *RcvUrgent(p)* is a predicate on state transitions. Unprimed occurrences of a variable represent its value in the starting state and primed occurrences represent its value in the ending state. The definition of *RcvUrgent(p)* has the form:

$$RcvUrgent(p) \triangleq$$
$$\exists\, i \in 1 .. Len(inq[p]) :$$
$$\wedge\ inq[p][i].type = \text{"Urgent"}$$
$$\wedge\ \forall j \in 1 .. (i-1) : inq[p][j].type \neq \text{"Urgent"}$$
$$\wedge\ inq' = [inq \text{ EXCEPT } ![p] = [j \in 1 .. (Len(inq[p]) - 1) \mapsto$$
$$\text{IF } j < i \text{ THEN } @[j] \text{ ELSE } @[j+1]]]$$
$$\wedge\ st' = [st \text{ EXCEPT } ![p] = \ldots]$$
$$\wedge\ \text{UNCHANGED } x$$

Module *ImSystem* defines a number of similar subactions. The next-state relation *Next* is defined to be the disjunction of subactions:

$$Next \ \triangleq\ \ldots \vee (\exists\, p \in Proc : RcvUrgent(p)) \vee \ldots$$

How many subactions like *RcvUrgent(p)* there are, and how large their definitions are, depend on the system. Here are the sizes (not counting comments) of a few actual specifications:

- A specification of sequential consistency [10] for a simple memory with just *read* and *write* operations is about two dozen lines.
- A simplified specification of the Alpha memory model [1] with the operations *read, uncached read, partial-word write, memory barrier, write memory barrier, load locked,* and *store conditional* is about 400 lines.
- High-level specifications of the cache coherence protocols for two large Alpha-based multiprocessors are about 1800 and 1900 lines.

3   Checking Models of a TLA$^+$ Specification

Traditional model checking works on finite-state specifications—that is, specifications with an *a priori* upper bound on the number of reachable states. The specification in our example module *ImSystem* is not finite-state because:

- The set *Proc* of processors can be arbitrarily large—even infinite.
- The number of states could depend on the unspecified parameter N.
- The sequences $inq[p]$ of messages may become arbitrarily long.

With TLC, one bounds the number of states by choosing a *model*. In our example, a model instantiates *Proc* with a set consisting of a fixed number of processors; it assigns a particular value to the constant N; and it bounds the length of the sequences $inq[p]$.

To use TLC to check our example, we can create a new module *MCImSystem* that EXTENDS the module *ImSystem* containing our specification. Module *MCImSystem* defines the predicate *Constr*, which asserts that each sequence $inq[p]$ has length at most 3:

$$Constr \triangleq \forall p \in Proc : Len(inq[p]) \leq 3$$

(We could declare a constant *MaxLen* to use instead of 3, and assign it a value along with the other constants *Proc* and N.)

The input to TLC is module *MCImSystem* and a configuration file that tells it the names of the initial condition (*Init*), the next-state relation (*Next*), and the constraint (*Constr*). The configuration file also declares values for the constants—for example, it might assert

$$Proc = \{p1, \ p2, \ p3\} \qquad N = 5$$

These declarations, together with the constraint, define the model that TLC tries to check.[4] Finally, the configuration file lists the names of one or more invariants—predicates that should be true of every reachable state.

TLC explores reachable states, looking for one in which (a) an invariant is not satisfied or (b) deadlock occurs—meaning that there is no possible next state. (Deadlock detection can be turned off.) The error report includes a minimal-length trace that leads from an initial state to the bad state.[5] TLC stops when it has examined all states reachable by traces that contain only states satisfying the constraint. (TLC may never terminate if this set of reachable states is not finite. In practice, it is easy to choose the constraint to ensure that the set is finite.)

In addition to the configuration file and module *MCImSystem*, TLC also uses the modules *ImSystem*, *Naturals*, and *Sequences* imported by *MCImSystem*. The TLA$^+$ *Naturals* module defines the natural numbers from scratch—essentially as an arbitrary set with a successor function satisfying Peano's axioms. A practical model checker will not compute 2+2 from such a definition. TLC allows any TLA$^+$ module to be overridden by a Java class (using Java's reflection) that provides efficient implementations of the operators and data structures defined

---

[4] TLC can also be used to do nonexhaustive checking by generating randomly chosen behaviors, in which case no constraint is needed.

[5] The trace is guaranteed to have minimal length only when TLC uses a single worker thread. We can easily modify TLC to maintain this guarantee when using multiple worker threads should nonminimality of the trace turn out to be a practical problem.

by that module. TLC provides Java classes for standard modules like *Naturals* and *Sequences*; a sophisticated user could write them for her own TLA$^+$ modules.

4 How TLC Works

TLC uses an explicit state representation instead of a symbolic one like a BDD because:

- Explicit state representations seem to work at least as well for the asynchronous systems that interest us [20].
- A symbolic representation would require additional restrictions on the class of TLA$^+$ specifications TLC could handle.
- It is difficult to keep a symbolic representation on disk.

TLC maintains two data structures: a set *seen* of all states known to be reachable, and a FIFO queue *sq* containing elements of *seen* whose successor states have not been examined. (Another implementation using different data structures is described in Section 6.) The elements of *sq* are actual states, while *seen* contains only the fingerprints of its states. TLC's fingerprints are 64-bit, probabilistically unique checksums [17]. For error reporting, an entry in *seen* also has a pointer to a predecessor state in *seen*. (The pointer is null for an initial state.)

TLC begins by generating and checking all possible states satisfying the initial predicate and setting *seen* and *sq* to contain exactly those states. In our example, there is one such state for each element of *xInitSet*.

TLC next rewrites the next-state relation as a disjunction of as many simple subactions as possible. In our example, the subactions include $RcvUrgent(p1)$, $RcvUrgent(p2)$, and $RcvUrgent(p3)$. (Recall that *Proc* equals $\{p1, p2, p3\}$ in the model.)

TLC then launches a set of worker threads, each of which repeatedly does the following. It removes the state s from the front of *sq*. For each subaction A, the worker generates every possible next state t such that the pair of states s, t satisfies A. To do this for action $RcvUrgent(p1)$, it finds, for each i in the set $1 .. Len(inq[p1])$, all possible values of the primed variables that satisfy the subaction's five conjuncts. (For this subaction, there is at most one i for which there exists a next state t, and that t is unique.) If there is no possible next state t for any subaction, a deadlock is reported. For each next state t that it generates, the worker does the following:

- Check if t is in *seen*.
- If it isn't, check if t satisfies the invariant.
- If it does, add t to *seen* (with a pointer to s).
- If t satisfies the constraint, add it to the end of *sq*.

An error is reported if a next state t is found that does not satisfy the invariant, or if s has no next state. In this case, TLC generates a trace ending in t, or in s if there is no next state t. Using the pointers in *seen*, TLC can generate a sequence of fingerprints of states. To generate the actual trace, TLC reruns the algorithm in the obvious goal-directed way.

A TLA$^+$ specification can have any initial predicate and next-state relation expressible in first-order logic and ZF set theory. Obviously, TLC cannot handle all such predicates. It must be able to compute the set of initial states and the set of possible next states from any given state. Space does not permit a description of the precise class of predicates TLC accepts. In practice, TLC seems able to handle specifications that describe actual systems, but not all abstract, high-level specifications. For example:

- It cannot handle either of the two specifications of sequential consistency in [8] because they are not written in TLA's "canonical form" with a single initial condition and next-state action. The specification *SeqDB2*, with its use of temporal universal quantification, lies well outside TLC's domain. The specification *SeqDB1* is the conjunction of two specifications in canonical form, which is easy to write in canonical form. (The initial condition or next-state action of a conjunction is the conjunction of the initial conditions or next-state actions, respectively.) TLC could then find simple errors by checking simple invariance properties. However, it could not verify arbitrary invariance properties because the specification is not machine-closed, meaning that the liveness property constrains the set of reachable states. It could therefore probably not find any subtle errors.

- It cannot not handle our original high-level specification of the Alpha memory model [1]. That specification uses a variable *Before* whose value is a relation on the set of all requests issued so far; and it defines a complicated predicate *IsGood(Before)* which essentially asserts that the results of those requests satisfy the Alpha memory requirements. Actions of the specification constrain the new value of *Before* with the conjunct

(∗) $(Before \subseteq Before') \wedge IsGood(Before')$

TLC cannot compute the possible new values of *Before* from this expression. However, the formula *IsGood(Before')* contained the conjunct

$Before' \in \text{SUBSET}\,(Req' \times Req')$

where *Req* is the set of possible sequences of requests, and SUBSET S is the set of all subsets of S. Moving this conjunct from *IsGood(Before')* to the beginning of formula (∗) allowed TLC to handle the specification. However, TLC finds possible next values of *Before* by first enumerating all the elements of SUBSET $(Req' \times Req')$, and there are an enormous number of them for any but the tiniest models. In a reasonable length of time, TLC can exhaustively explore only a model with two processors, each of which issues at most one request. Running TLC even on this tiny model led to the discovery of one error in the specification.

The inability to exhaustively check an abstract specification does not inhibit checking that a lower-level specification implements it. In that case, checking step-simulation just requires that each pair of states that satisfies the lower-level next-state relation also satisfies the higher-level one (under the refinement mapping). For the Alpha memory model, TLC can check this using the original next-state action, without having to rewrite the formula (∗).

- TLC has been applied to, and found errors in, one of the two specifications of cache coherence protocols for Alpha-based multiprocessors mentioned above. That specification was written by engineers with only a vague awareness of the model checker. The only TLC-related instruction they received was to use bounded quantification ($\exists x \in S : P$) rather than unbounded quantification ($\exists x : P$). We believe that TLC will be able handle the other cache coherence protocol specification as well, and we intend to make sure that it does.[6] That specification was written before the model checker was even conceived.

5   Representing States

Because TLA⁺ allows complex data structures, finding a normal form for representing states is nontrivial. The queue sq must contain the actual unexamined states, not just their fingerprints, and it can get very large [19]. We therefore felt that compactness of the normal form was important. The compact method of representing states that we devised could be used by other model checkers that allows complex types, and it is described here.

Our representation requires the user to write a *type invariant* containing a conjunct of the form $x \in T$ for every variable x, where T is a type. The types supported by TLC are based on atoms. An atom is an integer, a string, a primitive constant of the model (like $p1$, $p2$, and $p3$ in our example), or any other Java object with an equality method. A type is any set built from finite sets of atoms using most of the usual operators of set theory. For example, if S and T are types, then SUBSET S and $[S \rightarrow T]$, the set of functions from S to T, are types.

We first convert the value of each variable to a single number. We do this by defining, for each type T, a bijection C_T from T to the set of natural numbers less than $Cardinality(T)$. These bijections are defined inductively. We illustrate the definition by constructing C_T when T is the type $[Proc \rightarrow St]$ of the variable st in our example. In the model, the type of $Proc$ is $\{p1, p2, p3\}$, so we define C_{Proc} to be some mapping from $Proc$ to $0 \ldots 2$. An element f of $[Proc \rightarrow St]$ is represented as a triple. The jth element of this triple represents $f[C_{Proc}^{-1}(j)]$, which is an element of St. Therefore, f is represented by the triple $C_{St}(f[C_{Proc}^{-1}(0)]), C_{St}(f[C_{Proc}^{-1}(1)]), C_{St}(f[C_{Proc}^{-1}(2)])$. The value of $C_T(f)$ is the number represented in base $Cardinality(St)$ by these three digits.

In a similar fashion, if T is the Cartesian product $T_1 \times \ldots \times T_n$, we can define C_T in terms of the C_{T_i}. Since a state is just the Cartesian product of the values of the variables, this defines a representation of any state as a natural number. This representation uses the minimal number of bits. We use hash tables

---

[6] The specification uses the TLA⁺ action-composition operator, which is the only built-in nontemporal TLA⁺ operator that TLC does not yet implement.

and auxiliary data structures to compute the representation efficiently. The full details will appear elsewhere.

The compact representation did not provide as much benefit as we had expected for two reasons:

- Since the queue sq is kept on disk, the only benefit of a compact state representation is to reduce disk I/O. TLC succeeds so well in overlapping reading and writing of the queue with processing that reducing the amount of I/O saved little time.
- The method is optimal for representing any type-correct state, but in real specifications, the set of reachable states is likely to be much smaller than the set of type-correct ones. We found that the queue was not much larger when a simpler representation was used.

We therefore decided that the benefits of our compact representation did not outweigh the cost of computing it, and TLC now represents states with a simpler normal form that can be computed faster.

6 Using Disk

We have implemented two different versions of TLC. They both generate reachable states in the same way, but they use disk storage differently.

The first version is the one described in Section 4 that uses a set $seen$ of state fingerprints and a queue sq of states. The algorithm implementing the $seen$ set was designed and coded by Allan Heydon and Marc Najork. It represents $seen$ as the union of two disjoint sets of states, one kept as an in-memory hash table and the other as a sorted disk file having an in-memory index. To test if a state is in $seen$, TLC first checks the in-memory table. If the state is not there, TLC uses a binary search of the in-memory index to find the disk block that might contain it, reads that block, and uses binary search to see if the state is in the block. To add a state to $seen$, TLC adds it to the in-memory table. When the table is full, its contents are sorted and merged with the disk file, and the file's in-memory index is updated. Access to the disk file by multiple worker threads is protected by a readers-writers lock protocol.

The queue sq is implemented as a disk file whose first and last few thousand entries are kept in memory. This is a FIFO queue that uses one background thread to prefetch entries and another to write entries to the disk. Devoting a fraction of a processor to these background threads generally ensures that a worker thread never waits for disk I/O when accessing sq.

The second version of TLC uses three disk files: old, new, and $next$, with old and $next$ initially empty and new initially a sorted list of initial states. Model checking proceeds in rounds that consist of two steps:

1. TLC appends to the $next$ file the successors of the states in new. TLC then sorts $next$ and removes duplicate states.

2. TLC merges the *old* and *new* files to produce the next round's *old*. Any states in *next* that occur in this new *old* file are removed, and the resulting *next* file becomes the next round's *new*. TLC sets the next round's *next* file to be empty.

This algorithm results in a breadth-first traversal of the state space that reads and writes *old* once per level. As described thus far, it is the same as an algorithm of Roscoe [18], except he uses regions of virtual memory in place of files, letting the operating system handle the actual reading and writing of the disk. We improve the performance of this algorithm by implementing *next* with a disk file plus an in-memory cache, each entry of which is a state together with a *disk* bit. A newly generated state is added to the cache iff it is not already there, and the new entry's *disk* bit is set to 0. When the cache is full (the normal situation), an entry whose *disk* bit is 1 is evicted to make room. When a certain percentage of the entries have *disk* bits equal to 0, those entries are sorted and written to disk and their *disk* bits set to 1.[7] To reduce the amount of disk occupied by multiple copies of the same state in *next*, we incrementally merge the sets of states written to disk. This is done in a way that, in the worst case, keeps a little more than two copies of each state on disk, without increasing the time needed to sort *next*.

7   Experimental Results

TLC executed its first preliminary test in August of 1998 and has been used to debug specifications. The largest of these is the 1800 line TLA$^+$ specification of a cache coherence protocol for a new Compaq multiprocessor. This specification was written by the engineers in charge of testing. We have used TLC to find errors both in the specification and in a 1000-line invariant for use in a formal correctness proof. TLC has found about two dozen errors in the TLA$^+$ specification, two of which reflected errors in the actual RTL implementation and resulted in modifications to the RTL. (The other errors would presumably not have occurred had the TLA$^+$ specification been written by the design team rather than the testing team.) This specification has infinitely many reachable states. Checking it on a model with about 12M reachable states takes 7.5 hours on a two-processor 600MHz work station, and the state queue attains a maximum size of 250M bytes. The model with the most states that TLC has yet checked, which is for the 30-line specification *CCache* of [8], has over 60M reachable states and takes less than a day to check on a 600MHz uniprocessor work station.

8   Status and Future Work

By using a rich language such as TLA$^+$, we can have the engineers designing a system write a single TLA$^+$ specification that serves as a design document,

---

[7] The first version of TLC can also be improved in a similar fashion by adding a *disk* bit to the in-memory cache of the *seen* set.

as the basis for a mathematical correctness proof, and as input to TLC. By designing TLC to make explicit and disciplined use of disk, we have eliminated the dependency on main memory, which is the limiting factor of most model checkers. TLC is an industrial-strength tool that engineers are using to help design and debug their systems. TLC was released to other engineering groups in June of 1999. We hope to release TLC publicly in the fall of 1999.

In addition to improving performance and providing a better user interface, possible enhancements to TLC include:

- Checking that the output of a lower-level simulator is consistent with a specification.
- Checking liveness properties.
- Using partial-order methods and symmetry to reduce the set of states that must be explored.

We expect that the experience of engineers using TLC will teach us which of these are likely to have a significant payoff for our industrial users.

Acknowledgments

Homayoon Akhiani and Josh Scheid wrote the specification of the cache coherence protocol to which we are applying TLC. Mike Burrows suggested that TLC be disk based and recommended the second method of using the disk. Damien Doligez and Mark Tuttle are early users who provided valuable feedback. Sanjay Ghemawat implemented the high-performance Java runtime that we have been using. Jean-Charles Gregoire wrote the TLA$^+$ parser used by TLC. Mark Hayden is helping us improve the performance of TLC. Allan Heydon advised us on performance pitfalls in Java class libraries.

References

[1] Alpha Architecture Committee. *Alpha Architecture Reference Manual*. Digital Press, Boston, third edition, 1998.

[2] E. A. Ashcroft. Proving assertions about parallel programs. *Journal of Computer and System Sciences*, 10:110–135, February 1975.

[3] E.M. Clarke and E. A. Emerson. Design and synthesis of synchronization skeletons using branching time temporal logic. In *Workshop on Logics of Programs*, volume 131 of *LNCS*. Springer-Verlag, 1981.

[4] E.M. Clarke, E.A. Emerson, and A.P. Sistla. Automatic verification of finite-state concurrent systems using temporal logic. *ACM Transactions on Programming Languages and Systems*, 8(2), 1986.

[5] David L. Dill. The Murφ verification system. In *Computer Aided Verification. 8th International Conference*, pages 390–393, 1996.

[6] Z. Har'El and R. P. Kurshan. Software for analytical development of communication protocols. *AT&T Technical Journal*, 69(1):44–59, 1990.

[7] R. P. Kurshan and Leslie Lamport. Verification of a multiplier: 64 bits and beyond. In Costas Courcoubetis, editor, *Computer-Aided Verification*, volume 697 of *Lecture Notes in Computer Science*, pages 166–179, Berlin, June 1993. Springer-Verlag. Proceedings of the Fifth International Conference, CAV'93.

[8] Peter Ladkin, Leslie Lamport, Bryan Olivier, and Denis Roegel. Lazy caching in TLA. *Distributed Computing*, 12, 1999. To appear.

[9] Leslie Lamport. TLA—temporal logic of actions. At URL http://www.research. digital.com/SRC/tla/ on the World Wide Web. It can also be found by searching the Web for the 21-letter string formed by concatenating uid and lamporttlahomepage.

[10] Leslie Lamport. How to make a multiprocessor computer that correctly executes multiprocess programs. *IEEE Transactions on Computers*, C-28(9):690–691, September 1979.

[11] Leslie Lamport. Introduction to TLA. Technical Report 1994-001, Digital Equipment Corporation Systems Research Center, Palo Alto, CA, December 1994.

[12] Leslie Lamport. The temporal logic of actions. *ACM Transactions on Programming Languages and Systems*, 16(3):872–923, May 1994.

[13] Leslie Lamport. Specifying concurrent systems with tla$^+$. In Manfred. Broy and Ralf Steinbrüggen, editors, *Calculational System Design*, pages 183–247, Amsterdam, 1999. IOS Press.

[14] K. L. McMillan. *Symbolic Model Checking*. Kluwer, 1993.

[15] Susan Owicki and David Gries. Verifying properties of parallel programs: An axiomatic approach. *Communications of the ACM*, 19(5):279–284, May 1976.

[16] J. P. Queille and J. Sifakis. Specification and verification of concurrent systems in CESAR. In *Proc. of the 5th International Symposium on Programming*, volume 137 of *LNCS*, pages 337–350, 1981.

[17] M. O. Rabin. Fingerprinting by random polynomials. Technical Report TR-15-81, Center for Research in Computing Technology, Harvard University, 1981.

[18] A W Roscoe. Model-checking CSP. In *A Classical Mind: Essays in Honour of C A R Hoare*, International Series in Computer Science, chapter 21, pages 353–378. Prentice-Hall International, 1994.

[19] Ulrich Stern. *Algorithmic Techniques in Verification by Explicit State Enumeration*. PhD thesis, Technical University of Munich, 1997.

[20] Ulrich Stern and David L. Dill. Using magnetic disk instead of main memory in the Murφ verifier. In Alan J. Hu and Moshe Y. Vardi, editors, *Computer Aided Verification*, volume 1427 of *Lecture Notes in Computer Science*, pages 172–183, Berlin, June 1998. Springer-Verlag. 10th International Conference, CAV'98.

Efficient Decompositional Model Checking for Regular Timing Diagrams

Nina Amla[1], E. Allen Emerson[1], and Kedar S. Namjoshi[2]

[1] Department of Computer Sciences, University of Texas at Austin [* * *]
{namla,emerson}@cs.utexas.edu
URL: http://www.cs.utexas.edu/users/{namla,emerson}
[2] Bell Laboratories, Lucent Technologies
kedar@research.bell-labs.com
URL: http://cm.bell-labs.com/cm/cs/who/kedar

Abstract. Timing diagrams are widely used in industrial practice to express precedence and timing relationships amongst a collection of signals. This graphical notation is often more convenient than the use of temporal logic or automata. We introduce a class of timing diagrams called *Regular Timing Diagrams (RTD's)*. RTD's have a precise syntax, and a formal semantics that is simple and corresponds to common usage. Moreover, RTD's have an inherent compositional structure, which is exploited to construct an efficient algorithm for model checking a RTD with respect to a system description. The algorithm has time complexity that is linear in the system size and a small polynomial in the representation of the diagram. The algorithm can be easily used with symbolic (BDD-based) model checkers. We illustrate the workings of our algorithm with the verification of a simple master-slave system.

1 Introduction

The design of hardware systems includes the specification of timing behavior for circuit components. In industrial practice, this behavior is most often described graphically by timing diagrams. Timing diagrams are, however, often used informally and without a precise semantics, making it difficult to utilize them for the specification and verification of correct behavior. We address this issue by introducing the class of *Regular Timing Diagrams* (RTD's); which have a simple and precise semantics and an efficient, decompositional model checking algorithm. These diagrams describe changes of signal values over a finite time period, and precedence and timing dependencies between such events, such as "signal a rises within 5 time units of signal b falling" and "signal b is low when signal a rises". The time intervals are specified by integer constants, ensuring that the diagram defines a *regular* language.

[* * *] This work was supported in part by NSF grants CCR 941-5496, CCR 980-4736 and SRC Contract 98-DP-388.

A RTD, like the circuit it describes, may be either *asynchronous* or *synchronous*. A *synchronous* diagram includes one or more "clocks" with fixed periods; the time interval between any pair of events is determined up to the clock period. Asynchronous timing diagrams are used to specify handshaking protocols, like bus arbitration and memory access, while synchronous diagrams can specify timing requirements of clocked systems. The ordering between events is partial; such RTD's are called *ambiguous*. An unambiguous RTD has a total ordering on events (See Figure 1).

Since a RTD is defined for a finite time period, an important question that arises in defining the semantics is the manner in which an infinite computation satisfies a timing diagram? Fisler [13] considers two kinds of semantics: in the *invariant* semantics, the timing diagram must be satisfied at *every* state of a computation, while in the *basic iterative* semantics, the diagram must be satisfied iteratively, at points satisfying a precondition of the diagram. Our semantics is a reformulation of the basic iterative semantics, where we permit timing diagrams to be satisfied in an overlapping manner. For simplicity, in our current model, the precondition is a state property. In general, a precondition is a path property; it can be handled by introducing a monitor automaton for the property (see Section 2.2 for a discussion). This permits a system to satisfy diagrams that express the correctness of different aspects of its operation. For ambiguous diagrams, we further classify this semantics into a *weak* aspect, where a fresh linear ordering of the events is chosen for each satisfaction of the diagram, and a *strong* aspect, where a single linear order is chosen that applies to each satisfaction of the diagram.

The key observation that leads to efficient *model checking* [5, 22, 6] is that timing diagrams are compositional (conjunctive) in nature. This can be visualized informally as the waveforms acting independently and only interacting with other waveforms through a dependency. Rather than build the single, *monolithic* NFA (non-deterministic finite state automaton) or the temporal logic formula that corresponds to the entire diagram, we demonstrate that it is possible to decompose the diagram into properties of isolated waveforms and their interactions. This results in a conjunction of simpler properties that can be conveniently represented by a succinct ∀-automaton (∀FA) [21, 28]. A ∀FA (also known as "dual-run" or "universal" automaton) is a finite state automaton that accepts an input iff *every* run of the automaton along the input meets the acceptance criterion. ∀FA's can be exponentially more succinct than NFA's and naturally express properties that are conjunctive in nature.

Moreover, this conjunctivity can be exploited to verify smaller components of the timing diagram in isolation, thus avoiding the construction of the entire ∀-automaton. We present efficient algorithms that convert RTD's under the various semantics into ∀FA's that are in the worst case of size cubic in the size of the diagram and the largest time constant represented in unary (note that the unary size is exponential in the binary size). These constants are generally performance bounds and tend to be small; thus, we feel justified in claiming polynomial complexity. The use of ∀FA's permits the efficient use of the automata-theoretic language containment paradigm [29, 19, 20] to model checking. For a system M and RTD T, the verification check can be cast as $\mathcal{L}(M) \subseteq \mathcal{L}(\mathcal{A}_T)$, where $\mathcal{A}_T$ is the (small, polynomial size) $\forall$FA for the diagram T and $\mathcal{L}(X)$ denotes the language of X. This is equivalent to $\mathcal{L}(M) \cap \neg\mathcal{L}(\mathcal{A}_T) = \emptyset$. The complement language of a $\forall$FA is accepted by a NFA with identical structure but complemented acceptance condition. Hence, complementation (the $\neg\mathcal{L}(\mathcal{A}_T)$ term) is trivial, and the complexity of the model checking procedure is linear in the size of the structure and the size of the $\forall$FA $\mathcal{A}_T$. In addition, it is often possible to decompose $\mathcal{A}_T$ itself into a conjunction of smaller $\forall$FA's, which may be checked independently with M. It is also simple to produce a description of $\neg\mathcal{L}(\mathcal{A}_T)$ that can be input to a symbolic model checker. To illustrate our method, we show how the behavior of read and write transactions that is described by RTD's can be checked against a simple master-slave memory system.

We believe that this framework permits efficient model checking of timing specifications that are used in practice. Our review of industrial data books and discussions with engineers indicate that RTD's are sufficiently expressive for most industrial verification needs. With the exception of Fisler's work [13, 14], where the model checking algorithms have high complexity, other prior work considers timing diagram models that are at most as expressive as RTD's. The algorithm is linear in the structure size, polynomial in the number of diagram points and dependencies and in the unary size of the constants. The polynomial complexity of our decompositional algorithm is a significant improvement over the earlier monolithic approaches [13, 9], where the size may be exponential in the worst case. Not withstanding the Lichtenstein-Pnueli thesis [20], in practice, as one reaches the limits of applicability of symbolic model checking tools, the size of the specification is of importance. A detailed discussion of these points is in Section 5.

The rest of the paper proceeds as follows. In Section 2, we give a precise syntax and semantics for Regular Timing Diagrams. Section 3 outlines the algorithms that convert RTD's into $\forall$FA's and the model checking procedure. Section 4 describes how the algorithms are used with with the model checker VIS [3] for the verification of a master-slave system. We conclude with a discussion of related work in Section 5.

2 Regular Timing Diagrams - Syntax and Semantics

A Regular Timing Diagram (henceforth referred to as RTD or diagram) is specified by a number of finite waveforms defined over a set of "symbolic" values SV, and timed dependencies between points on the waveforms. The set of symbolic values includes 1 (High), 0 (Low) and X (unspecified). The set SV is ordered by $\sqsubseteq$, where $a \sqsubseteq b$ iff $a = X$ or $a = b$.

70

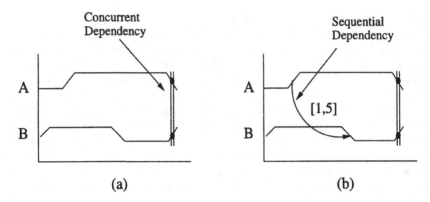

Fig. 1. (a) Ambiguous RTD (b) Unambiguous RTD

2.1 Regular Timing Diagrams: Syntax

Definition 1 (RTD) *A RTD is a tuple* (*WF, SD, CD*), *where*

- *WF is a finite set of waveforms over SV. Each waveform A of length n is a function* $A : [0, n) \rightarrow SV$. *A point of WF is a pair* (A, i) *where* $A \in WF$ *and* $i \in [0, length(A))$. $(A, 0)$ *is the* initial *point and* $(A, length(A) - 1)$ *is the* final *point of A.*
- *SD is the set of* sequential *dependencies on the points of WF. Each dependency is specified as* $(A, i) \overset{[a,b\rangle}{\rightarrow} (B, j)$, *where* $a \in N, b \in N \cup \{\infty\}, 1 \leq a$ *and* $a \leq b$. *(The "$\rangle$" bracket indicates either a closed or an open interval) For convenience,* $[k, \infty)$ *is often written as* $\geq k$, $[1, k]$ *as* $\leq k$ *and* $[k, k]$ *as* $= k$.
- *CD is a collection of mutually disjoint sets of points. Each set is called a* concurrent dependency. *The set of initial and final points of the diagram form predefined concurrent dependencies.*

Definition 2 (Event) *The events of a RTD* (*WF, SD, CD*) *are defined inductively as follows, where the rules are applied in the order shown.*

1. *For every waveform A in WF,* $(A, 0)$ *is an event.*
2. *For an event* (A, i) *with non-X value, the first change along waveform A to a non-X successor value* $A(j)$ *defines* (A, j) *as an event.*
3. *If* (A, i) *is a member of a concurrent dependency that contains an event, then* (A, i) *is an event.*
4. *If* (A, i) *is an event and* $(A, i) \overset{=k}{\Rightarrow} (B, j)$, *then* (B, j) *is an event.*

Notice that for any input string of vectors of signal values, every event has at most one position on the string. This "precise location" property of events is the key to our efficient model checking algorithm. For every event e, it is possible to construct a *DFA* we call *locator(e)* that accepts at the position on an input string where the event holds. This *DFA* essentially encodes the sequence of applications of the rules above that define the point e as an event.

A *symbolic point* of a RTD is either a concurrent dependency or a singleton set containing a point that is not in any concurrent dependency. The set of symbolic points is denoted by $\mathcal{SP}$. Informally, events in a symbolic point should occur simultaneously. The sequential dependencies of a RTD induce the following ordering relation $\prec$ on symbolic points: $p \prec q$ iff

- $(A,i) \in p$ and $(A,i+1) \in q$, for points $i,i+1$ of waveform A in *WF*, or
- there exist $e \in p$ and $f \in q$ such that $e \overset{\alpha}{\to} f$ is a sequential dependency.

The RTD syntax allows several definitions that run counter to intuition. For instance, dependencies may be cyclically related, or it may be possible that the location of a dependency is imprecise due to the presence of X (undetermined) parts of a waveform. These cases are ruled out by giving a notion of "well-formed" RTD's, which is defined below.

Definition 3 (Well-formed RTD) *A RTD is* well-formed *iff (i) every point of the RTD is an event and (ii) the transitive closure of $\prec$ ($\prec^+$) is not reflexive.*

The annotated RTD in Figure 2 can be expressed notationally as follows.

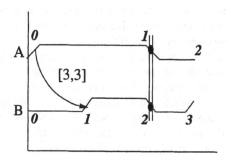

Fig. 2. Example: An Asynchronous RTD

$$WF : \{A, B\}$$
$$A : 0 \mapsto 1, 1 \mapsto 0, 2 \mapsto 0$$
$$B : 0 \mapsto 0, 1 \mapsto 1, 2 \mapsto 0, 3 \mapsto 1$$
$$SD : \{(A,0) \overset{[3,3]}{\to} (B,1)\}$$
$$CD : \{\{(A,0),(B,0)\}, \{(A,1),(B,2)\}, \{(A,2),(B,3)\}\}$$

There are four symbolic points in this RTD: $\{(B,1)\}$ together with the elements of *CD*.

2.2 Regular Timing Diagrams: Semantics

The semantics of a RTD is a set of infinite computations over *states*; each state is a vector indexed by the waveforms of the timing diagram. The set of states is denoted by Σ. The operator $\sqsubseteq$ defined earlier is extended to states as follows: $u \sqsubseteq w$ iff for each i, $u(i) \sqsubseteq w(i)$. A computation of the system to be verified consists of an infinite sequence of states from Σ. Since the syntax of a RTD describes only finite sequences of events, a key question is the appropriate extension to infinite computations.

The predefined initial and final concurrent dependencies can be viewed as the begin- and end- conditions of the finite sequence of events described by the RTD syntax; the initial concurrent dependency is a state predicate and the final concurrent dependency is a path predicate. For example, the begin-condition for the RTD in Figure 2 is $\langle A = 1, B = 0 \rangle$ and the end-condition is the locator for the concurrent dependency at the state $\langle A = 0, B = 1 \rangle$. As another example, if the diagram represents the behavior for a "memory-read" transaction, the begin- and end- conditions indicate the states that define the extent of this transaction. Clearly, this diagram should be checked only on the finite sub-computation that starts at a state satisfying the begin-condition and ends with a state satisfying the end-condition. It is sometimes necessary to make the begin-condition a *path* predicate; the path predicate identifies a sequence of states that indicate the start of a transaction. Such a path predicate can be handled in our current framework by constructing a "monitor" automaton that emits a signal whenever the path condition is satisfied; the presence of this signal, which is a state predicate, can be used as the begin-condition of the RTD.

One may thus consider an infinite sequence to satisfy a timing diagram iff the dependencies of the diagram are satisfied in each finite sub-sequence defined by the begin- and end- conditions. This statement, though, is still open to many interpretations, some of which are considered below. We first define what it means for a finite sequence of states to satisfy a timing diagram. Recall that the relation $\prec^+$ partially orders the set of symbolic points, $\mathcal{SP}$. In the following definitions $\mathcal{P}$ denotes the set of points in the diagram.

Definition 4 (Assignment) *An assignment π for a string σ of length n is a function $\pi : \mathcal{SP} \rightarrow [0, n)$, that is strictly monotonic w.r.t. $\prec$ ($p \prec q$ implies $\pi(p) < \pi(q)$) and maps the initial point of $\mathcal{SP}$ to 0.*

Two assignments $\pi : \mathcal{SP} \rightarrow [0, n)$ and $\xi : \mathcal{SP} \rightarrow [0, m)$ are *equivalent* iff they order symbolic points identically w.r.t. $<$ and $=$. Any assignment π induces the function $\hat{\pi} : \mathcal{P} \rightarrow [0, n)$ which maps a point (A, i) to k iff the (unique, by definition) symbolic point that includes (A, i) is mapped to k by π. From the definition of π, it follows that all points in a concurrent dependency are assigned a common position.

Definition 5 (RTD satisfaction) *A RTD $T = (WF, SD, CD)$ is satisfied by a finite sequence y over Σ^+ w.r.t. an assignment $\pi : \mathcal{SP} \rightarrow [0, |y|)$ (written as $y \models_\pi T$) iff the following conditions hold.*

- *Point consistency: For every point (A, i), if $\hat{\pi}((A, i)) = k$, then $A(i) \sqsubseteq y_k(A)$. Note that $y_j(A)$ denotes the value of waveform A at time j in y.*
- *Waveform consistency: Let $\hat{\pi}((A, i)) = k$ and $\hat{\pi}((A, i + 1)) = l$. For every $j \in [k, l)$, $A(i) \sqsubseteq y_j(A)$.*
- *Dependency consistency: For every sequential dependency $e \xrightarrow{[a,b)} f$, $(\hat{\pi}(f) - \hat{\pi}(e)) \in [a, b)$.*

For many systems, it is the case that the begin- condition for the timing diagram does not recur before the end- condition holds. For such systems, we may consider the following semantics. System computations may be described by the expression $(\Delta^+ \vee (\#\Delta^+ \$))^\omega$, where $\#$ and $\$$ are special vectors of Σ representing the satisfaction of the begin- and end- conditions respectively and $\Delta = \Sigma \backslash \{\#, \$\}$. The sequence of the form $\#\Delta^+\$$ is called a *transaction*.

Definition 6 (Weak Iterative Semantics) *An infinite sequence z satisfies a RTD T under the* weak iterative semantics *(written as $z \models_w T$) iff for every transaction $\#y\$$ on z, there exists an assignment π for which $\#y\$ \models_\pi T$.*

Definition 7 (Strong Iterative Semantics) *An infinite sequence z satisfies a RTD T under the* strong iterative semantics *(written as $z \models_s T$) iff there exists an assignment ξ such that for every transaction $\#y\$$ of z, there is an equivalent assignment π such that $\#y\$ \models_\pi T$.*

A notable class of systems where the assumption of non-overlapping transactions does not hold is those that involve some measure of pipelining. We may then consider the following generalization of the weak iterative semantics.

Definition 8 (Overlapping Semantics) *An infinite sequence z satisfies a RTD T under the* overlapping semantics *(written as $z \models_o T$) iff wherever $\#$ holds along z, there exists y such that $\#y\$$ is a prefix of the suffix computation from that point and for some assignment π, $\#y\$ \models_\pi T$.*

For the rest of the paper, we consider only the weak and strong iterative semantics in detail; the algorithm for the overlapping semantics is a slight modification of that for the weak iterative semantics and has the same complexity. We consider now an alternative formulation of Definition 5, which forms the basis for the decompositional algorithms for model checking. If $\#y\$$ satisfies the timing diagram, each event, by Definition 2 may be located precisely on the sequence. The key observation is that, since each dependency consists of precisely located events, it can be checked independently of the others. Let pt be the partial function that defines the location of events on a finite sequence.

Theorem 1 *For a RTD $T = (WF, SD, CD)$, and any finite transaction $z = \#y\$$, there exists an assignment π such that $z \models_\pi T$ iff each of the following conditions holds:*

- *Every event of T can be located on z and has a value consistent with that in T; i.e., pt is total, and if $pt(z, (A, i)) = k$ then $A(i) \sqsubseteq z_k(A)$.*

- Let $pt(z, (A, i)) = k$ and $pt(z, (A, i+1)) = l$. For every j in $[k, l)$, $A(i) \sqsubseteq z_j(A)$.
- For each sequential dependency $e \overset{[a,b)}{\to} f$, $(pt(z, f) - pt(z, e)) \in [a, b)$.
- For each pair of events e, f in a concurrent dependency, $pt(z, e) = pt(z, f)$.

Notice that the theorem essentially transforms the existential ($\exists$) condition of Definitions 6 through 8 into a universal ($\forall$) condition; this forms the basis for the decompositional check.

3 Decompositional Model Checking

Theorem 1 is fundamental to decomposing RTD's into a conjunction of properties of individual waveforms and ordering or timing restrictions on their interactions, which is the key to efficient model checking. In this section, we provide algorithms that translate a RTD under both strong and weak iterative semantics into a $\forall FA$. The basic iterative and overlapping semantics can be similarly handled. For clarity, we often describe the NFA for the complement language instead of the $\forall FA$.

Definition 9 ($\forall FA$) A $\forall FA$ on infinite strings $\mathcal{A} = (\Sigma, Q, \delta, q_0, \Phi)$ has a finite input alphabet Σ, finite state set Q, transition relation $\delta \subseteq Q \times \Sigma \times Q$, start state q_0 and acceptance condition Φ.

A run r of $\mathcal{A}$ on input x in Σ^ω is an infinite sequence of states of $\mathcal{A}$, where r_0 is an initial state, and for each i, $(r_i, x_i, r_{i+1}) \in \delta$. $\mathcal{A}$ accepts x by $\forall$-acceptance according to Φ iff every run r on x satisfies Φ. We define $\mathcal{L}(\mathcal{A})$ to be the set of strings accepted by $\mathcal{A}$; $\mathcal{L}_{NFA}$ by a $\exists$-acceptance and $\mathcal{L}_{NFA}$ by $\forall$-acceptance. In this paper, we consider Φ to be a Büchi acceptance condition. For any $\forall FA$ $\mathcal{A}$, let $\overline{\mathcal{A}}$ be the NFA with the same transition relation but complemented acceptance condition $\neg\Phi$.

Theorem 2 ([21, 28]) For any $\forall FA$ $\mathcal{A}$, $\neg\mathcal{L}_{\forall FA}(\mathcal{A}) = \mathcal{L}_{NFA}(\overline{\mathcal{A}})$.

3.1 RTD's under the weak iterative semantics

We describe here the NFA that accepts the complement of the weak-iterative language of a RTD $T = (WF, SD, CD)$. First, construct finite string automata for each waveform and dependency as follows:

- Waveform: If $(A, i+1)$ is defined in terms of (A, i), then $locator((A, i))$ is extended to ensure that the signal values up to the change of value that defines $(A, i+1)$ are above $A(i)$ in $\sqsubseteq$ order. Otherwise, $locator((A, i))$ is used to determine that the value at the position where (A, i) holds is above $A(i)$ in $\sqsubseteq$ order.
- Sequential dependency: For a sequential dependency $e \overset{[a,b)}{\to} f$, the automaton is a parallel composition of $locator(e)$ and $locator(f)$ that accepts iff the time between the acceptance of the locator DFA's is within $[a, b)$.

- Concurrent dependency: The ∀*FA* for a concurrent dependency C checks that for a fixed event e in C and every other event f in C, *locator*(e) and *locator*(f) accept at the same position on the input sequence.

The ω-*NFA* for the complement language operates as follows on an infinite input sequence: it nondeterministically "chooses" a transaction $\#y\$$ on the input, "chooses" which waveform or dependency fails to hold of the transaction, and accepts if the automaton for that entity (defined as given above) rejects. Notice that each automaton defined above is either a *DFA* or a ∀*FA*, both of which can be trivially complemented. The ∀*FA* obtained from this *NFA* by complementing the acceptance condition defines the language of the RTD under the weak iterative semantics. Denote this ∀*FA* by $\mathcal{A}_T$. For the diagram $T = (WF, SD, CD)$, let L be the size in unary of the largest constant in SD. Define $|T| = \#points + |SD| + |CD|$. The size of $\mathcal{A}_T$ is cubic in $|T|$ and L.

Theorem 3 (Correctness) *For any RTD T and $x \in \Sigma^\omega$, $x \models_w T$ iff $x \in L(\mathcal{A}_T)$. The size of $\mathcal{A}_T$ is polynomial in $|T|$ and the unary length of the largest constant in T.*

3.2 RTD's under the strong iterative semantics

Under the strong iterative semantics, every transaction on an input computation has to satisfy the RTD w.r.t. a single event ordering. The *NFA* for the complemented language accepts a computation iff

- Some transaction violates a waveform or dependency constraint. This is checked by the automaton defined for the weak-iterative semantics. Or,
- There is a transaction and a pair of events that occur in a different order from that in the first transaction. This is done by an automaton that "chooses" a pair of events unordered by $\prec^+$, executes the locator *DFA*'s for these events in parallel on the first transaction to determine their order, then "chooses" a subsequent transaction and executes the locator *DFA*'s of the same events on that transaction to determine the new order, and accepts if the orders differ.

Let $\mathcal{A}_T$ denote the ∀*FA* obtained from this *NFA* by complementing the acceptance condition. The size of $\mathcal{A}_T$ is cubic in $|T|$ and L for the first case; for the second, it is quadratic in $|T|$ and L with a multiplicative factor of the number of event pairs (which is bounded by $(\#points)^2$).

Theorem 4 (Correctness) *For any RTD T and $x \in \Sigma^\omega$, $x \models_s T$ iff $x \in L(\mathcal{A}_T)$. The size of the ∀*FA* $\mathcal{A}_T$ is polynomial in $|T|$ and L.*

3.3 Model Checking

The translation of a RTD to a small ∀*FA* implies that the language containment approach to model checking based on [29] gives an efficient algorithm. We need to check that $\mathcal{L}(M) \subseteq \mathcal{L}(\mathcal{A}_T)$, where M is the system to be verified and $\mathcal{A}_T$ is the $\forall FA$ for the RTD T. This is equivalent to $\mathcal{L}(M) \cap \neg\mathcal{L}(\mathcal{A}_T) = \emptyset$. Complementation (the $\neg\mathcal{L}(\mathcal{A}_T)$ term) is trivial for a $\forall FA$; the complemented automaton (a *NFA*) has the same structure but complemented acceptance condition. Hence, the emptiness check can be done in time linear in the size of the structure and a small polynomial in the size of T. The space complexity, by the results of [25], is logarithmic in the sizes of both M and T.

Theorem 5 *For a transition system M and a RTD T, the time complexity of model checking is linear in the size of M and a small polynomial in the size of T and the unary size of the largest constant in T.*

An alternative way of utilizing the $\forall FA$ construction is to note that, for the weak iterative semantics, the automaton essentially defines a language $(\Delta^+ \vee \#(\bigwedge_i L_i)\$)^\omega$, where the L_i's represent the languages of the dependencies. The lemma below shows that the ω-repetition distributes over the $\bigwedge_i$ in the following sense.

Lemma 1 *For finite-string languages L_i ($i \in [0,n)$) which are subsets of Δ^+, $(\Delta^+ \vee \#(\bigwedge_i L_i)\$)^\omega = \bigwedge_i (\Delta^+ \vee \#L_i\$)^\omega$.*

By this lemma, one can construct smaller ω-automata, one for each dependency, and check that the language of each has an empty intersection with $\mathcal{L}(M)$. This is often more efficient than the combined check, and may lead to quicker detection of any errors. We refer to this as the "decompositional" approach.

4 Applications

We demonstrate the use of these algorithms in the verification of a master-slave memory system using the model checker VIS, which is based on the automata-theoretic (language containment) approach to model checking. This example is small and is intended only as an illustration of how our algorithms may be used.

In the master-slave system (Figure 3), the master issues a read or a write instruction by asserting the corresponding line, and the slaves respond by accessing memory and performing the operation. The master chooses the instruction, puts the address on the address bus and then asserts the *req* signal. The slave whose tag matches the address awakens, services the request, then asserts the *ack* line on completion. Upon receiving the *ack* signal the master resets the *req* signal, causing the slave to reset the *ack* signal. Finally, the master resets the address and data buses. The memory read (Figure 4) and write cycles are specified as RTD's (interpreted under the weak iterative semantics).

The master-slave system was simplified by abstracting away some inessential details. First, the address bus was simplified to the tag of the slaves. Since VIS does not allow variables to be both input and output, the bidirectional data bus is represented as two 1-bit boolean variables, *Idata* and *Odata* that denote the input and output data buses respectively. The begin-condition for the read RTD

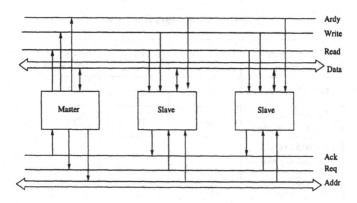

Fig. 3. Master-Slave Architecture

is the state that has *Ardy*, *Idata*, *Req*, *Ack* and *Write* being assigned 0 (low), the value of the address bus *Addr* is unknown and the *Read* signal is high. The end-condition for the read RTD is the state following the diagram where all the signals are low and *Addr* is *X*. Observe that during a read transaction the *write* signal remains unchanged and vice versa. This is a way of expressing negation in a timing diagram.

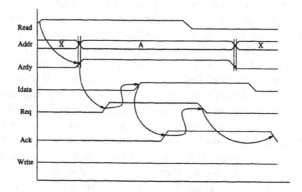

Fig. 4. RTD for the memory read cycle

The simplified master-slave system is represented in Verilog. For both RTD's, we created (as Verilog modules) both the complement of the $\forall FA$ and the complement *NFA* for individual dependencies and waveforms.

The language emptiness check passed for both the ambiguous read and write RTD translations. The results in Table 1 show that the decompositional procedure is indeed feasible and that the size of the system to be verified together

Table 1. Verification statistics

Design under verification	Number of BDD variables	Reachable state space	Number of BDD nodes
Master-Slave Design	61	109	275
Master-Slave and 1 waveform module	70	110	330
Master-Slave and 1 dependency module	70	390	532
Master-Slave and all read RTD modules	169	17796	2019

with a single dependency automaton may not be significantly larger, in terms of BDD variables, than the system itself.

5 Conclusions and Related Work

Several researchers have investigated timing diagrams and their use in automated verification. Boriello [2] proposes an approach to formalizing timing diagrams. Timing diagrams are described informally as regular expressions but no specific details or translation algorithms are given. Many other researchers [1, 26, 23, 4] have formalized timing diagrams and translated them to other formalisms (interval logics, trigger graphs etc.). Cerny et al. present a procedure [18] for verifying whether the finite behavior of a set of action diagrams (timing diagrams) is consistent; [17] uses constraint logic programming to check if a system satisfies finite action diagram specifications. Formal notions of timing diagrams have also proved to be useful in test generation and logic synthesis (cf. [27, 15, 12]).

Fisler [13, 14] proposes a timing diagram syntax and semantics that allows non-regular languages, and finds that these languages occur at all levels of the Chomsky hierarchy. The paper [14] provides a decision procedure that determines whether a regular language is contained in an unambiguous timing diagram language, and [13] provides an algorithm that translates a certain class of timing diagrams into CTL [5]. A key difference with our work is that this algorithm is restricted to a subset of unambiguous timing diagrams under the basic invariant semantics (our algorithms under both iterative and invariant semantics are defined for all types of diagrams). The regular containment procedure [14] has a high complexity (in PSPACE), while our algorithms have *polynomial* time complexity in the diagram size.

An important contribution in this area is the work done by Damm and colleagues at the University of Oldenburg on Symbolic Timing Diagrams (STD's) [9, 24, 8, 16, 7]. STD's may be compiled into first-order temporal logic formulae which are then used for model checking. STD's are extended in [11, 10] to

RTSTD's (Real-time STD's), where a translation into a timed propositional temporal logic TPTL is provided. Both these research efforts consider infinite languages and ambiguity. A key difference with our work lies in the fact that their translation is monolithic, in the sense that all dependencies are considered together; this can result in an exponential blowup in the size of the resulting formulae when the diagram is highly ambiguous. While it is possible to model check the first order temporal logic presented in [9, 10], the procedure is not very efficient.

This paper presents "regular" timing diagrams (RTD's), which have a simple syntax and precise, simple semantics that closely corresponds to common usage. From our discussions with engineers, we are led to believe that RTD's are expressive enough to represent many timing diagrams that arise in practice. As mentioned earlier, the algorithms proposed in this paper can also be used with synchronous RTD's.

Noteworthy contributions of this paper include polynomial time, decompositional algorithms for model checking timing diagram specifications, which are based on a decomposition of the RTD semantics into properties of each waveform and the way they interact. Such decompositions may also provide a way of composing RTD's and thereby building new RTD's hierarchically. Our algorithms generate a $\forall FA$ (NFA) corresponding to the RTD (the negation of the RTD). We can choose to use either the $\forall FA$ (by splitting it into smaller $\forall FA$'s) or its complement NFA in verifying that a system satisfies a RTD. These algorithms are a significant improvement over the earlier possibly exponential, monolithic translations. We have shown how our algorithms may be used in conjunction with a symbolic model checker such as VIS, to verify systems with specifications formulated as RTD's.

We are currently working on a tool that implements these translation and verification algorithms. We also intend to test the efficiency of our algorithms on industrial strength examples.

Acknowledgments We would like to thank Bob Kurshan, Kathi Fisler and Steve Keckler for helpful discussions and insightful comments.

References

1. C. Antoine and B. Le Goff. Timing Diagrams for Writing and Checking Logical and Behavioral Properties of Integrated Systems. In *Correct Hardware Design Methodologies*. Elsevier Sciences Publishers, 1992.

2. G. Borriello. Formalized Timing Diagrams. In *EDAC92*. IEEE Comput. Soc. Press, 1992.

3. R. Brayton, G. Hachtel, A. Sangiovanni-Vincentelli, F. Somenzi, A. Aziz, S. Cheng, S. Edwards, S. Khatri, Y. Kukimoto, A. Pardo, S. Qadeer, R. Ranjan, S. Sarwary, T. Shiple, G. Swamy, and T. Villa. VIS. In *FMCAD*, 1996.

4. V. Cingel. A Graph-based Method for Timing Diagrams Representation and Verification. In *Correct Hardware Design and Verification Methods*. Springer Verlag, 1993.

5. E. M. Clarke and E. A. Emerson. Design and Synthesis of Synchronization Skeletons using Branching Time Temporal Logic. In *Workshop on Logics of Programs*, volume 131. Springer Verlag, 1981.

6. E.M. Clarke, E.A. Emerson, and A.P. Sistla. Automatic Verification of Finite-State Concurrent Systems using Temporal Logic. *ACM Transactions on Programming Languages and Systems*, 8(2), 1986.

7. W. Damm and J. Helbig. Linking Visual Formalisms: A Compositional proof System for Statecharts cased on Symbolic Timing Diagrams. In E. R. Olderog, editor, *Programming Concepts, Methods and Calculi*. Elsevier Science B.V. (North Holland), 1994.

8. W. Damm, H. Hunger, P. Kelb, and R. Schlör. Using Graphical Specification Languages and Symbolic Model Checking in the Verification of a Production Cell. In C. Lewerentz and T. Lindner, editors, *Formal Development of Reactive Systems. Case Study Production Cell, LNCS 891*. Springer Verlag, 1994.

9. W. Damm, B. Josko, and Rainer Schlör. Specification and Verification of VHDL-based System-level Hardware Designs. In Egon Borger, editor, *Specification and Validation Methods*. Oxford University Press, 1994.

10. K. Feyerabend. Real-time Symbolic Timing Diagrams. Technical report, Department of Computer Science, Oldenburg University, September 1994.

11. K. Feyerabend and B. Josko. A Visual Formalism for Real Time Requirement Specifications. In *AMAST Workshop on Real-time systems and Concurrent and Distributed Software*. Springer Verlag, 1997.

12. K. Feyerabend and R. Schlör. Hardware synthesis from requirement specifications. In *EURO-DAC'96 with EURO-VHDL'96*. IEEE Computer Society Press, September 1996.

13. K. Fisler. *A Unified Approach to Hardware Verification Through a Heterogeneous Logic of Design Diagrams*. PhD thesis, Computer Science Department, Indiana University, August 1996.

14. K. Fisler. Containment of Regular Languages in Non-Regular Timing Diagrams Languages is Decidable. In *CAV*. Springer Verlag, 1997.

15. W. Grass, C. Grobe, S. Lenk, W. Tiedemann, C.D. Kloos, A. Marin, and T. Robles. Transformation of Timing Diagram Specifications into VHDL Code. In *Conference on Hardware Description Languages*, 1995.

16. J. Helbig, R. Schlör, W. Damm, G. Doehmen, and P. Kelb. VHDL/S - Integrating Statecharts, Timing diagrams, and VHDL. *Microprocessing and Microprogramming*, 1996.

17. F. Jin and E. Cerny. Verification of Real-Time Controllers against Timing Diagram Specifications using Constraint Logic Programming. In *IFIP EuroMICRO*, 1998.

18. K. Khordoc and E. Cerny. Semantics and Verification of Timing Diagrams with Linear Timing Constraints. *ACM Transactions on Design Automation of Electronic Systems*, 3(1), 1998.

19. R.P. Kurshan. *Computer-aided verification of coordinating processes: the Automata-theoretic approach*. Princeton University Press, 1994.

20. O. Lichtenstein and A. Pnueli. Checking that Finite State Concurrent Programs satisfy their Linear Specifications. In *POPL*, 1985.

21. Z. Manna and A. Pnueli. Specification and Verification of Concurrent Programs by ∀-Automata. In *POPL*, 1987.

22. J.P. Queille and J. Sifakis. Specification and Verification of Concurrent Systems in CESAR. In *Proc. of the 5th International Symposium on Programming*, volume 137 of *LNCS*, 1982.

23. Y.S.. Ramakrishna, P.M. Melliar-Smith, L.E. Moser, L.K. Dillon, and G. Kutty. Really Visual Temporal Reasoning. In *Real-Time Systems Symposium*. IEEE Publishers, 1993.
24. R. Schlör. A Prover for VHDL-based Hardware Design. In *Conference on Hardware Description Languages*, 1995.
25. A.P. Sistla, M. Vardi, and P. Wolper. The Complementation Problem for Büchi Automata with Applications to Temporal Logic. *TCS*, 49, 1987.
26. E.M. Thurner. Proving System Properties by means of Trigger-Graph and Petri Nets. In *EUROCAST*. Springer Verlag, 1996.
27. W.D. Tiedemann. Bus Protocol Conversion: from Timing Diagrams to State Machines. In *EUROCAST*. Springer Verlag, 1992.
28. M. Vardi. Verification of Concurrent Programs. In *POPL*, 1987.
29. M. Vardi and P.Wolper. An Automata-Theoretic approach to Automatic Program Verification. In *LICS*, 1986.

Vacuity Detection in Temporal Model Checking

Orna Kupferman[1] and Moshe Y. Vardi[2]*

[1] Hebrew University, The institute of Computer Science, Jerusalem 91904, Israel
Email: orna@cs.huji.ac.il, URL: http://www.cs.huji.ac.il/~orna
[2] Rice University, Department of Computer Science, Houston, TX 77251-1892, U.S.A.
Email: vardi@cs.rice.edu, URL: http://www.cs.rice.edu/~vardi

Abstract. One of the advantages of temporal-logic model-checking tools is their ability to accompany a negative answer to the correctness query by a counterexample to the satisfaction of the specification in the system. On the other hand, when the answer to the correctness query is positive, most model-checking tools provide no witness for the satisfaction of the specification. In the last few years there has been growing awareness to the importance of suspecting the system or the specification of containing an error also in the case model checking succeeds. The main justification of such suspects are possible errors in the modeling of the system or of the specification. Many such errors can be detected by further automatic reasoning about the system and the environment. In particular, Beer et al. described a method for the detection of *vacuous satisfaction* of temporal logic specifications and the generation of *interesting witnesses* for the satisfaction of specifications.

For example, verifying a system with respect to the specification $\varphi = AG(req \rightarrow AF grant)$ ("every request is eventually followed by a grant"), we say that φ is satisfied vacuously in systems in which requests are never sent. An interesting witness for the satisfaction of φ is then a computation that satisfies φ and contains a request. Beer et al. considered only specifications of a limited fragment of ACTL, and with a restricted interpretation of vacuity. In this paper we present a general method for detection of vacuity and generation of interesting witnesses for specifications in CTL^*. Our definition of vacuity is stronger, in the sense that we check whether all the subformulas of the specification affect its truth value in the system. In addition, we study the advantages and disadvantages of alternative definitions of vacuity, study the problem of generating linear witnesses and counterexamples for branching temporal logic specifications, and analyze the complexity of the problem.

1 Introduction

Temporal logics, which are modal logics geared towards the description of the temporal ordering of events, have been adopted as a powerful tool for specifying and verifying concurrent systems [Pnu81]. One of the most significant developments in this area is the discovery of algorithmic methods for verifying temporal-logic properties of *finite-state* systems [CE81,CES86,LP85,QS81,VW86a]. This derives its significance both from the

---

* Supported in part by the NSF grants CCR-9628400 and CCR-9700061, and by a grant from the Intel Corporation.

fact that many synchronization and communication protocols can be modeled as finite-state systems, as well as from the great ease of use of fully algorithmic methods. In temporal-logic *model checking*, we verify the correctness of a finite-state system with respect to a desired behavior by checking whether a labeled state-transition graph that models the system satisfies a temporal logic formula that specifies this behavior (for a survey, see [CGL93]).

Beyond being fully-automatic, an additional attraction of model-checking tools is their ability to accompany a negative answer to the correctness query by a counterexample to the satisfaction of the specification in the system. Thus, together with a negative answer, the model checker returns some erroneous execution of the system. These counterexamples are very important and they can be essential in detecting subtle errors in complex designs [CGMZ95]. On the other hand, when the answer to the correctness query is positive, most model-checking tools provide no witness for the satisfaction of the specification in the system. Since a positive answer means that the system is correct with respect to the specification, this at first seems like a reasonable policy. In the last few years, however, there has been growing awareness to the importance of suspecting the system (or the specification) of containing an error also in the case model checking succeeds. The main justification of such suspects are possible errors in the modeling of the system or of the behavior.

Early work on "suspecting a positive answer" concerns the fact that temporal logic formulas can suffer from antecedent failure [BB94]. For example, verifying a system with respect to the specification $\varphi = AG(req \rightarrow AF\,grant)$ ("every request is eventually followed by a grant"), one should distinguish between *vacuous satisfaction* of φ, which is immediate in systems in which requests are never sent, and non-vacuous satisfaction, in which φ's precondition is sometimes satisfied. Evidently, vacuous satisfaction suggests some unexpected properties of the system, namely the absence of behaviors in which the precondition was expected to be satisfied.

Several years of experience in practical formal verification have convinced the verification group in IBM Haifa Research Laboratory that vacuity is a serious problem [BBER97]. To quote from [BBER97]: "Our experience has shown that typically 20% of specifications pass vacuously during the first formal-verification runs of a new hardware design, and that vacuous passes always point to a real problem in either the design or its specification or environment." Often, it is possible to detect vacuity easily, by checking the system with respect to hand-written formulas that ensure the satisfaction of the preconditions in the specification [BB94,PP95]. Formally, we say that a formula φ' is a *witness formula* for the specification φ if a system M satisfies φ non-vacuously iff M satisfies both φ and φ'.[1] In the example above, it is not hard to see that a system satisfies φ non-vacuously iff it also satisfies $EF\,req$. Sometimes, however, the generation of witness formulas is not trivial, especially when we are interested in other types of vacuity passes, more involved than antecedent failure.

These observations led Beer et al. to develop a method for automatic generation of witness formulas [BBER97]. Witness formulas are then used for two tasks. First, for the original task of detecting vacuity, and second, for the generation of an *interesting*

---

[1] The notion of a witness formula that we use here is dual to the one used in [BBER97]. There, a system M satisfies φ vacuously iff M satisfies both φ and its witness φ'.

witness for the satisfaction of the specification in the system. A witness for the satisfaction of a specification in a system is a sub-system, usually a computation, that satisfies the specification. A witness is interesting if it satisfies the specification non-vacuously. For example, a computation in which both *req* and *grant* hold is an interesting witness for the satisfaction of φ above. An interesting witness gives the user a confirmation that his specification models correctly the desired behavior, and enables the user to study some nontrivial executions of the system. In order to generate an interesting witness for the specification φ, one only has to find a (not necessarily interesting) witness for the conjunction $\varphi \wedge \varphi'$ of the specification and its witness formula. This can be done using the counterexample mechanism of model-checking tools. Indeed, a computation is a witness for $\varphi \wedge \varphi'$ iff it is a counterexample to $\neg(\varphi \wedge \varphi')$.

While [BBER97] nicely set the basis for a methodology for detecting vacuity in temporal-logic specifications, the particular method described in [BBER97] is quite limited. The type of vacuity passes handled is indeed richer than antecedent failure, yet it is still very restricted. Beer et al. consider the subset w-ACTL of the universal fragment ACTL of CTL. The logic w-ACTL consists of all ACTL formulas in which all the (Boolean or temporal) binary operators have at least one operand that is a propositional formula. Many natural specifications cannot be expressed in w-ACTL. Beyond specifications that contain existential requirements, like $AGEF\,grant$ (and thus cannot be expressed in ACTL), this includes also universal specifications like $AG(AX\,grant \vee AX\neg grant)$, which ensures that the granting event do not distinguish between "brothers" (different successors of the same state) in the system, as we expect in systems with delayed updates (that is, when the reaction of the system to events occurs only in the successors of the position in which the event has occurred). The syntax of w-ACTL enables [BBER97] to associate with each specification, a single subformula (called *important subformula*) and the vacuity of passes of the specifications is then checked only with respect to this subformula. For example, in formulas like $AG(req \rightarrow AF\,grant)$, the algorithm in [BBER97] checks that *req* eventually holds in some path, yet it ignores the cases where $AF\,grant$ always holds. While, as claimed in [BBER97], the latter case is less interesting, it can still help in many scenarios. The restricted syntax of w-ACTL and the restriction to important subformulas led to efficient algorithms for detection of vacuity and generation of interesting witnesses.

In this paper we present a general method for detection of vacuity and generation of interesting witnesses for specifications in CTL* (and hence also LTL). Beyond the extension of the method in [BBER97] to highly expressive specification languages, our definition of vacuity is stronger, in the sense that we check whether all the subformulas of the specification affect its truth value in the system. In addition, we study the advantages and disadvantages of alternative definitions to vacuity, study the problem of generating linear witnesses and counterexamples for branching temporal logic specifications, and analyze the complexity of the problem.

From a computational point of view, we show that deciding whether a formula φ passes vacuously in a system M can be checked in time $O(C_M(|\varphi|) \cdot |\varphi|)$, where $C_M(|\varphi|)$ is the complexity of the model-checking problem for M and φ. Then, for φ in both LTL and CTL*, the problem of generating an interesting witness for φ in M (or deciding that no such witness exists) is PSPACE-complete. Both algorithms can be implemented symbolically on top of model checking tools like SMV and VIS. As explained in Section 4, part of the difficulty in generating an interesting witness comes from the fact that we insist on linear witnesses. When we consider worst-case complexity, the algorithm for generating interesting witnesses in [BBER97] is more efficient than ours (even when applied to w-ACTL formulas). Nevertheless, as explained in Section 4, for natural formulas, the performance of the algorithms coincides.

2 Temporal Logic

The logic *LTL* is a linear temporal logic. Formulas of LTL are built from a set AP of atomic proposition using the usual Boolean operators and the temporal operators X ("next time"), U ("until"), and $\tilde{U}$ ("duality of until"). We present here a positive normal form in which negation may be applied only to atomic propositions. Given a set AP, an LTL formula is defined as follows:

- **true, false**, p, or $\neg p$, for $p \in AP$.
- $\psi \vee \varphi$, $\psi \wedge \varphi$, $X\psi$, $\psi U \varphi$, or $\psi \tilde{U} \varphi$, where ψ and φ are LTL formulas.

We define the semantics of LTL with respect to a *computation* $\pi = \sigma_0, \sigma_1, \sigma_2, \ldots$, where for every $j \geq 0$, we have that σ_j is a subset of AP, denoting the set of atomic propositions that hold in the j'th position of π. We denote the suffix $\sigma_j, \sigma_{j+1}, \ldots$ of π by π^j. We use $\pi \models \psi$ to indicate that an LTL formula ψ holds in the path π. The relation $\models$ is inductively defined as follows:

- For all π, we have that $\pi \models$ **true** and $\pi \not\models$ **false**.
- For an atomic proposition $p \in AP$, we have $\pi \models p$ iff $p \in \sigma_0$ and $\pi \models \neg p$ iff $p \notin \sigma_0$.
- $\pi \models \psi \vee \varphi$ iff $\pi \models \psi$ or $\pi \models \varphi$.
- $\pi \models \psi \wedge \varphi$ iff $\pi \models \psi$ and $\pi \models \varphi$.
- $\pi \models X\psi$ iff $\pi^1 \models \psi$.
- $\pi \models \psi U \varphi$ iff there exists $k \geq 0$ such that $\pi^k \models \varphi$ and $\pi^i \models \psi$ for all $0 \leq i < k$.
- $\pi \models \psi \tilde{U} \varphi$ iff for every $k \geq 0$ for which $\pi^k \not\models \varphi$, there exists $0 \leq i < k$ such that $\pi^i \models \psi$.

We use the following abbreviations in writing formulas:

- $F\varphi = \text{true} U \varphi$ ("eventually").
- $G\varphi = \text{false} \tilde{U} \varphi$ ("always").

The logic CTL* is a branching temporal logic. A path quantifier, E ("for some path") or A ("for all paths"), can prefix an assertion composed of an arbitrary combination of linear time operators. There are two types of formulas in CTL*: *state formulas*, whose satisfaction is related to a specific state, and *path formulas*, whose satisfaction is related to a specific path. Formally, let AP be a set of atomic proposition names. A CTL* state formula (again, in a positive normal form) is either:

- **true, false**, p or $\neg p$, for $p \in AP$.

- $\psi \vee \varphi$ or $\psi \wedge \varphi$ where ψ and φ are CTL* state formulas.
- $E\psi$ or $A\psi$, where ψ is a CTL* path formula.

A CTL* path formula is either:

- A CTL* state formula.
- $\psi \vee \varphi$, $\psi \wedge \varphi$, $X\psi$, $\psi U \varphi$, or $\psi \tilde{U} \varphi$, where ψ and φ are CTL* path formulas.

The logic CTL* consists of the set of state formulas generated by the above rules. The logic *CTL* is a restricted subset of CTL*. In CTL, the temporal operators X, U, and $\tilde{U}$ must be immediately preceded by a path quantifier. Formally, it is the subset of CTL* obtained by restricting the path formulas to be $X\psi$, $\psi U \varphi$, or $\psi \tilde{U} \varphi$, where ψ and φ are CTL state formulas.

We denote the length of a formula φ by $|\varphi|$. When we consider subformulas of an LTL formula ψ, we refer to the syntactic subformulas of ψ, thus to path formulas. On the other hand, when we consider subformulas of a CTL* formula ψ, we refer to the state subformulas of ψ. Then, the *closure* $cl(\psi)$ of an LTL or a CTL* formula ψ is the set of all subformulas of ψ (including ψ but excluding **true** and **false**). For example, $cl(pU(Xq)) = \{pU(Xq), p, Xq, q\}$, and $cl(E(pU(AXq))) = \{E(pU(AXq)), p, AXq, q\}$. It is easy to see that the size of $cl(\psi)$ is linear in the size of ψ. We use $\varphi[\psi \leftarrow \xi]$ to denote the formula obtained from φ by replacing its subformula ψ by the formula ξ.

We define the semantics of CTL* (and its sublanguage CTL) with respect to *systems*. A system $M = \langle AP, W, R, w_0, L \rangle$ consists of a set AP of atomic propositions, a set W of states, a total transition relation $R \subseteq W \times W$, an initial state $w_0 \in W$, and a labeling function $L : W \to 2^{AP}$. A *computation* of a system is a sequence of states, $\pi = w_0, w_1, \ldots$ such that for every $i \geq 0$, we have that $\langle w_i, w_{i+1} \rangle \in R$. We define the *size* $|M|$, of a system M as above as $|W| + |R|$. We use $w \models \varphi$ to indicate that a state formula φ holds at state w (assuming an agreed fair module M). The relation $\models$ is inductively defined as follows (the relation $\pi \models \psi$ for a path formula ψ is the same as for ψ in LTL).

- For all w, we have that $w \models$ **true** and $w \not\models$ **false**.
- For an atomic proposition $p \in AP$, we have $w \models p$ iff $p \in L(w)$ and $w \models \neg p$ iff $p \notin L(w)$.
- $w \models \psi \vee \varphi$ iff $w \models \psi$ or $w \models \varphi$.
- $w \models \psi \wedge \varphi$ iff $w \models \psi$ and $w \models \varphi$.
- $w \models E\psi$ iff there exists a computation $\pi = w_0, w_1, \ldots$ such that $w_0 = w$ and $\pi \models \psi$.
- $w \models A\psi$ iff for all computations $\pi = w_0, w_1, \ldots$ such that $w_0 = w$, we have $\pi \models \psi$.
- $\pi \models \varphi$ for a computation $\pi = w_0, w_1, \ldots$ and a state formula φ iff $w_0 \models \varphi$.

A system M satisfies a formula φ iff φ holds in the initial state of M. The problem of determining whether M satisfies φ is the *model-checking* problem. For a particular temporal logic, a system M, and an integer n, we use $C_M(n)$ to denote the complexity of checking whether a formula of size n in the logic is satisfied in M.

3 Satisfying a Formula Vacuously

Intuitively, a system M satisfies a formula φ vacuously if M satisfies φ yet it does so in a non-interesting way, which is likely to point on some trouble with either M or φ. For example, a system in which *req* never occurs satisfies $AG(req \to AF\,grant)$ vacuously. In order to formalize this intuition, it is suggested in [BBER97] to use the following definition of when a subformula of φ affects its truth value in M.

Definition 1. [BBER97] *The subformula ψ of φ does not affect the truth value of φ in M (ψ does not affect φ in M, for short) if for every formula ξ, the system M satisfies $\varphi[\psi \leftarrow \xi]$ iff M satisfies φ.*

Note that one can talk about a subformula ψ affecting φ in M or about an *occurrence* of ψ affecting φ in M. As we shall see in Section 3.2, dealing with occurrences is much easier than dealing with subformulas. In the sequel, we assume for simplicity that all the subformulas of φ have single occurrences. (Equivalently, we could change the definition to talk about when a particular occurrence of ψ does not affect φ. All the results in the paper hold also for this alternative.)

As stated, Definition 1 is not effective, since it requires evaluation $\varphi[\psi \leftarrow \xi]$ for all formulas ξ. To deal with this difficulty, [BBER97] considers only a small class, called w-ACTL, of branching temporal logic formulas. In Theorem 1 bellow, we show that instead of checking the replacement of ψ by all formulas ξ, one can check only the replacement of ψ by the formulas **true** and **false**. For that, we first partition the subformulas of φ according to their *polarity* as follows. Every subformula ψ of φ may be either *positive in φ*, in the case it is in the scope of an even number of negations, or *negative in φ*, in the case it is in the scope of an odd number of negations (note that an antecedent of an implication is considered to be under negation)[2]. For a formula φ and a subformula ψ of φ, let $\varphi[\psi \leftarrow \bot]$ denote the formula obtained from φ by replacing ψ by **false**, in case ψ is positive in φ, and replacing ψ by **true**, in case ψ is negative in φ. Dually, $\varphi[\psi \leftarrow \top]$ replaces a positive ψ by **true** and replaces a negative ψ by **false**. We say that a Boolean or temporal operator $f(\xi_1)$ or $g(\xi_1, \xi_2)$ of temporal logic is *positively monotonic* if for every ξ_1 and ξ_2 we have that $f(\xi_1) \to f(\textbf{true})$ and $g(\xi_1, \xi_2) \to g(\textbf{true}, \xi_2) \wedge g(\xi_1, \textbf{true})$. Dually, the operator is *negatively monotonic* if for every ξ_1 and ξ_2 we have that $f(\xi_1) \to f(\textbf{false})$ and $g(\xi_1, \xi_2) \to g(\textbf{false}, \xi_2) \wedge g(\xi_1, \textbf{false})$. Since all the operators in $\text{CTL}^\star$ are positively monotonic, except for $\neg$, which is negatively monotonic, the following properties of positive and negative subformulas of φ can be easily proved by an induction on the structure of φ.

Lemma 1. *For a subformula ψ of φ and for every system M, if $M \models \varphi[\psi \leftarrow \bot]$, then for every formula ξ, we have $M \models \varphi[\psi \leftarrow \xi]$. Also, if $M \not\models \varphi[\psi \leftarrow \top]$, then for every formula ξ, we have $M \not\models \varphi[\psi \leftarrow \xi]$.*

---

[2] If we assume that the formula φ is given in a positive normal form, all the subformulas of φ, except maybe some propositions, are positive in φ. Since an assertion $\neg p$, for a proposition p, does not affect φ in M iff p does not affect φ in M, we can regard such assertions as atomic propositions, thus assume that all subformulas are positive in φ. In this section, however, we consider also formulas that are not in positive normal form, thus we refer to both positive and negative subformulas.

Lemma 1 implies that **true** and **false** are two "extreme" replacements for ψ in φ; thus instead of checking agreement on the satisfaction of φ with all replacements ξ, one may only consider these two extreme replacements. Theorem 1 below formalizes this intuition.

Theorem 1. *For every formula* φ, *a subformula* ψ *of* φ, *and a system* M, *the following are equivalent:*

(1) ψ *does not affect* φ *in* M.
(2) M *satisfies* $\varphi[\psi \leftarrow \textbf{true}]$ *iff* M *satisfies* $\varphi[\psi \leftarrow \textbf{false}]$.

Proof. Assume first that ψ does not affect φ in M. Then, in particular, $M \models \varphi[\psi \leftarrow \textbf{true}]$ iff $M \models \varphi$, and $M \models \varphi[\psi \leftarrow \textbf{false}]$ iff $M \models \varphi$. It follows that $M \models \varphi[\psi \leftarrow \textbf{true}]$ iff $M \models \varphi[\psi \leftarrow \textbf{false}]$. For the other direction, assume that M satisfies $\varphi[\psi \leftarrow \textbf{true}]$ iff M satisfies $\varphi[\psi \leftarrow \textbf{false}]$. Consider first the case ψ is positive in φ. We distinguish between two cases. First, if M satisfies $\varphi[\psi \leftarrow \textbf{false}]$, then, as ψ is positive in φ, it follows from Lemma 1 that for every formula ξ, we have $M \models \varphi[\psi \leftarrow \xi]$, and in particular $M \models \varphi$. Thus, ψ does not affect φ in M. Now, if M does not satisfy $\varphi[\psi \leftarrow \textbf{false}]$, we have that M does not satisfy $\varphi[\psi \leftarrow \textbf{true}]$ either. Then, as ψ is positive in φ, it follows from Lemma 1 that for every formula ξ, we have $M \not\models \varphi[\psi \leftarrow \xi]$, and in particular $M \not\models \varphi$. Thus, ψ does not affect φ in M. The case ψ is negative in φ is dual.

We can now define formally the notion of vacuous satisfaction:

Definition 2. [BBER97] *A system* M *satisfies a formula* φ *vacuously iff* $M \models \varphi$ *and there is some subformula* ψ *of* φ *such that* ψ *does not affect* φ *in* M.

Theorem 1 reduces the problem of vacuity detection to the problem of model checking of M with respect to the formulas $\varphi[\psi \leftarrow \textbf{true}]$ and $\varphi[\psi \leftarrow \textbf{false}]$ for all subformulas ψ of φ. In fact, by Lemma 1, whenever M satisfies φ, it also satisfies $\varphi[\psi \leftarrow \top]$ for all subformulas ψ of φ. Accordingly, M satisfies φ vacuously if $M \models \varphi$ and there is some subformula ψ of φ such that M satisfies $\varphi[\psi \leftarrow \bot]$. Since the number of subformulas of φ is bounded by $|\varphi|$, it follows that vacuity detection involves model checking of M with respect to at most $|\varphi|$ formulas, all smaller than φ. Hence the following theorem.

Theorem 2. *The problem of checking whether a system* M *satisfies a formula* φ *vacuously can be solved in time* $O(C_M(|\varphi|) \cdot |\varphi|)$.

3.1 Alternative definitions

In Definition 1, we require that for every ξ, the replacement of ψ by ξ does not affect the value of φ in M. One can also think about an alternative definition in which ψ does not affect φ in M if M satisfies φ iff for every formula ξ, we have that M satisfies $\varphi[\psi \leftarrow \xi]$. This alternative definition seems equivalent to Definition 1. Nevertheless, as we show below, the definitions are not equivalent and only Definition 1 agrees with our intuitive understanding of affecting a truth value. To see this, consider a system M

that has a single state with a self loop, in which p does not hold. Let $\varphi = \psi = p$. By the definition above, the formula ψ does not affect φ in M. Indeed, both sides of the iff condition in that definition do not hold: M does not satisfy p, and there is a formula ξ ($\xi = $ **false**) such that M does not satisfy $p[p \leftarrow \xi]$. Nevertheless, our intuition is that p does affect the truth value of p in M. This agrees with Definition 1. Indeed, there is a formula ξ ($\xi = $ **true**) such that $M \models p[p \leftarrow \xi]$ yet $M \not\models p$.

Note that the definition of when a system satisfies a formula vacuously is insensitive to the difference between the two definitions. Indeed, when $M \models \varphi$, both definitions require M to satisfy $\varphi[\psi \leftarrow \xi]$ for all ξ.

While Definition 1 cares about the satisfaction of φ in the initial state of M, and thus corresponds to *local* model checking, *global* model-checking algorithms calculate the set of all states that satisfy φ. Accordingly, if we use $M(\psi)$ to denote the set of states in M that satisfy ψ, one could also consider the alternative definition where ψ does not affect φ in M if for every formula ξ, $M(\varphi[\psi \leftarrow \xi]) = M(\varphi)$. The problem of this definition is that the replacement of ψ in ξ may change the set of states that satisfy φ in some non-interesting way, say with respect to non-reachable states. For example, consider a system M with one reachable state s with a self-loop, labeled $\{q\}$, and one non-reachable state s' with a self-loop, labeled $\emptyset$. The state s satisfies both $\varphi = AG(p \rightarrow q)$ and AGq. Thus, p does not affect φ in M according to Definition 1. On the other hand, while $M(\varphi) = \{s, s'\}$, we have that $M(AGq) = \{s\}$. Thus, p affects φ in M according to the global definition above. Since s' is not reachable, the fact that s' satisfies φ vacuously is not of real interest, thus we believe that the fact Definition 1 ignores such vacuous satisfaction meets our goals. It is easy, however, to extend all the results in the paper to handle also global vacuity. In particular, the corresponding variant of Lemma 1, namely $M(\varphi[\psi \leftarrow $ **false**$]) \subseteq M(\varphi) \subseteq M(\varphi[\psi \leftarrow $ **true**$])$ is valid for all ψ, thus global vacuous satisfaction of φ in M (and there are in fact several possible definitions here as well), can be detected in time $O(C_M(|\varphi|) \cdot |\varphi|)$.

3.2 Occurrences vs. subformulas

Recall that one can talk about a subformula ψ affecting φ in M or about an occurrence of ψ affecting φ in M. As we now show, the latter choice is computationally easier. Caring about whether a particular occurrence of ψ affects the value of φ in M, we assumed, for technical convenience, that all subformulas occur only once. Given ψ, φ, and M, Theorem 1 then suggests a simple solution for the problem of deciding whether ψ affects φ in M. Formally, the problem can be solved in time $O(C_M(|\varphi|))$. In particular, when φ is in CTL, the problem can be solved in time linear in M and φ [CES86]. When ψ has several occurrences, Theorem 1 is no longer valid. This is because different occurrences of ψ may have different polarities. We now show that in this case the problem of deciding whether ψ affects φ in M is most likely harder.

We say that ψ *affects* φ in M iff it is not the case that ψ does not affect φ in M. Thus, ψ affects φ in M iff there is a formula ξ such that either $M \models \varphi[\psi \leftarrow \xi]$ and $M \not\models \varphi$, or $M \not\models \varphi[\psi \leftarrow \xi]$ and $M \models \varphi$.

Theorem 3. *For φ in CTL, a subformula ψ of φ with multiple occurrences, and a system M, the problem of deciding whether ψ does not affect φ in M is co-NP-complete.*

Proof. We show that the complementary problem, of deciding whether ψ affects φ in M is NP-complete. To prove membership in NP, we first claim that if there is a formula ξ such that M does not agree on the satisfaction of φ and of $\varphi[\psi \leftarrow \xi]$, then there also exists such a formula ξ' of length $|M|$. Membership in NP then follows from fact that we can guess the formula ξ' above and check whether $M \models \varphi$ iff $M \models \varphi[\psi \leftarrow \xi']$. To prove hardness in NP, we do a reduction from SAT. Given $n \geq 0$, we define the Kripke structure $K_n = \langle \{q, r\}, \{0, \ldots, n+1\}, R, 0, L \rangle$, where $R = \{\langle 0, 1 \rangle, \langle 1, 2 \rangle, \ldots, \langle n, n+1 \rangle, \langle n+1, n+1 \rangle\}$, and L maps all states $i \in \{0, \ldots, n-1\} \cup \{n+1\}$ to $\emptyset$ and maps the state n to $\{r\}$. Thus, K_n is a chain of $n+2$ states none of which satisfies q, and only the state before the last one satisfies r. Giving a propositional formula θ over $p_0, \ldots, p_{n-1}$, let ψ be the CTL formula obtained from θ by replacing each occurrence of p_i by $(EX)^i q$. Then, let $\varphi = \psi \wedge (EX)^n q$. For example, if $\theta = (p_0 \vee p_1) \wedge (\neg p_1 \vee p_2)$, then $\varphi = (q \vee EXq) \wedge (\neg EXq \vee EXEXq) \wedge EXEXEXq$. Since no state of K_n satisfies q, the structure K_n does not satisfy φ. On the other hand, since, no matter what θ is, the only requirement that φ induces on the state n is to satisfy q, it is easy to see that there is a formula ξ such that $K_n \models \varphi[q \leftarrow \xi]$ iff θ is satisfiable: the formula ξ is induced by a satisfying assignment for θ and it holds at state i iff $i = n$ or p_i is assigned *true* in the satisfying assignment. Using the fact that n is the only state in which r holds, we can indeed "translate" each assignment to a corresponding ξ. In the example above, an assignment that assigns *true* to p_0 and p_2 induces the formula $\xi = r \vee EX(r \vee EXEXr)$. It follows that q affects φ in K_n iff θ is satisfiable.

4 Interesting Witnesses

When a good model-checking tool decides that a system M does not satisfy a required property φ, it returns a counterexample to the satisfaction of φ, namely, some erroneous execution of M. These counterexamples are very important and they help the user to detect problems in M or in φ. Most model-checking tools, however, provide no witness for the satisfaction of φ in M. Such a witness may be very helpful too, in particular when it describes an execution in which the formula is satisfied in an interesting way. In this section we discuss the generation of *interesting witnesses* to the satisfaction of LTL and CTL* formulas.

Definition 3. [BBER97] *Consider a system M and a formula φ such that $M \models \varphi$. A path π of M is an interesting witness for φ in M if π satisfies φ non-vacuously.*

The generation of an interesting witness involves two difficulties. The first is present in the case φ is a branching temporal logic formula and it involves the generation of a linear (rather than a branching) witness. This difficulty is analogous to the difficulty of constructing a linear counterexample in a system that violates a branching temporal logic formula. The second difficulty is present also when φ is a linear temporal logic formula and it involves the fact that all the subformulas of φ should affect the satisfaction of φ in the witness. Note that even when M satisfies φ non-vacuously, it may be that some paths of M satisfy φ vacuously. For example, a structure that satisfies $AG(req \rightarrow AF\,grant)$ non-vacuously may contain a path in which req never holds. Moreover, it may be that M satisfies φ non-vacuously, all the paths of M satisfy φ as well, yet no path of M is an interesting witness for φ. As an example, consider the formula above and a structure with two paths, one path that never satisfies req and a second path that always satisfies $grant$. To see another weakness of the definition of an interesting witness, consider the LTL formula $\varphi = G(req_1 \rightarrow F\,grant_1) \wedge G(req_2 \rightarrow grant_2)$. While a system M may satisfy φ non-vacuously and contain interesting witnesses for both $G(req_1 \rightarrow F\,grant_1)$ and $G(req_2 \rightarrow grant_2)$, the system M may not contain an interesting witness for φ, as both req_1 and req_2 are required to hold in such a witness. This difficulty arises since φ is a conjunction of two specifications, and it can be avoided by separating conjunctions to their conjuncts.

We start with the first difficulty. We say that a branching temporal logic formula φ is *linearly witnessable* if for every system M, if $M \models \varphi$ then M has a path π such that $\pi \models \varphi$. The following lemma follows immediately from the definition.

Lemma 2. *All formulas of the universal fragment* $ACTL^\star$ *of* $CTL^\star$ *are linearly witnessable, and so are all* $CTL^\star$ *formulas with a single existential path quantifier.*

It follows from Lemma 2 that if a formula has no existential path quantifiers, or has a single path quantifier, then it is linearly witnessable. This syntactic condition is a sufficient but not a necessary one. For example, the CTL formula $EXEFp$ is linearly witnessable, and so is the less natural formula $EXp \vee EX\neg p$. The latter example suggests that testing a formula for being linearly witnessable is at least as hard as the validity problem.

Theorem 4. *Given a CTL formula* φ, *deciding whether* φ *is linearly witnessable is in 2EXPTIME and is EXPTIME-hard.*

Proof. We start with the upper bound. We first claim that if there is a system M such that $M \models \varphi$ yet M has no path π such that $\pi \models \varphi$, then there also exists such an M with branching degree bounded by $|\varphi|$. The proof of the claim is similar to the proof of the bounded-degree property for CTL [Eme90]. Give φ, let $\mathcal{A}_\varphi$ be a nondeterministic Büchi tree automaton that accepts exactly all trees of branching degree at most $|\varphi|$ that satisfy φ [VW86b], and let $\mathcal{A}'_\varphi$ be nondeterministic Büchi word automaton that accepts exactly all words (i.e., trees of branching degree 1) that satisfy φ [VW94]. We expand $\mathcal{A}'_\varphi$ to a Büchi tree automaton $\mathcal{A}''_\varphi$ that accepts a tree iff the tree has a path accepted by $\mathcal{A}'_\varphi$ (in each state, $\mathcal{A}''_\varphi$ guesses a direction in which it follows $\mathcal{A}'_\varphi$). We prove that φ is linearly witnessable iff $\mathcal{L}(\mathcal{A}_\varphi) \subseteq \mathcal{L}(\mathcal{A}''_\varphi)$. Since the containment problem $\mathcal{L}(\mathcal{A}) \subseteq \mathcal{L}(\mathcal{A}')$ for Büchi tree automata can be solved in time that is polynomial in the size of $\mathcal{A}$ and exponential in the size of $\mathcal{A}'$ [EJ88,MS95], the 2EXPTIME upper bound follows. Assume first that φ is linearly witnessable, and let T be a tree in $\mathcal{L}(\mathcal{A}_\varphi)$. Then, T contains a path π such that π satisfies φ, implying that π is accepted by $\mathcal{A}'_\varphi$. Then, however, by the definition of $\mathcal{A}''_\varphi$, we have that T is also in $\mathcal{L}(\mathcal{A}''_\varphi)$. Assume now that φ is not linearly witnessable. then, by the bounded-degree property above, there is a system, and therefore also a tree T of branching degree at most $|\varphi|$ such that $T \models \varphi$ yet no path of T satisfies φ. Hence, while the tree T is in $\mathcal{L}(\mathcal{A}_\varphi)$, it is not accepted by $\mathcal{A}''_\varphi$, implying that $\mathcal{L}(\mathcal{A}_\varphi)$ is not contained in $\mathcal{L}(\mathcal{A}''_\varphi)$.

For an EXPTIME lower bound, we do a reduction from the satisfiability problem for CTL. Consider a formula φ over a set of atomic propositions that does not contain p and

q. We prove that φ is not satisfiable iff $\psi = \varphi \wedge EXp \wedge EXq$ is linearly witnessable. Clearly, if φ is not satisfiable, then so is ψ, which is therefore linearly witnessable. For the second direction, assume that φ is satisfiable, and consider a system M that satisfies φ. We define a system M' as follows. If the initial state of M has two or more successors, we label one of its successors by p and we label a different successor by q. If the initial state of M has only one successor, we duplicate it, and then proceed as above. It is easy to see that while M' satisfies ψ, no path of M' satisfies ψ, thus ψ is not linearly witnessable.

The gap between the upper and lower bounds in Theorem 4 is similar to gaps in related problems such as the complexity of determining whether a CTL* formula has an equivalent LTL formula (a 2EXPTIME upper bound and an EXPTIME lower bound [KV98b]), the complexity of determining whether an LTL formula has an equivalent alternation-free μ-calculus formula (an EXPSPACE upper bound and a PSPACE lower bound [KV98a]), and several more problems. Essentially, in all the problems above we check the equivalence between a set of trees that satisfy $A\varphi$, for an LTL formula φ, and a set of trees that is defined directly by some branching-time formalism. The best known translation of $A\varphi$ to a tree automaton involves a doubly-exponential blow up. This is because the nondeterministic automaton for φ, whose size is exponential in $|\varphi|$, needs to be determinized before its expansion into a tree automaton, or, alternatively (as in the proof above), the nondeterministic tree automaton for $E\neg\varphi$ needs to be complemented. The doubly-exponential size of the tree automaton then leads to EXPSPACE and 2EXPTIME upper bounds. On the other hand, typical EXPSPACE and 2EXPTIME lower-bound proofs for temporal logic [VS85,KV95] require the use of temporal logic formulas that do not fit into the restricted syntax that is present in the problems above (e.g., formulas of the form $A\varphi^d \rightarrow \varphi$ for some CTL* formula φ).

The generation of interesting witnesses in [BBER97] goes through a search for a counterexample for a "witnessing formula". This generation succeeds only for witnesses formulas for which a linear counterexample exists. It is claimed in [BBER97] that almost all interesting CTL formulas indeed have linear counterexamples. We say that a branching temporal logic formula is *linearly counterable* iff for every system M, if $M \not\models \varphi$ then M has a path π such that $\pi \not\models \varphi$. The following theorem, which characterizes linearly counterable formulas, follows immediately from the definitions of linearly witnessable and linearly counterable.

Theorem 5. *For a branching temporal logic formula φ, we have that φ is linearly counterable iff $\neg\varphi$ is linearly witnessable.*

Note that a formula φ may be both linearly witnessable and linearly counterable (in which case $\neg\varphi$ is both linearly witnessable and linearly counterable as well). The formulas AGp and EFq, for example, fall in this category. In fact, by Lemma 2, all formulas with at most one universal and one existential path quantifiers are both linearly witnessable and linearly counterable.

In the context of model checking, however, a particular system M is given, and while φ may not be linearly witnessable, it may still have a linear witness in M. We say that φ is *linearly witnessable in M* if $M \models \varphi$ implies that M has a path π such that $\pi \models \varphi$. In order to check whether φ is linearly witnessable in M, we first need the following notation. For a branching temporal logic formula φ in a positive normal form, let φ^d be the LTL formula obtained from φ by eliminating its path quantifiers. For example, $(AGEFp)^d = GFp$. By [CD88], φ has an equivalent LTL formula iff φ is equivalent to $A\varphi^d$.

Theorem 6. *For a branching temporal logic formula φ and a system M, we have that $M \not\models A\varphi^d$ iff M has a path π such that $\pi \not\models \varphi$.*

Proof. Assume first that $M \not\models A\varphi^d$. Then, M has a path π such that $\pi \not\models \varphi^d$. Since the branching degree of π is 1, the path π does not satisfy φ either. For the other direction, assume that M has a path π such that $\pi \not\models \varphi$. Since the branching degree of π is 1, the path π does not satisfy φ^d either. Hence, $M \not\models A\varphi^d$.

Theorem 7. *For a CTL* formula φ and a system M, deciding whether φ is linearly witnessable in M is PSPACE-complete.*

Proof. Replacing the formula φ in Theorem 6 by the formula $\neg\varphi$, we get that $M \not\models A(\neg\varphi)^d$ iff M has a path π such that $\pi \models \varphi$. It follows that φ is linearly witnessable in M iff $M \models \varphi \rightarrow E\varphi^d$. Membership in PSPACE then follows from CTL* model-checking complexity [EL87]. Given a system M and an ACTL formula φ, it is shown in [KV98b] that the model-checking problem $M \models A\varphi^d \rightarrow \varphi$ is PSPACE-complete. Equivalently, given a system M and an ECTL formula φ, the model-checking problem $M \models \varphi \rightarrow E\varphi^d$ is PSPACE-complete. Since φ is linearly witnessable in M iff $M \models \varphi \rightarrow E\varphi^d$, hardness in PSPACE follows (in fact, already for φ in ECTL).

In practice, we are interested in generating a linear witness (and thus in the question of linear witnessability) only in systems M that satisfy φ. Note that the proof of Theorem 7 shows that deciding whether φ is linearly witnessable in M is PSPACE-complete already for M as above.

We now study the second difficulty: finding an interesting linear witness. Recall that the generation of interesting witnesses in [BBER97] goes through a search for a counterexample for a witnessing formula. The definition of the witnessing formula in [BBER97] crucially depends on the restricted syntax of w-ACTL. Below we generate a witnessing formula for general branching or linear temporal logic formulas. Given a formula φ (in either LTL or CTL*), we define

$$witness(\varphi) = \varphi \wedge \bigwedge_{\psi \in cl(\varphi)} \neg\varphi[\psi \leftarrow \bot].$$

Note that the length of $witness(\varphi)$ is quadratic in $|\varphi|$. Intuitively, a path π satisfies $witness(\varphi)$ if π satisfies φ and in addition, π does not satisfy the formula $\varphi[\psi \leftarrow \bot]$ for all the subformulas ψ of φ. Thus, all subformulas of φ affect its value in π.

Theorem 8. *For a formula φ and a system M, a counter example for $\neg witness(\varphi)$ in M is an interesting witness for φ in M.*

Proof. Let π be a counterexample for $\neg witness(\varphi)$ in M. Then, π satisfies $witness(\varphi)$. As such, π satisfies φ, yet for all subformulas ψ of φ, the path π does not satisfy the formula $\varphi[\psi \leftarrow \bot]$. It follows that all subformulas ψ of φ affect φ in π, hence π satisfies φ non-vacuously.

Theorem 9. *For an LTL or a* CTL* *formula* φ *and a system* M, *an interesting witness for* φ *in* M *can be generated in polynomial space. Deciding whether such a witness exists is PSPACE-complete.*

Proof. By Theorem 8, one can generate an interesting witness π for φ in M by generating a counterexample for $\neg witness(\varphi)$ in M. When φ is an LTL formula, so is $\neg witness(\varphi)$, and the generation of π can be done by generating a path in the intersection of M and a Büchi word automaton for $\neg witness(\varphi)$. Membership in PSPACE then follows from the fact that the automaton for an LTL formula ξ is of size exponential in ξ [VW94], and the generation of a path in the intersection of the automaton with M can be done on-the-fly and nondeterministically in space that is logarithmic in the sizes of M and the automaton. When φ is a CTL* formula, we know, as discussed in the proof of Theorem 6, that a counterexample in M for $A(\neg witness(\varphi))^d$ is also a counterexample for $\neg witness(\varphi)$ in M. Thus, the generation can proceed as in the case of LTL formulas, replacing $\neg witness(\varphi)$ by $(\neg witness(\varphi))^d$. In both cases, the lower bound follows by a reduction from LTL model checking [SC85].

The lower bound in Theorem 9 implies that the generation of interesting witnesses may require, at the worst case, space that is polynomial in the length of the specification, which in practice means that it may require time that is exponential in the length of the specification. On the other hand, the method in [BBER97] requires only linear time. The comparison of the two approaches from a complexity-theoretic point of view is actually a special case of the traditional comparison between LTL and CTL model-checking complexity. Indeed, while the generation in [BBER97] goes through the counterexample mechanism for CTL formulas [CGMZ95], ours go through the counterexample mechanism for LTL formulas, which uses an automata-theoretic reduction (exponential in the worst case) to CTL counterexample generation [VW86a]. Our experience with this comparison teaches us that, in practice, standard LTL model checkers perform nicely on most formulas. In fact, for formulas that can be expressed in both LTL and CTL, LTL model-checking tools often proceeds essentially as CTL model-checking tools. Intuitively, both model checkers proceed according to the semantics of the formula and are insensitive to the syntax in which the formula is given (for a detailed analysis and comparison of the two verification paradigms see [KV98b]). Experimental results of LTL and CTL model checking of common specifications support our observation and show no advantage to the branching paradigm [Cla97,BRS99]. In addition, once LTL model checking (or generation of counterexamples) is reduced to detection of a fair computation in the product of the system and the automaton for the negation of the specification, such a detection can be performed using CTL model-checking tools, thus our method can be implemented symbolically on top of model checkers such as SMV or VIS.

5   Discussion

We presented a general method for detection of vacuity and generation of interesting witnesses for specifications in CTL*. The results in the paper can be easily extended to handle systems with *fairness* conditions. A typical fairness condition for a system $M = \langle AP, W, R, w_0, L \rangle$ is a tuple $\langle F_1, \ldots, F_k \rangle$ of subsets of W. Such a condition means that we restrict attention to computations that visit each F_i infinitely often [Fra86]. It is known that model-checking algorithms extend to systems with such fairness conditions [CES86,VW86a]. Since our method is based on the model-checking algorithm, it can therefore be easily extended to handle fairness. Also, being based on the model-checking algorithm, our method is fully automatic, and all the common heuristics for coping with the state-explosion problem are applicable to it. As with model checking, the discouraging complexity bounds for the problems discussed in the paper do rarely appear in practice. An interesting open question is how to find interesting witnesses of minimal length (cf. [CGMZ95]).

Vacuity check is only one approach to challenge the correctness of the verification process. We mention here two recent related approaches. An approach that is closely related to vacuity is taken in the process of constraint validation in the verification tool FormalCheck [Kur98]. In order to validate a set of constraints about the environment, the constraints are converted into specifications and are checked with respect to a model of the environment. Sometimes, however, there is no model of the environment, and instead, FormalCheck proceeds with some heuristic sanity checks for constraint validation. This includes a search for enabling conditions that are never enabled, and a replacement of all or some of the constraints by **false**. A different approach is described in [KGG99], where the authors extend the notion of *coverage* from testing to model checking. Given a specification and its implementation, bisimulation is used in order to check whether the specification covers the entire functionality performed by the implementation. If the answer is negative, the specification is suspected for not being sufficiently restrictive.

Acknowledgments We thank Shoham Ben-David and Yaakov Crammer for helpful comments on a previous version of this work.

References

[BB94]   D. Beaty and R. Bryant. Formally verifying a microprocessor using a simulation methodology. In *Proc. 31st DAC*, pp. 596–602. IEEE Computer Society, 1994.

[BBER97]   I. Beer, S. Ben-David, C. Eisner, and Y. Rodeh. Efficient detection of vacuity in ACTL formulas. In *Proc. 9th CAV*, LNCS 1254, pp. 279–290, 1997.

[BRS99]   R. Bloem, K. Ravi, and F. Somenzi. Efficient decision prcedures for model checking of linear time logic properties. In *Proc. 11th CAV*, LNCS, 1999.

[CD88]   E.M. Clarke and I.A. Draghicescu. Expressibility results for linear-time and branching-time logics. In *Proc. Workshop on Linear Time, Branching Time, and Partial Order in Logics and Models for Concurrency, LNCS* 354, pp. 428–437, 1988.

[CE81]   E.M. Clarke and E.A. Emerson. Design and synthesis of synchronization skeletons using branching time temporal logic. In *Proc. Workshop on Logic of Programs, LNCS* 131, pp. 52–71, 1981.

[CES86]   E.M. Clarke, E.A. Emerson, and A.P. Sistla. Automatic verification of finite-state concurrent systems using temporal logic specifications. *ACM Transactions on Programming Languages and Systems*, 8(2):244–263, 1986.

[CGL93]   E.M. Clarke, O. Grumberg, and D. Long. Verification tools for finite-state concurrent systems. In *Decade of Concurrency – Reflections and Perspectives (Proceedings of REX School), LNCS* 803, pp. 124–175, 1993.

[CGMZ95] E.M. Clarke, O. Grumberg, K.L. McMillan, and X. Zhao. Efficient generation of counterexamples and witnesses in symbolic model checking. In *Proc. 32nd DAC*, pp. 427–432. IEEE Computer Society, 1995.

[Cla97] E. Clarke. Private communication, 1997.

[EJ88] E.A. Emerson and C. Jutla. The complexity of tree automata and logics of programs. In *Proc. 29th FOCS*, pp. 368–377, White Plains, 1988.

[EL87] E.A. Emerson and C.-L. Lei. Modalities for model checking: Branching time logic strikes back. *Science of Computer Programming*, 8:275–306, 1987.

[Eme90] E.A. Emerson. Temporal and modal logic. *Handbook of Theoretical Computer Science*, pp. 997–1072, 1990.

[Fra86] N. Francez. *Fairness*. Texts and Monographs in Computer Science. Springer-Verlag, 1986.

[KGG99] S. Katz, D. Geist, and O. Grumberg. "Have I written enough properties ?" a method of comparison between specification and implementation. In *10th CHARME*, LNCS, 1999.

[Kur98] R.P. Kurshan. *FormalCheck User's Manual*. Cadence Design, Inc., 1998.

[KV95] O. Kupferman and M.Y. Vardi. On the complexity of branching modular model checking. In *Proc. 6th CONCUR, LNCS* 962, pp. 408–422, 1995.

[KV98a] O. Kupferman and M.Y. Vardi. Freedom, weakness, and determinism: from linear-time to branching-time. In *Proc. 13th LICS*, pp. 81–92, 1998.

[KV98b] O. Kupferman and M.Y. Vardi. Relating linear and branching model checking. In *PROCOMET*, pp. 304 – 326, 1998. Chapman & Hall.

[LP85] O. Lichtenstein and A. Pnueli. Checking that finite state concurrent programs satisfy their linear specification. In *Proc. 12th POPL*, pp. 97–107, 1985.

[MS95] D.E. Muller and P.E. Schupp. Simulating alternating tree automata by nondeterministic automata: New results and new proofs of theorems of Rabin, McNaughton and Safra. *Theoretical Computer Science*, 141:69–107, 1995.

[Pnu81] A. Pnueli. The temporal semantics of concurrent programs. *Theoretical Computer Science*, 13:45–60, 1981.

[PP95] B. Plessier and C. Pixley. Formal verification of a commercial serial bus interface. In *Proc. of 14th Annual IEEE International Phoenix Conf. on Computers and Communications*, pp. 378–382, March 1995.

[QS81] J.P. Queille and J. Sifakis. Specification and verification of concurrent systems in Cesar. In *Proc. 5th International Symp. on Programming*, pp. 337–351, LNCS 137, 1981.

[SC85] A.P. Sistla and E.M. Clarke. The complexity of propositional linear temporal logic. *Journal ACM*, 32:733–749, 1985.

[VS85] M.Y. Vardi and L. Stockmeyer. Improved upper and lower bounds for modal logics of programs. In *Proc 17th STOC*, pp. 240–251, 1985.

[VW86a] M.Y. Vardi and P. Wolper. An automata-theoretic approach to automatic program verification. In *Proc. 1st LICS*, pp. 322–331, 1986.

[VW86b] M.Y. Vardi and P. Wolper. Automata-theoretic techniques for modal logics of programs. *Journal of Computer and System Science*, 32(2):182–221, April 1986.

[VW94] M.Y. Vardi and P. Wolper. Reasoning about infinite computations. *Information and Computation*, 115(1):1–37, November 1994.

Using Symbolic Model Checking to Verify the Railway Stations of Hoorn-Kersenboogerd and Heerhugowaard

Cindy Eisner

IBM Haifa Research Laboratory
Matam Advanced Technology Center
Haifa, 31905 Israel
eisner@il.ibm.com

Abstract. Stålmarck's proof procedure is a method of tautology checking that has been used to verify railway interlocking software. Recently, it has been proposed [SS98] that the method has potential to increase the capacity of formal verification tools for hardware. In this paper, we examine this potential in light of an experiment in the opposite direction: the application of symbolic model checking to railway interlocking software previously verified with Stålmarck's method. We show that these railway systems share important characteristics which distinguish them from most hardware designs, and that these differences raise some doubts about the applicability of Stålmarck's method to hardware verification.

1 Introduction

Stålmarck's proof procedure is a method of tautology checking that has been used to verify railway interlocking software [GKV94, Fok95]. Based on the observation that these systems are hardware-like, it has been suggested that this method may have the potential to increase the capacity of formal verification tools for hardware [SS98]. Indeed, Biere, Cimatti, Clarke and Zhu [BCCZ98] have built a symbolic model checker in which boolean decision procedures like Stålmarck's method replace BDDs.

In this paper, this potential is examined in light of an experiment in the opposite direction: the application of symbolic model checking to railway interlocking software[1]. It is shown that the two railway stations commonly cited as successful applications of Stålmarck's method are *robust*: most properties required of them hold for all states in the state space rather than holding for the reachable states only. It is also shown that these models exhibit *locality*, in that only a small number of inputs toggle in any one counter-example to a non-valid formula. Finally, we show that the properties checked for these models belong to a subset of CTL formulas we call AGAX formulas. We show how the characteristics of robustness and locality cause these models to be particularly suitable to symbolic model checking of AGAX formulas, and speculate that robustness also aids the application of Stålmarck's method to these types of formulas. Finally, we note that most hardware systems do not exhibit robustness, which raises doubts about the applicability of Stålmarck's method to hardware.

---

[1] The models of stations Hoorn-Kersenboogerd and Heerhugowaard used in this paper are the property of Holland RailConsult and are used with permission.

The remainder of this paper is structured as follows. Section 2 covers the basics of symbolic model checking. Section 3 describes the structure of railway interlocking software in the language VLC, and the structure of properties required of such software. Section 4 discusses the application of symbolic model checking to the verification of such software, and defines AGAX formulas, which are typical of those used to verify VLC models. Section 5 defines robustness and locality, analyzes why it is "easy" to model check AGAX formula in robust systems, and shows how locality can be used to aid counter-example generation for false formulas. Section 6 presents experimental results of the application of symbolic model checking to the stations at Hoorn-Kersenboogerd and Heerhugowaard. Section 7 concludes with some speculation regarding the application of Stålmarck's method to robust systems, and casts doubt on the applicability of Stålmarck's method to hardware verification.

2 Preliminaries

CTL, or Computation Tree Logic [CE81], is a temporal logic useful for reasoning about the ongoing behavior of reactive systems, and is the logic used by the symbolic model checker SMV [McM93]. In CTL, temporal operators occur in pairs consisting of A or E, followed by F, G, U, or X, as follows:

1. Every atomic proposition is a CTL formula, and
2. If f and g are CTL formulas, then so are $\neg f, (f \wedge g), AXf, EXf, A[fUg], E[fUg]$

The remaining operators are viewed as abbreviations of the above, as follows: $f \vee g = \neg(\neg f \wedge \neg g), AFg = A[trueUg], EFg = E[trueUg], AGf = \neg E[trueU\neg f]$ and $EGf = \neg A[trueU\neg f]$.

The semantics of a CTL formula is defined with respect to a model M. A model is a quadruple (S, S_0, R, L), where S is a finite set of states, $S_0 \subseteq S$ is a set of initial states, $R \subseteq S \times S$ is the transition relation, and L is the valuation, a function mapping each state with a set of atomic propositions true in that state. We require that there is at least one transition from every state. A computation path of a model M is an infinite sequence of states $(s_0, s_1, s_2, \cdots)$ such that $R(s_i, s_{i+1})$ is true for every i.

The notation $M, s \models f$ means that the formula f is true in state s of model M. The notation $M \models f$ is equivalent to $\forall s \in S_0 \ M, s \models f$. The semantics of a CTL formula is defined as follows:

$M, s \models p \Longleftrightarrow p \in L(s)$, where p is an atomic proposition
$M, s \models \neg f \Longleftrightarrow M, s \not\models f$
$M, s \models f \wedge g \Longleftrightarrow M, s \models f$ and $M, s \models g$
$M, s_i \models AX \ f \Longleftrightarrow$ for all paths $(s_i, s_{i+1}, ...), M, s_{i+1} \models f$
$M, s_i \models EX \ f \Longleftrightarrow$ for some path $(s_i, s_{i+1}, ...), M, s_{i+1} \models f$
$M, s_i \models A[fUg] \Longleftrightarrow$ for all paths $(s_i, s_{i+1}, ...), \exists k \geq i$ such that $M, s_k \models g$, and $\forall j$ such that $i \leq j < k, M, s_j \models f$
$M, s_i \models E[fUg] \Longleftrightarrow$ for some path $(s_i, s_{i+1}, ...), \exists k \geq i$ such that $M, s_k \models g$ and $\forall j$ such that $i \leq j < k, M, s_j \models f$

Emerson and Clarke [CE81b] have shown that the operators of CTL can be characterized as fixed points. This is the basis of CTL model checking. In symbolic CTL model checking, fixed points are frequently expensive to calculate, and their calculation is one of the sources of state space explosion. In the sequel, we will show a way to avoid this explosion for the railway stations of Hoorn-Kersenboogerd and Heerhugowaard.

3 Railway interlocking software in the language VLC

The language VLC is described in full by Groote, Koorn and van Vlijmen [GKV94]. Briefly, a VLC program describes a reactive system which continually executes control cycles. In each control cycle a set of inputs is latched, which means that their value cannot change during the control cycle. Then, the next state value of the internal variables and outputs is calculated. The outputs are transmitted simultaneously to the outside world at the end of the calculation. Finally, the internal variables and outputs are latched.

For the most part, a VLC program consists of a group of boolean equations describing the next state function of the internal variables and outputs, where '+' indicates boolean or, '*' indicates boolean and, and ".N." indicates boolean not. A time delay may be associated with a boolean equation, which means that the assignment is executed only if the right hand side of the equation has been true for the number of control cycles indicated by the time delay.

Following is an example given by [GKV94]:

```
1 DIRECT INPUT SECTION
2 I
3 OUTPUT SECTION
4 U
5 CODE SYSTEM SECTION
6 CURRENT RESULT SECTION
7 R
8 SELF-LATCHED PARAMETER SECTION
9 V
10 TIMER EXPRESSION RESULT SECTION
11 Q
12 BOOLEAN EQUATION SECTION
13 APPLICATION = Example
14 TIME DELAY = 2 SECONDS BOOL Q = I
15 BOOL R = Q + V
16 BOOL V = Q * .N.R
17 BOOL U = V
18 END BOOLEAN EQUATION SECTION
```

Fig. 1. An example VLC program

The boolean assignment to Q on line 14 in Figure 1 is time delayed by 2 seconds (where one second indicates one control cycle). This means that Q will get the value

$TRUE$ only if I is $TRUE$ and was also $TRUE$ in the previous 2 control cycles. The boolean assignment to R on line 15 will use the new value of Q when calculating $Q \lor R$, because it appears after the assignment to Q in the sequential order of the program statements. Similarly, the boolean assignment to V on line 16 will use the new values of Q and R to calculate $Q \land \neg R$ and the boolean assignment to U on line 17 will use the new value of V.

Properties verified by Groote et al [GKV94] and Fokkink [Fok95] are propositional formulas using current state variables and past state variables, as follows: if V is a current state variable, then V_{-j-1} indicates its value one cycle in the past, V_{-j-2} indicates its value 2 cycles in the past, etc.

4 The application of symbolic model checking to VLC code

In order to apply symbolic model checking to VLC code, it is necessary to translate VLC to the input language of the symbolic model checker. Here we describe the translation to the language EDL, a dialect of SMV [McM93] accepted by the symbolic model checker RuleBase [BBEL96].

The explanations which follow use the "Little Yard" of [GKV94] in VLC (Figure 2) and EDL (Figure 3).

```
1 DIRECT INPUT SECTION
2 I
3 OUTPUT SECTION
4 Pr Pn A B C
5 CODE SYSTEM SECTION
6 CmdA CmdB CmdC Cmdr
7 CURRENT RESULT SECTION
8 E
9 SELF-LATCHED PARAMETER SECTION
10 TIMER EXPRESSION RESULT SECTION
11 P
12 BOOLEAN EQUATION SECTION
13 APPLICATION = LY
14 BOOL E = A * B * C + A * B + A * C + B * C
15 TIME DELAY = 1 SECONDS BOOL P = I
16 BOOL Pr = .N.A * .N.B * .N.C * Cmdr
17 BOOL Pn = .N.Pr
18 BOOL A = CmdA * .N.CmdB * .N.CmdC * .N.E * P * Pn
19 BOOL B = CmdB * .N.CmdA * .N.CmdC * .N.E * P * Pr
20 BOOL C = CmdC * .N.CmdA * .N.CmdB * .N.E * P * Pn
21 END BOOLEAN EQUATION SECTION
```

Fig. 2. Program "Little Yard" in VLC

```
1 – inputs
2 var I, CmdA, CmdB, CmdC, Cmdr: boolean;
3 – next state variables
4 var Pr_out, Pn_out, A_out, B_out, C_out, E_out, P_out: boolean;
5 assign next(Pr_out) := Pr;
6 assign next(Pn_out) := Pn;
7 assign next(A_out) := A;
8 assign next(B_out) := B;
9 assign next(C_out) := C;
10 assign next(E_out) := E;
11 assign next(P_out) := P;
12 – counters
13 var P_count(0): boolean;
14 assign init(P_count(0)) := 0;
15 next(P_count(0)) := if !(P_temp) then 0 else P_count_inc(0) endif;
16 define P_count_inc(0) := if P_count(0)=1 then 1 else P_count(0)+1 endif;
17 – current state symbols
18 define E := A_out&B_out&C_out | A_out&B_out | A_out&C_out | B_out&C_out;
19 define P := P_temp & (P_count(0)=1);
20 define P_temp := I;
21 define Pr := !A_out & !B_out & !C_out & Cmdr;
22 define Pn := !Pr;
23 define A := CmdA & !CmdB & !CmdC & !E & P & Pn;
24 define B := CmdB & !CmdA & !CmdC & !E & P & Pr;
25 define C := CmdC & !CmdA & !CmdB & !E & P & Pn;
```

Fig. 3. Program "Little Yard" in EDL

4.1 Current vs. latched state variables

There is one important semantic difference between VLC code and EDL that is immediately obvious in the syntax: VLC is sequential in nature, i.e., the order of the statements matters (as in most software languages), while EDL code is parallel - there is no importance to the order of statements (as in most hardware languages). Therefore, the translation must take into account the position of each VLC statement as follows: if the variable V is used in an expression which appears before the assignment of a new value to V, the old, latched value of V should be used in the calculation. If the variable V is used in an expression which appears after the assignment of a new value to V, the new value of V should be used in the calculation.

Thus, in the VLC statement on line 14 of Figure 2, the signals A and B and C refer to the values the previous cycle, because no new values have yet been set. The equivalent EDL statement is that on line 18 in Figure 3 in which we must explicitly state (by the use of the $*\_out$ variables) that the assignment to E uses the previous values. On the other hand, in the VLC statement on line 17 of Figure 2 the signal Pr refers to the value the current cycle, because Pr receives a new value in a statement appearing above this statement in the code. This translates into the EDL statement of line 22 in Figure 3 in which we explicitly state (by the use of the current state symbol Pr) that the assignment to Pn uses the current value of Pr.

4.2 Time delay statements

The time delay statements of VLC are translated into temporary variables which count up to the delay, and whose value is tested to have reached the delay before assignment to the signal being assigned. Thus, the time delayed boolean assignment of line 15 in Figure 2 is translated into the EDL counter of lines 13-16 in Figure 3 plus the assignment of line 19. In this small example, a one-bit counter is used, but of course it is usually the case that a wider counter is needed.

4.3 Translation of propositional formulas to CTL

The formulas used by [GKV94] and [Fok95] have the property that they are translatable into a subset of CTL formulas described below. We will see below that this aids the model checking process for the railway stations in question.

Definition 1 (nested-AX formula). *A nested-AX formula is defined as follows:*

1. *Every propositional formula is a nested-AX formula*
2. *If f is a nested-AX formula, then AX f is a nested-AX formula*
3. *If p is a propositional formula and f is a nested-AX formula, then $p \rightarrow AX f$ is a nested-AX formula*

Definition 2 (AGAX formula). *An AGAX formula is defined as follows: if f is a nested-AX formula, then AG f is an AGAX formula*

Definition 3 (Depth). *The depth of an AGAX formula is the number of AX operators it contains.*

The propositional formulas used by [GKV94] and [Fok95] as described in section 3 translate into AGAX formulas in CTL as follows. The current state variables and past state variables are "shifted into the future" by as many cycles as are needed to get rid of the past. For instance, the propositional formula

$$\neg(74\_R\_ACO\_J\_1 \vee 74\_R\_ACO) \rightarrow 68\_R\_ACO \tag{1}$$

is equivalent to the propsitional formula

$$\neg 74\_R\_ACO\_J\_1 \rightarrow (\neg 74\_R\_ACO \rightarrow 68\_R\_ACO) \tag{2}$$

We shift the past state variable 74_R_ACO_J_1 to the current state variable 74_R_ACO by dropping the _J_1, and the current state variables to next state variables by adding an AX. The result is the equivalent CTL formula

$$AG(\neg 74\_R\_ACO \rightarrow AX(\neg 74\_R\_ACO \rightarrow 68\_R\_ACO)) \tag{3}$$

4.4 Motivation for the remainder of this paper

Using the translations described above, the models of stations Hoorn-Kersenboogerd and Heerhugowaard were converted into the input format of RuleBase. The station Hoorn-Kersenboogerd required approximately 200 variables after reduction and was checked relatively easily. However, station Heerhugowaard, which required approximately 600 variables after reduction, showed surprising results. Despite the large size, some formulas were checked easily. However, other formulas suffered from state space explosion during model checking, despite the fact that they induced the exact same model as the easily checked formulas. An investigation into the reasons for this difference led to the work described in this paper.

5 Robustness and locality in symbolic model checking

In this section we will define robustness and locality, and show how these properties aid the model checking of AGAX formulas.

5.1 Robustness

A formula AGf is true in a model if the formula f is true in all reachable states of the model. Informally, a system is robust with respect to a formula AGf if f is true for all states of the model and not only the reachables, or if f is false in the model.

Definition 4 (Robust). *Let $M = (S, S_0, R, L)$ and $M' = (S, S, R, L)$. A model M is robust with respect to a specification f if $M \models f \rightarrow M' \models f$.*

It was observed in the experiments described below that the railway models checked were robust with respect to almost all (47 out of 51) formulas. In the sequel, we will call a model robust when we assume that it is robust with respect to most of the formulas we require to hold.

5.2 Model checking AGAX formulas in robust models

Consider the process of model checking the following AGAX formula in a robust model:

$$AG(a \rightarrow AXb) \tag{4}$$

First, we negate b to get $\neg b$. Then, we take a backward symbolic step to find the set of states which model $EX\neg b$. Finally, we intersect the result with the set of states which model a to get the set of states BAD which models $a \wedge EX\neg b$. There are two cases. Either the intersection of BAD with the reachables is non-empty, and then the original formula is false in the model, or the intersection is empty, and the original formula is true in the model. In order to decide, the model checking algorithm will take backward symbolic steps until either an initial state is seen, or a fixed point is reached.

However, if the model is robust with respect to the formula and the original formula is valid, the set BAD is the empty set and the fixed point calculation will be trivial. In fact, the process of model checking any model which is robust with respect to a valid AGAX formula will consist of a number of backward symbolic steps equal to the depth of the AGAX formula, plus one trivial fixed point calculation.

For a false AGAX formula, the set BAD will not be empty, and the backward symbolic steps will not be trivial. However, the fact that BAD is not empty is itself indicative that the formula is false.

At this point we have an explanation for the behavior observed in Section 4.4. The question was why some formulas model checked easily while others, which induced the same reduction, suffered from state space explosion. The answer is that the formulas which model checked easily were true formulas, with respect to which the model was robust. Thus, the set BAD was empty, and the fixed point calculation was trivial. The formulas which suffered from state space explosion were false formulas, for which the set BAD was not empty.

For a model which is designed to be robust, we have discovered a decision procedure which is faster than full symbolic model checking: simply compare set BAD with the empty set. However, we have still not solved the problem of generating a counter-example for false formulas. We would like to be able to generate a counter-example while avoiding state space explosion. This is the subject of the next subsection.

5.3   Generating counter-examples

One way to generate a counter-example is to notice that in a model which is designed to be robust, it is not necessary to show a counter-example in the original model M. Rather, it is enough for a counter-example to show that the model is not robust, that is, that the formula is false in model M' of definition 4. This can be done without any changes to the symbolic model checker itself, as follows.

Definition 5 (Non-deterministic inputs method). *Code the initial states of the model to be the set of all states.*

In EDL, this is accomplished by coding $assign\ init(v) := \{0, 1\}$; for every state variable v in the model.

Now, if the formula is false, the set BAD will be non-empty as before. However, the path from an initial state to BAD will be trivial. If the formula is true, the set BAD will be empty as before, and the fixed point calculation will be trivial.

Possibly (indeed, most probably) the counter-example generated according to the non-deterministic inputs method will not start in a true initial state of model M. However, if the model is intended to be robust, the counter-example generated in this manner is sufficient to show that it is not so, and is therefore useful. In the sequel, we use the following definitions to distinguish between a counter-example generated from model M, and one generated from model M'.

Definition 6 (True counter-example). *A true counter-example is a counter-example which starts in some initial state.*

Definition 7 (Bogus counter-example). *A bogus counter-example is a counter-example which starts in some state which is not an initial state.*

Until we have generated it, there is no useful bound on the length of a true counter-example for most CTL formulas. However, it is easy to see that there exists a bogus counter-example to an AGAX formula with length exactly equal to the depth of the formula plus one. For instance, the formula

$$AGp \qquad\qquad (5)$$

where p is a propositional formula, has depth 0. A true counter-example will consist of a path from an initial state to some state in which $\neg p$, while a bogus counter-example can consist simply of a state in which $\neg p$, and thus will have length 1. The formula

$$AG(p \rightarrow AXq) \qquad\qquad (6)$$

where p and q are propositional formulas, has depth 1. A true counter-example will consist of a path from an initial state to some state in which p holds, followed by a state in which $\neg q$, while a bogus counter-example can consist simply of a state in which p holds followed by a state in which $\neg q$, and thus will have length 2.

In the case that a true counter-example in model M is desired, we can make use of locality to generate it as described below.

5.4 Locality

The definition of locality is informal, and leads to a heuristic method of searching for a true counter-example for models which exhibit it.

Definition 8 (Locality). *A model has locality if for every false formula, there exists a counter-example in which most inputs have constant value.*

It was observed in the experiments described below that all counter-examples generated for false formulas exhibited locality. Intuitively, this makes sense for railway models, because the properties which prevent trains from crashing or derailing must be local properties: dependent on only the behavior of "close" tracks or signals. In other words, to push it to the extreme: the fact that two trains do not crash in Amsterdam should not be dependent on the state of a signal in Istanbul. Thus, if a counter-example exists, there should also exist a counter-example in which signals on "far" tracks are quiet.

We have seen previously that if a system is robust, no fixed point calculations are needed to decide on the validity of an AGAX formula. Further, if we are willing to accept a counter-example which is bogus, we can also generate a counter-example with less symbolic steps than needed for a true counter-example. Now we will show how to make use of locality to generate a true counter-example for a robust system when the symbolic steps cause state space explosion.

The method heuristically searches for the inputs which are not needed in the counter-example. It then sets these to 0 and uses pre-model checking reductions [BBEL96] to reduce the size of the model. The method is as follows:

Set half of the inputs to 0, check the formula. Probably the result finds that the formula is true. This is inconclusive, so free up some of the inputs and rerun. If the run does not terminate quickly (i.e., starts a fixed point calculation), then the formula is false even with the inputs chosen set to 0. Choose some more inputs to set to 0 from the half not previously chosen, and so on, until the model is small enough to complete a true counter-example generation.

6 Experimental Results

The experimental results presented below support the observations made above regarding model checking robust models, and demonstrate the effectiveness of the proposed heuristic for counter-example generation of robust models which exhibit locality.

The railway stations were verified in four modes:

1. the properties are checked for all states of the system
2. the properties are checked only for the reachable states of the system
3. the properties are checked for all states of the system with the aid of some simple invariants which are separately checked
4. the properties are checked only for the reachable states of the system with the aid of some simple invariants which are separately checked

The first mode assumes that the systems are robust, while the second conforms more closely to the way the VLC code is used in practice. The third and fourth modes were used as a way to deal with the size problems of the station at Heerhugowaard [Fok99].

6.1 Station Hoorn-Kersenboogerd

Four invariants used by mode 3 were checked in mode 1 with a run time of 9 seconds and memory usage of 33 MB[2]. Thus, the invariants themselves were found to be robust.

Forty-seven formulas were checked for station Hoorn-Kersenboogerd, including all the formulas described in [Fok95], as well as some additional formulas [Gro98]. The formulas included known true and known false formulas which were checked in modes 1, 2 and 3 as described above using the model checker RuleBase [BBEL96]. Results are shown per rule in Table 1 below, where a rule is a group of formulas.

Notice that there is little if any difference in run time between modes 1 and 2 for rules containing only true formulas. This is because the additional fixed point calculation needed by mode 2 is trivial for formulas which are true in mode 1. The additional fixed point calculation can be significant in the case of a false formula. This is especially evident in rules spoor1 and page11. Finally, in all cases, the use of invariants significantly speeded the model checking.

The model is robust with respect to all formulas with the exception of 4 formulas of rule page11. All results agreed with the expected results of [Gro98, Fok95], with one exception. RuleBase found one formula of rule page7 to be false which was found to be true by [Fok95]. After some additional investigation, this formula was confirmed as false by the current version of the tool used in the original work [Gro99]. This formula is now under investigation by Holland RailConsult. This leads to an important conclusion

---

[2] All statistics shown are for an IBM RS/6000 workstation model 140.

Table 1. Run time and memory usage for Station Hoorn-Kersenboogerd

Rule	Vars	mode 1			mode 2			mode 3		
		run time	memory	false	run time	memory	false	run time	memory	false
spoor1	209	64 s	35 MB	x	766 s	51 MB	x	5 s	36 MB	x
spoor2	209	1 s	31 MB		1 s	31 MB		1 s	32 MB	
spoor3	209	1 s	31 MB		1 s	31 MB		1 s	32 MB	
spoor4	209	1 s	31 MB		1 s	31 MB		1 s	32 MB	
page6	209	28 s	29 MB		37 s	30 MB		1 s	32 MB	
page7	215	468 s	80 MB	x	574 s	198 MB	x	152 s	80 MB	x
page8	215	114 s	42 MB		112 s	47 MB		28 s	43 MB	
page9	215	121 s	43 MB		88 s	45 MB		28 s	44 MB	
page10	215	115 s	43 MB		70 s	51 MB		29 s	46 MB	
page11	215	423 s	43 MB	x	42316 s	216 MB		37 s	36 MB	
page12	215	74 s	33 MB		33 s	31 MB		22 s	34 MB	
page13	215	75 s	33 MB		33 s	31 MB		21 s	34 MB	
page14	209	28 s	29 MB		37 s	31 MB		1 s	32 MB	

of this experiment: for safety critical systems, it is necessary to apply two independent methods of formal verification.

The longest true counter-example was of length 126 cycles.

6.2 Station Heerhugowaard

Forty-five invariants were checked in mode 1 with a run time of 1-2 minutes per invariant. Thus, the invariants for station Heerhugowaard were found to be robust.

Four formulas from [Gro98] were checked for this station. The formulas were checked in modes 1, 2, 3 and 4 as described above using the model checker RuleBase [BBEL96]. Results are shown per rule in Table 2 below, where a rule is a group of formulas.

Table 2. Run time and memory usage for Station Heerhugowaard

Rule	Vars	mode 1			mode 2			mode 3			mode 4		
		run time	memory	false	run time	memory	false	run time	memory	false	run time	memory	false
hh2	594	522 s	162 MB	x	*space out*			778 s	166 MB	x	*space out*		
hh3	594	337 s	157 MB		280 s	144 MB		263 s	149 MB		263 s	151 MB	
hh4	594	416 s	247 MB		287 s	149 MB		268 s	149 MB		277 s	149 MB	
hh5	588	*space out*			*space out*			*space out*			*space out*		

For this station, as for station Hoorn-Kersenboogerd of the previous section, it is apparent that for rules containing only true formulas (hh3 and hh4), there is little difference in run times between modes 1 and 2, because the fixed point calculations are trivial for true formulas in a robust model. In station Heerhugowaard the difficulty of the fixed point calculations for false formulas is evident. Neither of the rules containing false formulas could complete when a fixed point calculation was needed.

The heuristic method described in Section 5.4 was used to create true counter-examples for rules hh2 and hh5 (note that bogus counter-examples were created easily for them, and should be enough to debug a robust model in the usual case). Results are shown in Table 3 below.

Table 3. Run time and memory usage for heuristic generation of counter-example

Rule	Vars (original)	Vars (after heur.)	heur. iterations	run time	memory
hh2	594	189	7	2370 s	89 MB
hh5	588	165	10	51 s	55 MB

For these rules, results agreed with those described in [Gro98].
The length of the longest true counter-example was 124 cycles.

7 Conclusions and future directions

We have seen that symbolic model checking of AGAX formulas can avoid fixed point calculations by making use of the robustness of a model. This includes true formulas and bogus counter-examples for false formulas. In addition, we have seen how locality can be used to heuristically reduce models, thus enabling true counter-example generation even for very large (almost 600 state variables) models.

The use of robustness is not limited to model checking, however. It is also applicable to the use of Stålmarck's method, and can be used to limit the unfolding of the model to the depth of the formula. Indeed, for the work described in [GKV94] and [Fok95], this is exactly what was done [Fok99]. The following calculation shows that it was done out of necessity: The station Heerhugowaard has 4789 triples per cycle. If robustness had not been used, 4789 * 124 = 593836 triples would have been needed to generate the counter-example of length 124. This is above the size of the largest formula reported by [SS98], which was 350000.

Although [BCCZ98] report good results for some hardware models typically difficult for BDD-based methods, for instance multipliers, the above calculations cast doubt on the applicability of Stålmarck's method for verification of typical control intensive hardware. This is because control intensive hardware does not usually exhibit robustness, and a quick calculation indicates that well in excess of 350000 triples would be needed to create the counter-examples we see in our hardware. The intuition that Stålmarck's method performs best for robust models is strengthened by the authors of [BCCZ98] themselves, who, in a new paper [BCRZ99] propose a methodology based on first weeding out the easy cases: those formulas for which the system is robust.

These results are not conclusive, of course. Future work involves experiments in applying Stålmarck's method to large hardware models. Finally, we have seen that for safety critical systems, (at least!) two methods of formal verification should be used.

Acknowledgements

Thanks are due to Jan Friso Groote for supplying the railway models used in this paper, and to Holland RailConsult for agreeing to such. Thank you to Wan Fokkink for supplying the invariants which enabled the verification of station Heerhugowaard. The help of both Jan Friso Groote and Wan Fokkink in investigating the discrepancy of section 6.1 was greatly appreciated. Thank you to Sharon Keidar for implementation of optimizations to RuleBase which solved the size problems for these models. Thank you to Shoham Ben-David for careful review and important comments.

References

[BBEL96] I. Beer, S. Ben-David, C. Eisner, A. Landver, "RuleBase: an Industry-Oriented Formal Verification Tool", in Proc. 33rd Design Automation Conference 1996, pp. 655-660.

[BCCZ98] A. Biere, A. Cimatti, E. Clarke, Y. Zhu, "Symbolic Model Checking without BDDs", in TACAS '99, to appear.

[BCRZ99] A. Biere, E. Clarke, R. Raimi, Y. Zhu, "Verifying Safety Properties of a PowerPC Microprocessor Using Symbolic Model Checking without BDDs", submitted, CAV '99.

[CE81] E.M. Clarke and E.A. Emerson, "Design and synthesis of synchronization skeletons using Branching Time Temporal Logic", in Proc. Workshop on Logics of Programs, Lecture Notes in Computer Science, Vol. 131 (Springer, Berlin, 1981) pp. 52-71.

[CE81b] E.M. Clarke and E.A. Emerson, "Characterizing Properties of Parallel Programs as Fixed-point", in Seventh International Colloquium on Automata, Languages, and Programming, Volume 85 of LNCS, 1981.

[CG+95] E. Clarke, O. Grumberg, K. McMillan, X. Zhao, "Efficient Generation of Counterexamples and Witnesses in Symbolic Model Checking", Design Automation Conference 1995, pp. 427-432.

[Fok95] W.J. Fokkink, "Safety criteria for Hoorn-Kersenboogerd Railway Station", Logic Group Preprint Series 135, Utrecht University 1995.

[Fok99] W.J. Fokkink, personal communication to C. Eisner.

[GKV94] J.F. Groote, J.W.C. Koorn and S.F.M. van Vlijmen: "The Safety Guaranteeing System at Station Hoorn-Kersenboogerd." Technical Report 121, Logic Group Preprint Series, Utrecht Univ., 1994.

[Gro98] J.F. Groote, personal communication to C. Eisner.

[Gro99] J.F. Groote, personal communication to C. Eisner.

[McM93] K.L. McMillan, "Symbolic Model Checking", Kluwer Academic Publishers, 1993.

[SS98] M. Sheeran and G. Stålmarck, "A Tutorial on Stålmarck's Proof Procedure for Propositional Logic", in Second International Conference on Formal Methods in Computer-Aided Design, FMCAD '98, Volume 1522 of LNCS, 1998, pp. 82-99.

Practical Application of Formal Verification Techniques on a Frame Mux/Demux Chip from Nortel Semiconductors

Y. Xu[1], E. Cerny[2], A. Silburt[1], A. Coady[1], Y. Liu[1], P. Pownall[1]

[1]Nortel Semiconductors, Ottawa, Ontario, Canada
{xuying, silburt, acoady, yingliu, philp}@nortelnetworks.com

[2]Dépt. IRO, Université de Montréal, Montréal, Québec, Canada
cerny@iro.umontreal.ca

Abstract. We describe the application of model checking using FormalCheck to an industrial RTL design. It was used as a complement to classical simulation on portions of the chip that involved complex interactions and were difficult to verify by simulation. We also identify certain circuit structures that for a certain type of queries lend themselves to manual model reductions which were not detected by the automatic reduction algorithm. These reductions were instrumental in allowing us to complete the formal verification of the design and to detect two design errors that would have been hard to detect by simulation. We also provide a technique to estimate the length of a random simulation needed to detect a particular design error with a given probability; this length can be used as a measure of its difficulty.

1 Introduction

The rapid increase in the complexity of microelectronics systems and the degree of integration of subsystems and systems on one silicon chip make their design and verification increasingly difficult. The amount of Verilog or VHDL RTL code written to describe a system for the synthesis tools is large, but the simulation test bench code written to verify that the chip satisfies its specification is much larger, and growing [1]. This is a direct consequence of the desire to achieve short time to market while producing an error free design, and due to the growing size of the chips to be verified. Many techniques and tools are being introduced on the market to improve the situation, to cut down on the amount of code and/or to improve the verification coverage of the design. The list includes:

Simulation oriented:

- Improved verification productivity through better test-bench construction environments [8].
- Reduced number of testbench cases by using random input sequences [6, 15].
- Improved observability through automatically generated property checkers [6].

- Improved coverage evaluation through classical software code-coverage techniques [e.g., 9] or based on state / transition information of the network of finite-state machines underlying the design [6].

(Semi-)Formal methods:

- Symbolic model checking based on verifying a temporal logic specification of properties [13, 5] or using language inclusion tests of ω-automata [4, 10].
- Partial state-space exploration (simulation and symbolic exploration combined) [7].
- Theorem proving at higher levels of design abstraction [2, 14].

Although simulation remains the main verification workhorse, formal verification (model checking in particular) using commercially available tools is making slow inroads when verifying complex interactions between design modules.

In this paper, we describe our experience in formal verification using FormalCheck [4, 10] (a model checker based on language inclusion test of ω-automata) from Lucent Technologies as applied to a large ASIC design. We show that a judicious deployment of FormalCheck on a small subset of modules was efficient and highly beneficial to the improvement of the quality of a Frame Multiplexer/Demultiplexer chip (called FMD in what follows).

The principal problem in the application of a model checker to a large design is the so-called *state explosion*: the representation of the state space that has to be maintained by the tool during verification growing beyond the available computer memory. The result is that the verification cannot be completed. To counter this problem, one should concentrate on smaller subsystems that are difficult to verify by simulation. Usually these are complex control structures implemented using a number of cooperating state machines. In addition, one should consider scaling down any datapaths, register arrays, and / or constraining the input space. Both of these techniques may change the behavior of the design or limit the state-space exploration, thus making the verification result less useful or even invalid. A better solution is to apply model reduction or abstraction techniques. These can simplify the model representation by introducing complete nondeterminism on state variables in those parts of the design that should not have influence on the properties to be verified. If the property holds even after the reduction, then it is guaranteed to hold on the original design. If not, then either a lesser reduction must be used or, there is a design error which then should be confirmed by simulating the error trace (counter-example) produced by the model checker. In our verification work, we applied model checking to Context Switch, which is a critical subsystem of FMD. We scaled down the number of channels, and developed a reduction technique that allowed us to successfully complete the verification when combined with the automatic model reductions carried out by Formal Check. In the process, two important design errors were discovered. The contributions of this paper can thus be summarized as follows:

- A description of the verification approach: isolation of the modules, creation of a model of the use environment, and formulation of properties.

- Identification of model structures that can be considerably reduced by converting the majority of state variables into primary inputs. This makes the next-state transition function of these state variables completely non-deterministic. Yet, the automatic reduction algorithm as implemented in FormalCheck (Cospan [4]) were not able to detect this reduction possibility. By combining the power of our manual reductions with the automatic ones, the Context Switch subsystem was efficiently verified in 512Mb of memory.

We also describe a measure of "hardness" of a design error detected by a model checker. It consists of computing an estimate on the length of a random input sequence that would detect the particular error with a given probability (e.g., a measure of how long it would take to detect the error by random simulation).

The paper is organized as follows: In Section 2 we briefly describe our design and the verification environment. In Section 3 we introduce the FMD design and the overall verification strategy. In Section 4 we concentrate on the Context Switching subsystem that was deemed to require verification by model checking, the classes of properties verified, and the design errors detected. Then, in Section 5, we discuss a model reduction needed for the verification to complete. In Section 6 we conclude the paper. The appendix contains a description of a method for computing the length of a random simulation sequence to detect a design error previously identified by a model checker.

2   The design verification environment

The RTL design of a chip like the FMD is carried out by a team of designers according to the specification of the product. The verification of its component blocks is done by simulation by the respective designers before chip-level integration. The interaction of the blocks and the conformance of the design to the specification is verified by simulation of the whole chip once the RTL code is ready. This simulation follows a test plan that identifies all the features whose functionality must conform to the specification.

It is well known that features involving complex interactions between internal design modules are hard to verify by simulation, and that model checking can be effective in such cases. This was exactly the case of certain parts of the FMD design. We therefore deployed FormalCheck to those portions of the design that would have required long simulation sequences and yet with no guarantee on the completeness of verification.

In the following section we describe briefly the FMD design and the portion verified using FormalCheck.

3   The design of the Frame Multiplexer/Demultiplexer

The FMD chip is part of a system used in multiplexing/demultiplexing framed data between various channels and a SONET line. It consists of some 250k gates. The architecture contains a datapath over which the data frames travel and are processed, under the control of cooperating state machines. Since the frame and mux/demux parameters may depend on the specific connection and channel number (0 to 27), each time a piece of information is to be processed, its "context" information is retrieved from a Context Memory, the data and the context are updated and then written back to memory. The accesses to the memory are thus shared by the datapath and by the controlling processor. The controlling processor sets up the initial context information and retrieves status information from internal registers and the memory asynchronously relative to the datapath operations. There are thus concurrent accesses to the Context Memory by the Datapath and the Microprocessor.

The arbitration between the accesses and the retrieval / storing of the context information is handled by a Context Switch subsystem. A Datapath access always takes precedence, and thus a Microprocessor access is held off until a free cycle exists in which the Datapath is not accessing the Context Memory. Furthermore, a Microprocessor access can piggyback on a Datapath access. For power consumption and efficiency concerns, the Context Memory is subdivided into 6 memory blocks according to the type of information and frequency of access, hence, there are 6 Context Memory controllers. There are also 28 Facility Data Link (FDL) FIFOs in the Context Switch (one FIFO for each Datapath channel). The control structure in the Context Switch (Figure 1) is complex, and it also contains queues for the requests and a 3-stage pipeline for the Read-Modify-Write processing sequence on the data. To minimize

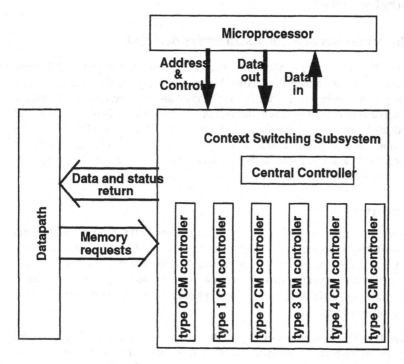

Fig. 1. Bock diagram of the Context Switch subsystem of FMD

power consumption, many of the pipeline registers are only virtual, i.e., they do not physically appear in the design. This is possible, because the data flow pattern is such that in some cases buffering over two cycles is sufficient. This overall complexity makes the verification of the Context Switch quite difficult by simulation, because it is hard to imagine all the input combinations that force different orders of events inside the Context Switch.

It follows that the design blocks of the Context Switching Subsystem were identified as the natural candidates for formal verification. We isolated them, defined their operating environment (non-deterministic state machines and constraints on the inputs) and then verified their operation using a set of properties derived from the global test plan, from the intimate understanding of how the Context Switch should operate based on the chip-level specification, and from RTL code reviews. In the following sections we describe the typical structure of the properties that were verified and then discuss the necessary model reductions for the verification to be completed on a Ultra Sparc 10 workstation with 512Mb of memory. The Context Switching blocks represent some 20k gates described in about 3500 lines of RTL Verilog code, not counting comments. There are some 840 state variable bits. Depending on the property and the reductions achieved, the reachable state space consists of 10^{12} to 10^{20} states (as reported by FormalCheck).

4 The properties and the design errors detected

A total of 110 queries were verified using FormalCheck on the Context Switch, some queries consisted of a number of properties. We primarily targeted the mutual exclusion of accesses to the context memories by the Datapath and the Microprocessor, and the proper emission of control signals to the memory. We did not verify the correctness of the contents stored in the memory, since the memory is (logically) correct by construction and the physical data connections to it were verified by traditional simulation test benches. In addition, the various information fields in the memory were also tested by simulation.

4.1 Properties

Due to the similarities among the memory controllers in the receive and transmit context switching blocks, the 110 queries can be subdivided into 4 classes discussed below. We also constructed (in Verilog) a non-deterministic model of the user that mimics the normal operating environment in which the Context Switch is utilized. In what follows, the underlined words are keywords of the FormalCheck Property specification user interface.

1. Unicity of the Microprocessor accesses to the receive-side Context Memory. This query consists of 9 properties which check that only one of the possible Microprocessor accesses is valid in any clock cycle. These are all safety properties of the form "*never condition*", where *condition* is a boolean formula over the control signals involved in the accesses. It describes the possible contention situations.

2. A Datapath write to the Context Memory must be preceded by a read 2 clock cycles earlier. The address value itself to the Context Memory is not verified. Again, this safety property has the form "*never condition*", except that *condition* now involves delayed control values. These delays were either already present in the design or they were obtained using state machines built from within the user interface of Formal Check using if-then-else constructs and state variables.

3. When the Microprocessor requests access to a data select register, then the 7th bit in the address must be 1. This safety property has the form[1] "*always condition => conclusion*", where *condition* characterizes the particular type of access and *conclusion* verifies that the 7th bit is actually 1.

4. All Microprocessor read or write requests to the context memories are eventually acknowledged. This is a liveness property verified under the fairness constraint that there is no Datapath accessing the memory for 4 consecutive clock cycles. It has the form "*After condition Eventually conclusion*", where *condition* describes the signal values representing a request, and *conclusion* describes the acknowledgment. The fairness constraint is constructed with the help of additional state variables implemented from the user interface of FormalCheck.

In the following we mention two properties that have a structure similar to those we have described, but which each had a major impact on the verification. Either because it lead us to the identification of circuit structures amenable to manual reductions (Property A), thus allowing us to complete the verification of many other properties, or because it detected important and difficult design errors (Property B).

Property A: The Datapath FDL FIFO consumer (head) and producer (tail) indexes are updated correctly on each such queue. The size of each FIFO is 128 1-byte entries. For example, when the Microprocessor reads a byte from the FIFO of channel 3, then the consumer index should be increased by one. The component properties of this query have a similar structure as (4) above, however, it is mentioned here because:

a) The original property definition had to be broken up into two parts, so as to implement the verification over the modulo 128 incrementation of the indexes: We first consider the case when the index is in the range of 0-126, and then the case when the index is changing from 127 to 0.

b) As we mentioned earlier, there are 28 FIFOs accessed using a common path. The selection of which FIFO is to be incremented is based on information contained in the request. The verification could not complete because of the number and the size of the FIFOs. In spite of the fact that the property would select a specific queue, the automatic reductions could not identify a sufficiently large number of state variables to reduce. Yet, as will be seen in the next section, the structure of the model and the property allows more powerful reductions which, if carried out manually, allow completing the verification of this kind of property on the design.

---

1. Implication "A => B" is replaced by "(not A) or B" to specify the proposition to FormalCheck.

Property B: When a valid Microprocessor read request to the Context Memory is acknowledged, the read address (channel number) issued to the memory 2 cycles earlier is the same as the channel number specified by the Microprocessor.

Table 1 shows the CPU time, the memory size, the total state space, and the reachable states for the queries involving Properties A and B. Notice that Property A was verified after the reduction in only some 30 minutes of CPU time, while without the reductions it could not complete at all. The execution time of Property B is typical of the longest verification times we encountered. Either it would complete within about 90 minutes, or it would run out of physical memory and start heavily paging (thrashing), since model checking (like simulation) has relatively poor spacial locality in memory accesses. We would thus periodically observe the status of the verification process using the Unix *ps* command. Normally, on an unloaded workstation, the CPU utilization would be above 80%. When it runs out of memory and starts thrashing, the CPU utilization by the process drops below 5% and it is time to terminate the process (unless one wishes to wait almost indefinitely or until the process runs out of the allowed virtual address space).

Table 1. Verification Statistics

	CPU time (minutes)	Memory (MB)	total states	reachable states
Property A	34.4	27.40	8.44e14	1.95e12
Property B	85.7	167.72	1.77e21	8.70e19

4.2 Error detection

As mentioned earlier, due to the concurrent accesses to shared resources, the number of possible behaviors that must be verified in the Context Switch is very large. It is thus difficult to carry out the verification by simulation. FormalCheck implicitly enumerates all the possibilities which allowed us to detect two difficult design errors using Property B. These errors were found after all the planned simulations of the entire FMD RTL design had been executed. The errors were related to different issues in the implementation, but both exhibited the same behavior: when the Microprocessor reads the Context Memory, it could receive data from a wrong address.

More specifically, the counter-example[1] generated by Formal Check that leads to the first design error confirmed that the particular input sequence would have been difficult to foresee in a simulation scenario: When the Microprocessor requests to write to a read-only address or to access an invalid address, it is acknowledged which allows it to initiate a new request. At the same time, however, the Microprocessor-initiated

---

1. It is an input sequence computed by the model checker that shows what inputs must be applied to reach a point where the property is violated, i.e., leading to the design error.

memory read signal may incorrectly be activated in this case. Therefore, if there is a valid Microprocessor read to the Context Memory immediately following the first access, it would receive the (default) data obtained as a result of the preceding incorrect read.

The sequence leading to the second error was even more complex and difficult to stimulate. It now involved an interaction between the Datapath and the Microprocessor accesses to the Context Memory and the internal registers of the Context Switch. Again, incorrect data was returned to the Microprocessor, but this time it was due to an error in the complicated control circuitry that handles updating of the virtual pipeline registers: The registers for addresses and data as related to the Datapath requests were not updated in the same clock cycle. A Microprocessor access to these registers that would occur between the two parts of the update could read an incoherent pair of address-data values.

A question is often asked when a design error is detected using a new kind of a verification tool. How hard would it be to detect this error using simulation? We could measure the difficulty of detection by estimating the chance of hitting the "right" sequence in a simulation testbench. This is impossible with directed testbenches that depend on the ingenuity of the designers, but, as mentioned in Section 1, random simulation may be used to verify designs without having to devise specific input sequences. Therefore, we propose to measure the detection difficulty by estimating the length of the input sequence needed to stimulate the design error with a given probability. This is similar to the estimation of random test sequences for stuck-at-0/1 faults, except that we do not have a simple fault model here.

In the Appendix we outline a method that allows us to make such estimation in some cases, based on the counter-example generated by the model checker. Using that method, we have calculated that to detect the first error with the probability of 99.99%, we would need some 24199 input patterns. This does not look like a formidable length of a sequence to simulate, however, it requires setting up the random testbench and formulating the property checkers in the form of finite-state machines implemented as Verilog modules. Last and not least, the random simulation could only detect this error with a certain probability, but not with absolute certainty, while the model checker guarantees that the design does indeed satisfy the property, for any input sequence.

The second error is even more complex and harder to detect by simulation, as it requires a particular coordination between the Datapath and the Microprocessor accesses. Unfortunately, here the counter-example sequence is not unique and we could not compute the estimation (we discuss possible ways around this problem in the appendix). Still, due to the more complex coordination of the events that lead to the error, it is likely that the length of a random input sequence would be much longer than in the first case, and, in addition the test bench would have to reliably predict the correct results at this level in the design environment. This would be difficult based on a more abstract behavioral model of the whole chip as the reference for comparison. A model checker reduces this investment in test-bench development at the block level.

5 Model reductions

Due to the state explosion problem, FormalCheck could not complete the verification of properties that refer to channel numbers and compare the resulting values (like Property A) even when the blocks related to the Context Switch were isolated from the chip design, the number of channels was scaled down from 28 to 8, and the automatic iterated reduction algorithm was used. We found, however, that effective manual reductions[1] could be carried out on certain state variables, after which the verification of the properties took only about 30 minutes. Naturally, in the case of a violation of a stated property, like the one which identified the design errors mentioned above, we had to make certain that it was not a false negative answer to the query. We achieved that by examining the counter-example and the Verilog code, and then confirming the error by a Verilog simulation of the counter-example sequence. In both cases where we obtained a negative answer, the simulation confirmed the presence of a design error, i.e., no false negative.

The type of reductions that were not detected by the automatic reduction algorithm and which we manually applied were similar in many of the designs that we verified. The structure of the designs is quite common in dataprocessing circuits, in which the appropriate processing context (set of registers, memory data, etc.) of an arriving datum is selected based on a data descriptor, processing is carried out, and the datum is sent out, while possibly updating the context information.

The situation can be summarized by the structure shown in Figure 2 that corresponds to the following Verilog code extracted from the Context Switch model:

```
module littleFifoDesign(fifo_num, fifo_rd, rst, clock, fifocsmr_old);

    input fifo_num; // select context by FIFO number entered on this input; this is
        the data descriptor (0 or 1)
    input fifo_rd; // increment data if 1; represents the data processing operation
    input rst; // reset
    input clock;
    output fifocsmr_old; // output FIFO index corresponding to the fifo number

    reg [6:0] fifocsmr_R0;   // 7-bit FIFO consumer index 0
    reg [6:0] fifocsmr_R1; // 7-bit FIFO consumer index 1
    wire [6:0] fifocsmr_new, fifocsmr_old; // interconnections

    always @(posedge clock)
    begin
```

---

1. The manual reduction consists of replacing the state variable by a free (unconstrained) primary input that in effect converts the next-state transition function into a completely non-deterministic one, meaning that the next-state value can be any. The reachable state-space remains the same or is even larger, however its symbolic representation and those of the next-state functions become trivial, thus reducing the amount of memory needed by the model checker.

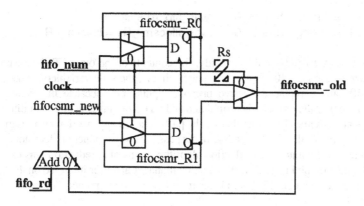

Fig. 2. Example of FIFO circuit structure for manual reduction

```
if (rst == 1'b0)
begin
{fifocsmr_R0, fifocsmr_R1} <= {2{7'b0000000}};
end
else
begin
  case (fifo_num)
    1'b0: fifocsmr_R0 <= fifocsmr_new;
    1'b1: fifocsmr_R1 <= fifocsmr_new;
  endcase
  end
end
assign  fifocsmr_old  = fifo_num == 1'b0 ?
               fifocsmr_R0 : fifocsmr_R1;
assign fifocsmr_new = fifocsmr_old + {6'b000000, fifo_rd};
endmodule
```

The property to be checked verifies that the FIFO consumer indexes are updated correctly:

Property:fifo_index_Without_Rs

Always:(was_fifo_num == 1 && was_fifo_rd == 1 &&

was_fifocsmr_R1 <= 126) => (fifocsmr_R1 == was_fifocsmr_R1 + 1)

The signals was_fifo_num, was_fifo_rd and was_fifocsmr_R1 are the signals fifo_num, fifo_rd and fifocsmr_R1 delayed by one clock cycle using environment state machines defined either in the Verilog code or through the FormalCheck user interface as follows:

if (clock == rising && rst != 0) was_fifo_num = fifo_num;

```
if (clock == rising && rst != 0)    was_fifo_rd = fifo_rd;
if (clock == rising && rst != 0)    was_fifocsmr_R1 = fifocsmr_R1;
```

To update the fifocsmr_R1 correctly, the contents of the other FIFO consumer index register (fifocsmr_R0) should not matter, that is, the state variable(s) related to fifocsmr_R0 can be replaced by free primary inputs. Yet, because this reduction depends on the way the property is formulated and on the circuit structure including the additional state variables for remembering the past values, it appears that every register value may depend on all the other ones and this possible model reduction does not seem to be detected by the automatic reduction algorithm. A manual reduction was carried out with the help of the FormalCheck reduction manager and user interface on 49 state variable bits - the 7 channels FIFO indexes (of 7 bits each) not related to the selection to be verified.

Note that to decrease the size of the model without the above reductions, we could have constrained the primary inputs such that fifo_num = 1'b1. However, this would have eliminated the presence of the other selection(s) from the input stream, thus potentially hiding a design error that manifests itself only in the presence of different consecutive operations. For example, consider the case where a register **Rs** (as shown shaded in Figure 2) is added by mistake in the design. If we constrain fifo_num = 1'b1 and initialize Rs to 1 during reset, then the property would still pass without detecting the error. However, it is detected under the manual reduction of R0.

Finally, the property stated above can also be reformulated without the use of the explicit state variables was_fifo_num, was_fifo_rd and was_fifocsmr_R1 to memorize the preceding values as follows:

```
After:    ( fifo_num == 1 &&
           fifo_rd == 1 &&
           fifocsmr_R1 == 0 &&
           rst == 1 &&
           clock == rising )
Always: ( fifocsmr_R1 == 1 )
Unless after:( clock == rising)
```

and enumerated over all values (0 .. 127) of fifocsmr_R1.

Using this formulation which is much longer to manipulate due to the large number of instances of the property (enumeration) the automatic reduction algorithm finds the proper reductions automatically. This seems to indicate that the presence of the additional state variables can introduce false dependencies between state variables as far as the reduction algorithm is concerned, and thus disallows it from abstracting fifocsmr_R0. Since the use of the state variables makes the formulation of properties much easier in many cases, we are investigating how to improve reduction algorithms to take situations like this into account. In the meantime we complement the automatic reductions by manual ones, based on our knowledge of the design, as indicated earlier.

6   Conclusions

We described the verification of a critical portion of a large ASIC design using a commercial model checker. Model checking was applied to a subsystem that is difficult to verify by simulation. It allowed us to detect two very hard design errors that were not detected by the simulation suites. To complete the model checking in the available workstation memory, we had to apply manual model reductions. In the process we identified circuit structures that lend themselves to manual reductions for certain specific properties. Once such reductions were carried out, all our queries were efficiently verified in 512Mb of memory.

To illustrate how difficult the detected errors are, we computed estimates on the length of a random input sequence needed to detect the error with a certain probability. The computation requires constructing a recognizer of a detection sequence which can be in some cases obtained from the counter-example generated by the model checker.

One of the difficult tasks of carrying out the formal verification was to isolate the RTL code related to the subsystem of interest and to construct an appropriate model of the environment, that is, to identify the minimal input constraints. With what we have learned about model reductions, we hope that it might be actually possible, with some improvements in the model-checking technology and the design/verification methodology, to verify important queries on the entire chip model. This would lead to higher verification quality and better understanding of the division of coverage of the chip-level test plan features by simulation and by model checking.

References

1. A. Silburt Invited Lecture: ASIC/System Hardware Verification at Nortel: A View from the Trenches. *Proceeding of the 9th IFIP WG10.5 Advanced Research Working Conference on Correct Hardware Design and Verification Methods (CHARME'97)*, October,1997, Montreal Canada.

2. A.J. Camilleri. A role for Theorem Proving in Multi-processor Design. Proceeding of *the 10th International Conference on Computer Aided Verification (CAV'98)*,Vancouver, BC Canada, June/July 1998.

3. D.L.Dill. What's Between Simulation and Formal Verification. Proceeding of *the 35th Design Automation Conference (DAC'98)*, San Francisco, CA, USA, June 1998

4. FormalCheck User's Guide. Bell labs Design Automation, Lucent Technologies, V1.1, 1997

5. K. L. McMillan. *Symbolic model checking - an approach to the state explosion problem*. Ph.D. thesis, SCS, Carnegie Mellon University, 1992. (See also Cadence Design Systems, Inc. www pages.)

6. 0-In Design Automation, Inc., "Whitebox verification," http://www.0-in.com/tools.

7. 0-In Design Automation, Inc., "0-In search," http://www.0-in.com/tools.

8. System Science, Inc. (Synopsys, Inc.), "The VERA verification system," http://www.systems.com/products/vera.

9. TransEDA, Inc., "VeriSure- Verilog code coverage tool," http://www.transeda.com/products.

10. Lucent Technologies,"FormalCheck model checker," http:/www.bell-labs.com/org/blda/product.formal.html.

11. P. Bratley, B.L. Fox, L.E. Schrage, *A Guide to Simulation*, 2nd Edition, Springer-Verlag, New York, 1986.

12. L. Devroye, *Non-Uniform Random Variate Generation*, Springer-Verlag, New York, 1986.

13. CheckOff User Guide, Siemens Nixdorf Informations Systemen AG & Abstract Hardware Limited, January, 1996.

14. Lambda user's manual, Abstract Hardware limited, 1995.

15. Verisity Design, Inc. Specman Data Sheet, http://www.verisity.com.

A. APPENDIX: Estimating the length of a random input sequence

We shall show how one can quantify the hardness of a design error by estimating the length of a random input sequence that would detect the error. The problem could be stated as follows:

Suppose that a synchronous RTL design has input ports on which one of $n \geq 1$ symbols can occur in each clock cycle, independently of the choice made in the preceding cycles. Assume also that the probability distribution function of the symbols on the input in a clock cycle is uniform, i.e., a symbol is chosen with probability $p = 1/n$. Let a design error E be detected by Formal Check such that the only input sequence that detects it consists of m consecutive symbols on the inputs, regardless what preceded this Detection Sequence (DS). The question is: given that the error is to be detected with probability P_d, what is the minimal length L of a random simulation sequence (independent choice at each clock cycles, uniform choice from n symbols) such that the detection sequence appears on the input at least once, i.e, that the error is detected.

A.1 An analytical solution

The proposed method can actually be used in a more complex setting than the one stated in the problem definition above, but we shall illustrate it on the simplified case. Construct a finite state automaton that recognizes the first occurrence of the DS in a sequence. For the problem above, such a recognizer is shown in Figure 3. The symbol U represents the set of the n symbols $|U| = n$, and A_i is the subset of symbols that can occur (one of them) in position i in the DS. Starting in state 0, the automaton will reach state m if it detects DS. If now we replace the symbols labeling the transitions by the probabilities of occurrence of these symbols, we shall obtain a discrete time Markov chain, with initial state 0. We can then compute the probability of reaching state m in at most $L > n$ transitions. Let p_i be the probability that one of the symbols from A_i occurs. The Markov chain corresponding to the automaton in Figure 3 is shown in Figure 4.

The probability state transition matrix of this chain is $T = [p_{ij}]$, where p_{ij} is the probability of reaching state i from state j in one step (transition). For example, entry $(0,0)$ is $p_{00} = 1-p_1$. The sum of the elements in any one column must be equal to one, since there is a transition from any state for all the input symbols. Let $S_0 = [1, 0, ..., 0]^T$, $|S_0| = m+1$, be the probability of being in one of the states at time 0, i.e., in this case, we start from state 0, hence its entry in S_0 is 1, and the other m entries are 0. The

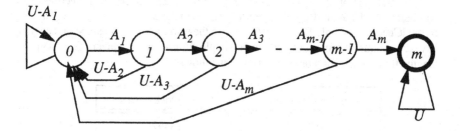

Fig. 3. A simple recognizer automaton, initial state is 0, state m is acceptin

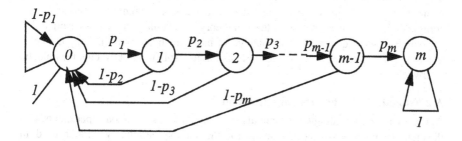

Fig. 4. Discrete time Markov chain for automaton in Figure 3

vector of probabilities of being in one of the $m+1$ states after L transitions is given by $P_L = T^L S_0$. However, the accepting state m (the one in which the detection sequence has been observed somewhere in past) has a self-loop transition, i.e., once it is reached in $k < L$ transitions it will stay there; therefore, if we take the last element (m) of P_L then this element contains the probability of having reached state m in up to L transitions. That is, the probability of detecting the error in a random sequence of length L is given by

$$P_L(m) = (T^L S_0)_m .$$

For a regular structure (e.g., when $p_i = p_j$), there may be a closed form solution for $P_L(m)$, but in general, given a P_d, we can compute $T^k S_0$ numerically until $P_L(m) \geq$

P_d. The number of iterations k then is the minimum length L of a sequence to produce DS on the input of the design with probability P_d.

In the case of the first design error reported under Property B in Section 4.2, it was determined that the 4-cycle sequence reported in the counter-example generated by FormalCheck is the only DS that can provoke the error, and that it can be preceded by any other input. It thus fitted the above simple form of a recognizer. The computed lengths for two detection probabilities are as follows:

Detection Probability	Nb. of clock cycles
0.999	18492
0.9999	24199

This assumes that the vectors were applied at the inputs of the blocks; longer sequences would be needed if applied from the primary inputs of the chip as discussed in Section 4.2.

In more complex situations where there are potentially many different detection sequences, the construction of the appropriate Markov chain may be difficult, since only one such sequence is reported in the counter-example. However, FormalCheck being based on language inclusion test of ω-automata, computes implicitly the complete DS recognizer, hence, at least in theory, one could compute the estimate on the length of the sequence. Unfortunately, for most practical cases this may not be of much use, because the size of the corresponding explicit Markov chain would be overwhelming.

A.2 Estimation of L by random simulation

Another way of estimating L is to actually perform random simulation experiments with different (uncorrelated) starting seeds for the random number generator used in producing the inputs. There are two problems with this approach:
- We wanted to avoid random simulation runs in the first place.
- The simulation runs can be quite long.

These problems could potentially be alleviated by using special techniques to bias the input sequences (knowing the error) to produce good estimators even with short random simulation sequences [11, 12]. If, however, the sequence recognizer is complex (i.e., not just a single detection sequence) then determining the appropriate bias formula may be difficult.

Fortunately, the need for such measures of error "hardness" may only be required until formal methods find their firm place in the normal design flow. This will happen when we can easily identify features in the test plan that can be verified most efficiently using formal methods, and when designers become more aware of such novel verification techniques.

Efficient Verification of Timed Automata Using Dense and Discrete Time Semantics[*]

Marius Bozga, Oded Maler, Stavros Tripakis

VERIMAG, Centre Equation, 2, av. de Vignate, 38610 Gières, France
{bozga,maler,tripakis}@imag.fr

Abstract. In this paper we argue that the semantic issues of *discrete* vs. *dense* time should be separated as much as possible from the pragmatics of state-space representation. Contrary to some misconceptions, the discrete semantics is not inherently bound to use state-explosive techniques any more than the dense one. In fact, discrete timed automata can be analyzed using any representation scheme (such as DBM) used for dense time, and *in addition* can benefit from enumerative and symbolic techniques (such as BDDs) which are not naturally applicable to dense time. DBMs, on the other hand, can still be used more efficiently by taking into account the *activity* of clocks, to eliminate redundancy.

To support these claims we report experimental results obtained using an extension of Kronos with BDDs and variable-dimension DBMs where we verified the asynchronous chip STARI, a FIFO buffer which provides for skew-tolerant communication between two synchronous systems. Using discrete time and BDDs we were able to prove correctness of a STARI implementation with 18 stages (55 clocks), better than what has been achieved using other techniques. The verification results carry over to the dense semantics.

Using variable-dimension DBMs we have managed to verify STARI for up to 8 stages (27 clocks). In fact, our analysis shows that at most one third of the clocks are active at any reachable state, and about one fourth of the clocks are active in 90% of the reachable states.

1 Introduction

The analysis of discrete systems such as programs or digital circuits, while taking into account the temporal uncertainty associated with transition delays, is a very challenging and important task. In [MP95] and elsewhere, it has been demonstrated that reasonable models of digital circuits with uncertain delay bounds can be translated systematically into timed automata [AD94], which can then be analyzed using various verification tools. However, this remains a theoretical possibility as long as the performance bottleneck for timed verification remains (see the discussion in [BMPY97] as well as [TKB97]). During the last decade the

---

[*] This work was partially supported by the European Community Esprit-LTR Project 26270 VHS (Verification of Hybrid systems) and the French-Israeli collaboration project 970MAEFUT5 (Hybrid Models of Industrial Plants).

VERIMAG laboratory has been engaged in development of the timing analyzer KRONOS [DOTY96] and in attempts to improve its performance using various techniques.

Timed automata operating on the dense real time axis constitute an instance of *hybrid automata* and their analysis confronts researchers with problems usually not encountered in "classical" finite-state verification. These problems, such as Zeno's paradox (the possibility of making infinitely-many steps in a bounded interval) or the representation of an uncountable number of states, which are related to the foundations of mathematics, sometime give a false impression on the essense of timing analysis. We argue that this essence does not depend on the time domain, and that the difference between dense and discrete time semantics is "epsilon", so to speak.

We demonstrate these claims by applying discrete and dense techniques to a non-trivial case-study where we obtain the best performance results achieved so far for timed automata. More precisely, we report the application of two techniques: the BDD-based verification using the discrete time semantics [ABK+97] [BMPY97], and the "standard" DBM-based method using variable-sized matrices based on clock activity analysis [DY96], to a real hardware design, the STARI chip due to M. Greenstreet [Gre97]. This chip is an asynchronous realization of a FIFO buffer, composed of a sequence of stages, each consisting of two Muller C-elements and one NOR gate. According to the principles laid out in [MP95], and similarly to [TB97], each such stage is modeled as a product of 3 timed automata, each with 4 states and one clock. The (skewed) transmitter and receiver are modeled as two timed automata using a shared clock.

We have modeled the intended behavior of the FIFO buffer operationally as an automaton, and were able to verify that an 18-stage implementation (55 clocks) indeed realizes the specification. These are, to the best of our knowledge, among the *best performance results* for timed automata verification, and they show that some real circuits behave better than artificial examples of the kind we used in [BMPY97].

The rest of the paper is organized as follows. In section 2 we give a very informal survey of timed automata, their verification techniques and their discrete and dense semantics. In section 3 we describe STARI and its desired properties, which are then modeled using timed automata in section 4. The performance results are reported and analyzed in section 5, and future work is discussed at the end.

2 Verification Using Timed Automata

2.1 Timed Automata

Timed automata can represent systems in which actions take some unknown, but bounded, amount of time to complete, in a rigorous and verifiable manner. They are essentially automata operating on the continuous time scale, employing auxiliary continuous variables called clocks. These clocks, while in a given state, keep on increasing with time. Their values, when they cross certain thresholds can enable some transitions and also force the automaton to leave a state. Temporal uncertainty is modeled as the possibility to choose between staying in a state and taking a transition during an interval $[l, u]$.

Since their introduction in 1990 [AD94], Timed Automata (TA) attracted a lot of attention from the verification community, mainly for the following reasons:

1. They constitute a computational model in which one can faithfully represent many real-world situations where timing constraints interfere with discrete transitions.
2. In spite of the fact that their state-space is non-countable, their reachability problems (which are the essence of any verification problem) are decidable. Several verification tools such as Kronos [DOTY96], Timed Cospan [AK96] and Uppaal [LPY97] have been implemented, featuring various verification and synthesis algorithms which explore, this way or another, the state-space of TA.

The use of a non-countable state-space, $Q \times X$ where $|Q| = m$ and $X = [0, k]^d$, along with dense time, excludes immediately any verification method which is based on explicit enumeration of states and trajectories. All existing TA verification algorithms are based, either explicitly or implicitly, on the region graph construction [AD94]: an equivalence relation is defined on X, unifying clock configurations from which the future behaviors are essentially identical (i.e. two clock valuations $\mathbf{x}$ and $\mathbf{x}'$ are equivalent, $\mathbf{x} \sim \mathbf{x}'$, if the same sequences of discrete transitions are possible from $\mathbf{x}$ and from $\mathbf{x}'$). It turns out that in TA this relation is of finite index, and the quotient graph of a TA, modulo $\sim$, is a finite-state automaton with a combination of discrete transitions and abstract "time-passage" transitions, indicating the temporal evolution of clock values from one equivalence class to another.

Verification tools for TA, either construct first the region automaton (whose size is $O(mk^d d!)$) and then use standard discrete verification algorithms, or calculate the reachable configurations successively while representing them as unions of polyhedra of certain restricted form. These "symbolic states", which are generated by the equivalences classes of $\sim$, are polyhedra which can be represented by combinations of inequalities of the form $x_i \prec c$ or $x_i - x_j \prec c$ where x_i, x_j are clock variables, $\prec$ is either $<$ or $\leq$, and c an integer constant in $\{0, 1, \ldots, k-1\}$. For convex regions, there is a canonical form based on an irredundant set of inequalities, which can be efficiently represented using an $O(n^2)$-sized integer matrix, known as the *difference bounds matrix* (DBM). The main computational activity in TA verification is the storage and manipulation of sets of these matrices during fixed-point computations. The major bottleneck is due to the fact that the number and size of DBMs grows exponentially with the number of clocks and the size of k (roughly, the size of the largest constant in the TA after normalization). Moreover the representation of non-convex polyhedra as unions of convex ones is not unique. There have been many attempts to break the computational bottleneck associated with the manipulation of DBMs, such as [WD94,H93,B96,DY96,LLPY97,DT98], to mention a few, and to be able to verify larger timed automata. One approach, [DY96], is based on the observation that not all clocks are active at any configuration (see also [SV96]). A clock x_i is inactive in a configuration $(q, \mathbf{x})$ if it is reset to zero before any future test of its value. In that case its value can be eliminated from the state description of the system. Consequently one can use variable-sized DBMs restricted to the relevant clocks in every region of the TA. In section 5 we will report the results of clock activity analysis of STARI and the performance of the variable-sized DBMs.

2.2 The Joy of Discrete Time

There is, however, an alternative semantics for TA based on discrete (and in fact, integer) time, which has already been discussed in early works about real-time logics (see the survey [AH92]). According to this view, time steps are multiples of a constant, and at every moment the automaton might choose between incrementing time or making a discrete transition. Consider the fragment of a 2-clock timed automaton depicted at the left of Figure 1. The automaton can stay in the state and let the time progress (i.e. let the values of x_1 and x_2 grow with derivative 1) as long as $x_1 \leq u$. As soon as x_1 reaches l (we assume $l \leq u$) it can take a transition to another state and reset x_1 to zero. By restricting the time domain to the integers, the staying conditions ("invariants") in every state are replaced by "idle" transitions as in the right of Figure 1.

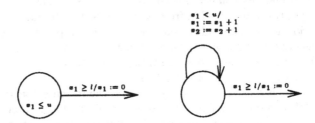

Fig. 1. A timed automaton and its discrete time interpretation.

Under this interpretation clocks are nothing but *bounded integer variables*, whose values are incremented simultaneously by time transitions and some of them are reset to zero by certain discrete transitions. Such systems are *finite-state*, but some components of the state-space, namely the clocks, have additional structure (addition and linear-ordering of clock values), which can be exploited by verification algorithms. In particular, any representation scheme for the dense semantics which is based on clock inequalities can be specialized for the discrete semantics. Since on discrete order, a strict inequality of the form $x_i < c$ can be written as the non-strict inequality $x_i \leq c - 1$, discrete regions can be expressed using exclusively non-strict inequalities. Hence even DBM-based methods can be tuned to work better on discrete time since the space of DBMs is smaller. A typical step in the iterative calculation of reachable states is depicted in Figure 2 for the dense (left) and discrete (right) semantics.

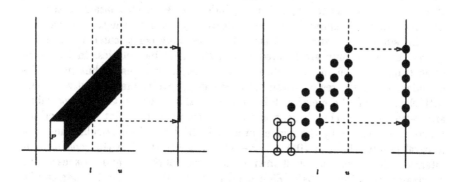

Fig. 2. Calculation of reachable configurations, starting from the initial set P, for the dense and discrete timed automata of Figure 1.

In addition to these methods one can take advantage of the finite-state nature of discrete TA and apply techniques which *cannot* be applied directly to dense time. One possibility is to push clocks values into states and transform the TA into a finite automaton (either off-line or on-the-fly). This provides for depth-first traversal of the state-space, as well as other search regimes. An alternative approach is the one advocated in [ABK+97,BMPY97] in which the clocks values are encoded in binary and subsets of them are written as BDDs. The advantage of this approach is that it gives a *canonical representation* for any subset (convex or not) of the state-space, and that it combines naturally with BDD-based representation of the control states. Most of this paper is a report of one success story of this approach, where a non-trivial system with 55 clocks has been verified. However, before that there is one little point to be clarified: although discrete time verification is inherently more efficient than dense time, it is certainly less expressive, and one might want to know what is sacrificed in order to improve performance. Our answer, based on the results in [HMP92,AMP98], which we explain below is: "not much".

2.3 Why Discrete Time Suffices

Consider two clock valuations $\mathbf{x}$ and $\mathbf{x}'$ sharing the same open unit square $S = (c_1, c_1 + 1, c_2, c_2 + 1)$ (Figure 3-(a)). Clearly, every inequality of the form $x_i \prec c$, $i \in \{1, 2\}$ and $\prec \in \{<, \leq, >, \geq\}$ is satisfied by $\mathbf{x}$ iff it is satisfied by $\mathbf{x}'$. Hence a transition that can be taken at $\mathbf{x}$, leading to a new valuation $\mathbf{y}$, iff it can be taken at $\mathbf{x}'$ leading to a point $\mathbf{y}'$ on the same square as $\mathbf{y}$. For the same reasons time can progress at $\mathbf{x}$ iff it can do so at $\mathbf{x}'$, however unless the order among the fractional parts of x_1 and x_2 is the same in $\mathbf{x}$ and $\mathbf{x}'$ they might reach different squares as time goes by. Only if they belong to the same triangular subset of X, namely a set of the form $\{x : \langle x_1 \rangle \prec \langle x_2 \rangle\}$ where $\langle x_i \rangle$ denotes the fractional part of x_i, they will meet the same squares during time evolution (Figure 3-(b)). Combining these facts we obtain the equivalence relation on X which guarantees that all the members of an equivalence class can exhibit essentially the same behaviors.

This simple (and simplified) story becomes more complicated if transition guards and invariants are allowed to contain strict inequalities. In that case some transitions might be enabled in the interior of a region but not in its boundaries, and the region graph becomes a more involved mathematical object with elements of all dimensionalities from 0 to n. If, however, timing constraints are restricted to be closed (i.e. non-strict) every boundary point satisfies *all* the constraints satisfied by the regions in its neighborhood. In particular the set of integral points, the grid $\{0, 1, \ldots, k-1\}^n$ "covers" all X in the sense that it intersects the boundaries of all open full-dimensional regions and satisfies all the constraints that they satisfy (Figure 3-(c)). Hence these integral points can be taken as *representatives* and all the (qualitative) trajectories starting from them cover all the possible behaviors of the system.

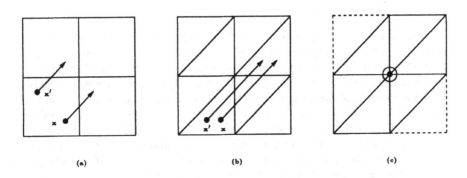

Fig. 3. a) Two points satisfying the same constraints; b) Two equivalent points; c) An integer point satisfying all the constraints satisfied by its six neighboring regions.

To be more precise, a discrete run might be a slight variation of some of the dense runs it represents: it may sometimes have few transitions taken simultaneously while in the dense run these transitions are separated by small amount of time. Nevertheless, the following results [HMP92,AMP98] underlie the soundness of discrete verification:

Theorem 1 (Emptiness of Closed TA). *The set of dense behaviors of a closed TA is non-empty iff it contains a discrete run.*

Combining this with the fact that untimed properties can be expressed as TAs in which all constraints are *true* (which is closed), we have:

Corollary 1 (Discrete Time Verification). *A closed TA satisfies an untimed property φ iff its discrete time version satisfies φ.*

If the TA is not closed, its closure is an over-approximation and a satisfaction of any linear-time property by the closure implies satisfiability by the original automaton.

3 STARI Description

STARI (Self-Timed At Receiver's Input) [Gre97] is a novel approach to high-bandwidth communication which combines synchronous and self-timed design techniques. Generally speaking, a transmitter communicates synchronously with a receiver through an asynchronous FIFO buffer (see Figure 4). The FIFO makes the system tolerant to time-varying skew between the transmitter and receiver clocks. An internal handshake protocol using acknowledgments prevents data loss or duplication inside the queue.

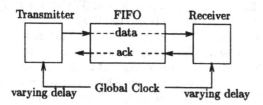

Fig. 4. The STARI overview.

The functioning of STARI is based on a rather intuitive idea. The FIFO must be initialized to be half-full. During each period of the clock one value is inserted to the FIFO by the transmitter and one value is removed by the receiver. Due to the complementary nature of these actions no control is required to prevent queue underflow or overflow. Short-term fluctuations in the clock rates of the transmitter and the receiver are handled by inserting or removing, more items to or from the queue.

Following the STARI model proposed by Tasiran and Brayton in [TB97], which differs slightly from the original description in [Gre97], we represent the boolean values *true* and *false* by *dual rail* encoding (see Figure 5). An auxiliary *empty* value is needed to distinguish between the case of two consecutive identical values and the case of one value maintained during more than one clock cycle. The transmitter is constrained to send sequences of *true* and *false* where each two occurrences of these values are separated by an occurrence of *empty*. The STARI chip consists of a linear array of n identical stages, each capable of storing a data value X.

The following two properties need to be proved to ensure the correct operation of the STARI circuit:

X	true	false	empty
X.t	1	0	0
X.f	0	1	0

Fig. 5. Dual rail encoding.

- Each data value output by the transmitter must be inserted in the FIFO before the next one.
- A new value must be output by the FIFO before each acknowledgment from the receiver

We specify the desired behavior of an n-stage STARI as an ideal FIFO buffer combined with a receiver and a transmitter respecting the abovementioned convention (see Figure 6). Note that in this specification, every transition is labeled with a pair of *put* and *get* actions, with the intended meaning that they can occur in *any order* including simultaneously. The goal of the verification is to show that if we hide the internal operations of STARI, the realizable sequences of *put*'s and *get*'s conform with this specification.

The operation principle of a stage k can be summarized as follows: *it may copy its predecessor value ($X_k := X_{k-1}$) when its successor has already copied (and acknowledged) its current value ($X_k = X_{k+1}$).* Using the dual rail encoding of data values, such a behavior can be achieved using two Muller C-elements that hold the $X.t$ and $X.f$ components, and one NOR gate for computing the acknowledgment (see Figure 7).

A Muller C-element works as follows: when the two inputs become identical, after some delay the output takes on their value, otherwise the output maintains its previous value. Consider, for example, a situation where stages k and $k+1$ hold the *empty* value, stage $k-1$ the *true* value and $Ack_{k+1} = 0$. When Ack_{k+1} becomes 1, the C-element for $X_k.f$ remains unchanged at 0 because its inputs are different (i.e. $Ack_{k+1} = 1$, $X_{k-1}.f = 0$). However, both the inputs of the C-element for $X_k.t$ are equal to 1 ($Ack_{k+1} = X_{k-1}.t = 1$), and after some delay, it will switch to 1. This way the *true* value has been copied from stage $k-1$ to stage k.

4 Modeling STARI by Timed Automata

The correct functioning of STARI depends on the timing characteristics of the gates (the time it takes, say, for a C-element to switch) and its relation with the central clock period and the skew between the receiver and transmitter. We model the uncertainty concerning the delay associated with gates using the bi-bounded delay model, that is, we associate with every gate an interval $[l, u]$ indicating the lower and upper bounds for its switching delay (see [L89], [BS94], [MP95] and [AMP98] for the exact definitions).

Following [MP95] we can model any logical gate with a delay $[l, u]$ using a timed automaton with 4 states (0-stable, 0-excited, 1-stable and 1-excited) and one clock. In particular, each stage of STARI is modeled by the three timed automata of Figures 8, 9 and 10.

Let us look at the automaton of Figure 8 which models the $X.t$ component of the kth stage. Its state is characterized by two boolean variables $X_k.t$, $x_k.t$, the former stores the gate output and the latter stores the gate internal value, i.e. the value to which the gate "wants" to go after the delay. The stable states are those in which $X_k.t = x_k.t$. The conditions for staying and leaving stable states are complementary and do not depend on clock values: for example, the automaton leaves state $(0,0)$ and goes to the unstable state $(0,1)$ exactly when both its inputs are 1. During this transition the clock variable $C_k.t$ is reset to zero. The automaton can stay at $(0,1)$ as long as $C_k.t < u_C$ and can change its output and stabilize in $(1,1)$ as soon as $C_k.t \geq l_C$, where $[l_C, u_C]$ is the delay interval associated with a C-element. The automaton for the $X.f$ component (Figure 9) is exactly the same (with different inputs) and the automaton for the NOR gate (Figure 10) is similarly characterized by two boolean variables Ack_k, ack_k, a clock variable $C_k.a$ and a delay bounded by $[l_N, u_N]$. This means that an n-stage STARI can be translated into $3n$ automata with $3n$ clocks and $6n$ boolean variables.

In addition to the automata for modeling the stages, we need three other automata for modeling the transmitter, the reciever and their clock cycle. The global *clock cycle* is modeled by a simple timed automaton using one clock variable C. Whenever C reaches the cycle size p it is reset to zero. (see Figure 11).

The *transmitter* is modeled as a 3-state automaton (Figure 11). At each clock cycle it puts a value at the input ports of the first stage ($X_0.t$ and $X_0.f$), according to the convention that every pair of data items is separated by an *empty* item. Moreover, the transmission can be done with some skew with respect to the clock cycle, bounded by the s_T constant, that is, the actual time of transmission can be anywhere in the interval $[p - s_T, p]$.

The *receiver* is a 1-state automaton (see Figure 11) which reads the current output value (i.e. $X_n.t$ and $X_n.f$) and acknowledges the reception by modifying Ack_{n+1} according to whether or not X_n is empty. As in the transmitter, a skew bounded by s_R is allowed.

Note that the receiver and transmitter skews *cannot accumulate* during successive cycles. They always range in an interval depending on the (perfect) global clock cycle. However, each one can vary non-deterministically from one cycle to another. This is more general than assuming a fixed skew given in advance, or a fixed skew chosen at start-up from a given interval. The transitions of these automata are annotated by action names such as *put* and *get* whose role is explanatory – they have no effect on the functioning of the system.

5 Verification Results and Performance Analysis

5.1 Discrete Time and BDDs

We have modeled various instants of STARI, each with a different number of stages. For each instance we have composed the timed automata and then min-

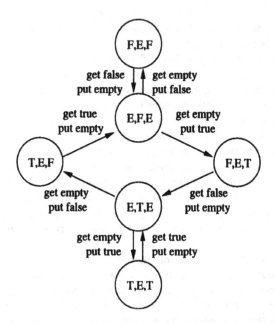

Fig. 6. The specification of an ideal 3-stage buffer. The states correspond to the buffer contents.

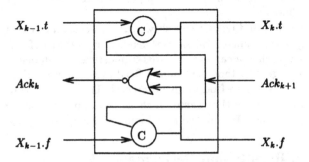

Fig. 7. Stage k of STARI.

135

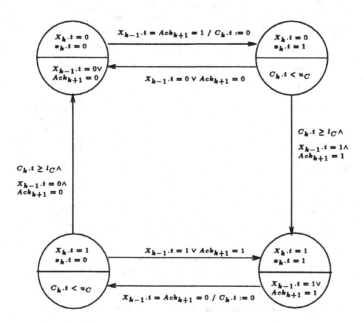

Fig. 8. The timed automaton for the C-element $X_k.t$.

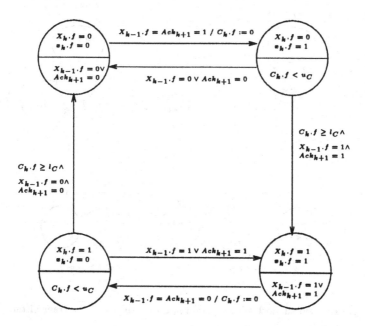

Fig. 9. The timed automaton for the C-element $X_k.f$.

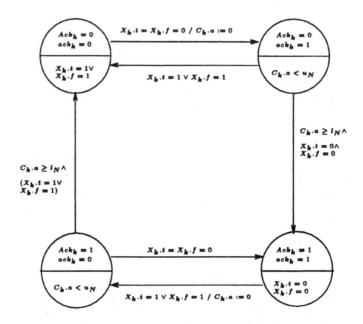

Fig. 10. The timed automaton for the NOR gate Ack_k.

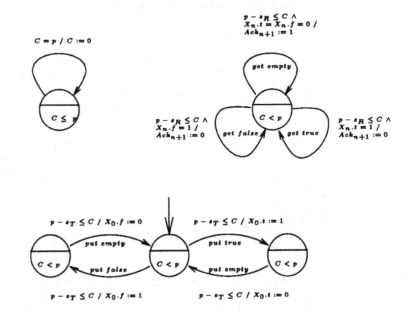

Fig. 11. A global clock with a period p, the receiver and the transmitter.

imized them by hiding the unobservable transitions. In Figure 12 one can see the automaton obtained for three stages, where in addition to the *put* and *get* actions, we left also the *tick* action which indicates the end of the global clock cycle. After hiding the *tick* we obtain a realization of the ideal FIFO as specified in Figure 6.

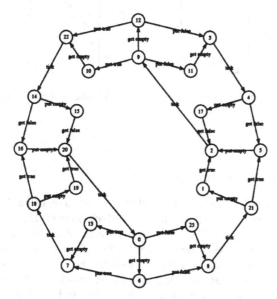

Fig. 12. A three-stage realizations of Stari with internal actions hidden.

We used the following concrete parameters to model the chip: $[l_C, u_C] = [l_N, u_N] = [2, 4]$ for gate delays, $p = 12$ for the clock cycle and $s_T = s_R = 2$ for the maximal allowed skew of the transmitter and the receiver.

The BDD implementation is based on the model-checker SMI [B97] and it uses the CUDD package [S95]. The 18-stage STARI is modeled by a network of timed automata summing up to 55 clocks. It uses 286 BDD variables to encode the clocks and the states. The reachable state space is of order of 10^{15} states.

We were able to prove that each STARI model with $n \leq 18$ stages, right initialized with m distinct values ($m \sim n/2$) *simulates* an ideal buffer of size m. Moreover, we verified that the transition graphs of the implementation and the specification are equivalent with respect to the *branching* bisimulation [vGW89], if we consider only the reading and writing to be observable. The equivalence is verified symbolically using the method described in [FKM93]. The time and the memory needed[1] to perform this verification are presented in Figure 13.

---

[1] All the results reported here were obtained on a Pentium II with 512 MB of memory.

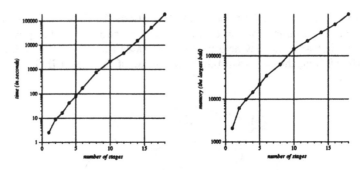

Fig. 13. Time and memory consumption for STARI verification using BDDs.

5.2 Variable-Dimension DBMs

We have also verified STARI, interpreted over dense-time, using the DBM representation of Kronos and the forward-analysis technique of [DT98]. To overcome the explosion associated with the size and number of DMBs we have used the techniques of [DY96,DT98], based on the notion of active and inactive clocks.

As one can see in Figure 8, the basic building block which is used to model a timed gate is a four-state automaton with one clock which is active only in the unstable states. So a-priori, each clock is active in half of the global control states. However, in real designs, especially when there is some natural order in which information flows in the circuit, the average over the reachable states of the number of active clocks can be much smaller.

The information concerning clock activity has been extracted automatically from the TA description and,using the variable-dimension DBM library of KRONOS, we were able to verify STARI with up to 8 stages (27 clocks). The main reason for the relative inferiority compared to the BDD approach is the large size of the discrete state-space (2^{24}): using DBMs, discrete variables are enumerated, whereas using discrete time and BDDs all variables (including clocks) are handled uniformly, which results in more compact representation. Future techniques, combining BDDs for the state variables and DBMs for the clocks might improve performance significantly.

Figure 14 shows the time performance and the number of symbolic states (discrete variables plus DBM) generated for number of stages. We have also measured the number of active clocks in each symbolic state and the results confirm our expectations that only a small fraction of clocks are active at any time. For instance, in the case of 8 stages, out of 27 clocks at most 9 were active, and this in less than 4% of the total number of DBMs generated (see diagram on the right of Figure 15). In more than 85% of the symbolic states, only 6 to 8 clocks were active. The distributions have the same shape for other STARI configurations.

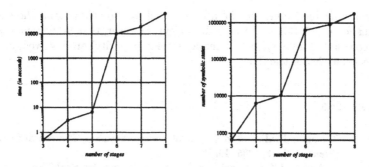

Fig. 14. Experimental results for STARI verification using DBMs.

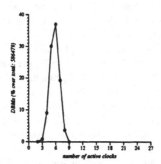

Fig. 15. The distribution of active clocks in an 8-stage STARI.

6 Discussion

Our performance resuts are significantly better than those reported by Tasiran and Brayton [TB97], from whom we have adopted the model. They prove, using techniques developed in [TAKB96], that every stage can be abstracted into a timed automaton having 5 states and only *one* clock. Using this abstract model and the tool Timed-Cospan they were able to verify an 8-stage buffer, while using the detailed model they could not verify more than 3 stages. Another attempt to verify STARI was reported by Belluomini and Myers [BM98] who model the circuit using a variant of timed Petri nets which they verify using the tool POSET which employs partial-order methods. The largest example they reported was of 10 stages. Yoneda and Ryu [YR99] improve these results significantly using circuit-specific heuristics.

We have demonstrated that a rather large example can be verified by tools based on timed automata, and we hope that this will contribute to the wide adoption of timed automata as a model for quantitative timing analysis. Our results indicate that in certain cases, discretized BDD-based approaches outperform other techniques. In the future we will try to characterize the class of systems for which this is the case. It is clear, however, that in examples where large constants (or equivalently, smaller time granularity) are involved, discrete time becomes less attractive.

References

[AD94]   R. Alur and D.L. Dill, A Theory of Timed Automata, *Theoretical Computer Science* 126, 183–235, 1994.

[AH92]   R. Alur and T.A. Henzinger, Logics and Models for Real-Time: A survey, J.W. de Bakker et al (Eds.), *Real-Time: Theory in Practice*, LNCS 600, 74-106, Springer, 1992.

[AK96]   R. Alur, and R.P. Kurshan, Timing Analysis in COSPAN, in R. Alur, T.A. Henzinger and E. Sontag (Eds.), *Hybrid Systems III*, LNCS 1066, 220-231, Springer, 1996.

[ABK+97]   E. Asarin, M. Bozga, A. Kerbrat, O. Maler, A. Pnueli and A. Rasse, Data-Structures for the Verification of Timed Automata, in O. Maler (Ed.), *Proc. HART'97*, LNCS 1201, 346-360, Springer, 1997.

[AMP98]   E. Asarin, O. Maler and A. Pnueli, On the Discretization of Delays in Timed Automata and Digital Circuits, in R. de Simone and D. Sangiorgi (Eds), *Proc. Concur'98*, LNCS 1466, 470-484, Springer, 1998.

[B96]   F. Balarin, Approximate Reachability Analysis of Timed Automata, *Proc. RTSS'96*, 52-61, IEEE, 1996.

[B97]   M. Bozga, SMI: An Open Toolbox for Symbolic Protocol Verification, Technical Report 97-10, Verimag, 1997.
           http://www.imag.fr/VERIMAG/DIST_SYS/SMI/

[BM98]   W. Belluomini and C.J. Myers, Verification of Timed Systems Using POSETs, in A.J. Hu and M.Y. Vardi (Eds.), *Proc. CAV'98*, 403-415, LNCS 1427, Springer, 1997.

[BMPY97]   M. Bozga, O. Maler, A. Pnueli, S. Yovine, Some Progress in the Symbolic Verification of Timed Automata, in O. Grumberg (Ed.) *Proc. CAV'97*, 179-190, LNCS 1254, Springer, 1997.

[BS94]   J.A. Brzozowski and C-J.H. Seger, *Asynchronous Circuits*, Springer, 1994.

[DOTY96]   C. Daws, A. Olivero, S. Tripakis, and S. Yovine, The Tool KRONOS, in R. Alur, T.A. Henzinger and E. Sontag (Eds.), *Hybrid Systems III*, LNCS 1066, 208-219, Springer, 1996.

[DT98]   C. Daws and S. Tripakis, Model checking of Real-time Reachability Properties using Abstractions, *Proc. TACAS'98*, LNCS 1384, 1998.

[DY96]   C. Daws and S. Yovine, Reducing the Number of Clock Variables of Timed Automata, *Proc. RTSS'96*, 73-81, IEEE, 1996.

[D89]   D.L. Dill, Timing Assumptions and Verification of Finite-State Concurrent Systems, in J. Sifakis (Ed.), *Automatic Verification Methods for Finite State Systems*, LNCS 407, 197-212, Springer, 1989.

[FKM93]   J.C. Fernandez, A. Kerbrat, and L. Mounier, Symbolic Equivalence Checking, In C. Courcoubetis (Ed.), *Proc. CAV'93*, LNCS 697, Springer, 1993.

[Gre97]   M. R. Greenstreet, STARI: Skew Tolerant Communication, to appear in *IEEE Transactions on Computers*, 1997.

[H93]   N. Halbwachs, Delay Analysis in Synchronous Programs, in C. Courcoubetis (Ed.), *Proc. CAV'93*, LNCS 697, 333-346, Springer, 1993.

[HMP92]   T. Henzinger, Z. Manna, and A. Pnueli. What Good are Digital Clocks?, in W. Kuich (Ed.), *Proc. ICALP'92*, LNCS 623, 545-558, Springer, 1992.

[LLPY97] K. Larsen, F. Larsson, P. Pettersson and W. Yi, Efficient Verification of Real-Time Systems: Compact Data Structure and State-Space Reduction, *Proc. RTSS'98*, 14-24, 1997.

[LPY97]   K.G. Larsen, P. Pettersson and W. Yi, UPPAAL in a Nutshell, *Software Tools for Technology Transfer* 1/2, 1997.

[L89]   H.R. Lewis, Finite-state Analysis of Asynchronous Circuits with Bounded Temporal Uncertainty, TR15-89, Harvard University, 1989.

[MP95]   O. Maler and A. Pnueli, Timing Analysis of Asynchronous Circuits using Timed Automata, in P.E. Camurati, H. Eveking (Eds.), *Proc. CHARME'95*, LNCS 987, 189-205, Springer, 1995.

[S95]   F. Somenzi, CUDD: CU Decision Diagram Package, 1995.

[SV96]   J. Springintveld and F.W. Vaandrager, Minimizable Timed Automata, in B. Jonsson and J. Parrow (Eds.), *Proc. FTRTFT'96*, LNCS 1135, 130-147, Springer, 1996.

[TAKB96] S. Tasiran R. Alur, R.P. Kurshan and R. Brayton, Verifying Abstractions of Timed Systems, in *Proc. CONCUR'96*, 546-562, Springer, 1996.

[TB97]   S. Tasiran and R.K. Brayton, STARI: A Case Study in Compositional and Hierarchical Timing Verification, in O. Grumberg (Ed.) *Proc. CAV'97*, 191-201, LNCS 1254, Springer, 1997.

[TKB97]   S. Tasiran, Y. Kukimoto and R.K. Brayton, Computing Delay with Coupling using Timed Automata, *Proc. TAU'97*, 1997.

[vGW89]   R. J. van Glabbeek and W. P. Weijland, Branching-Time and Abstraction in Bisimulation Semantics (extended abstract), CS R8911, Centrum voor Wiskunde en Informatica, Amsterdam, 1989.

[WD94]   H. Wong-Toi and D.L. Dill, Approximations for Verifying Timing Properties, in T. Rus and C. Rattray (Eds.), *Theories and Experiences for Real-Time System Development*, World Scientific Publishing, 1994.

[YR99]   T. Yoneda and H. Ryu, Timed Trace Theoretic Verification Using Partial Order Reductions, in *Proc. Async'99*, 108-121, IEEE Press, 1999.

From Asymmetry to Full Symmetry: New Techniques for Symmetry Reduction in Model Checking *

E. Allen Emerson and Richard J. Trefler

Department of Computer Sciences and Computer Engineering Research Center
University of Texas, Austin, TX, 78712, USA

Abstract. It is often the case that systems are "nearly symmetric"; they exhibit symmetry in a part of their description but are, nevertheless, globally asymmetric. We formalize several notions of near symmetry and show how to obtain the benefits of symmetry reduction when applied to asymmetric systems which are nearly symmetric. We show that for some nearly symmetric systems it is possible to perform symmetry reduction and obtain a bisimilar (up to permutation) symmetry reduced system. Using a more general notion of "sub-symmetry" we show how to generate a reduced structure that is simulated (up to permutation) by the original asymmetric program.

In the symbolic model checking paradigm, representing the symmetry reduced quotient structure entails representing the BDD for the orbit relation. Unfortunately, for many important symmetry groups, including the full symmetry group, this BDD is provably always intractably large, of size exponential in the number of bits in the state space. In contrast, under the assumption of full symmetry, we show that it is possible to reduce a textual program description of a symmetric system to a textual program description of the symmetry reduced system. This obviates the need for building the BDD representation of the orbit relation on the program states under the symmetry group. We establish that the BDD representing the reduced program is provably small, essentially polynomial in the number of bits in the state space of the original program.

1 Introduction

Model checking [CE81] (cf. [QS82] [LP85] [CES86]) is an algorithmic method for determining whether a finite state system, M, satisfies a temporal logic formula, f. Lichtenstein and Pnueli [LP85] have argued that in practice the complexity of model checking will be dominated by $|M|$, the size of M. Unfortunately, $|M|$ may be exponentially larger than the textual description of M. For example, a system comprised of n identical processes running in parallel, each of which has 3 local states, may have 3^n reachable states.

---

* The authors' work was supported in part by NSF grant CCR-980-4736 and SRC contract 98-DP-388.

Symmetry reduction is an abstraction technique which endeavors to substantially ameliorate this state explosion problem by exploiting the fact that many computer systems are symmetric in their design and implementation (cf. [JR91] [ID96] [ES96] [CE+96] [HI+95] [MAV96] [ES97] [GS97] [ET98] [AHI98]). Such symmetry can be seen to be a form of redundancy from the standpoint of model checking temporal logic formulae. The state graph, M, of many synchronization and coordination protocols which are the parallel composition of n processes identical up to renaming, often exhibits considerable symmetry. For example, the mutual exclusion protocol contains states (C_1, T_2) and (T_1, C_2) representing the states where process 1 is in its critical section and process 2 is attempting to reach its critical section and vice versa. These two states are related by the permutation (1 2) which drives process index 1 to 2 and 2 to 1; in general the permutation (1 2) when applied systematically to the states and transitions of M results in M again, that is (1 2) is an automorphism of M. Aggregating states which are equivalent up to permutation factors out the symmetry of a system and model checking is then performed on the symmetry reduced structure – a substantial, often exponential, savings can be achieved.

While symmetry reduction methods offer great potential, there are several obstacles to its more widespread application. Firstly, it is often the case that protocols are not symmetric; they may contain a high degree of symmetry in some part of their design but their global behavior is asymmetric. This can occur, for instance, in systems with processes which are identical up to renaming and the assignment of priorities. The readers–writers protocol, a refinement of the mutual exclusion protocol, is one such example. In the mutual exclusion algorithm the two processes competing for access to their critical sections are given equal priority; in the readers–writers protocol the writer is given priority. While the global state graph of the readers–writers protocol is asymmetric, it is symmetric in every aspect except the transition from the state where both processes are attempting to access their critical sections.

Secondly, BDD [Br86] based symbolic representation of the symmetry relation used in forming symmetry reduced quotient structures can be proved to be of exponential size. From this it has been argued that symmetry and symbolic representation [Mc92] [BC+92] do not combine profitably [CE+96].

We describe solutions to both these problems in this paper. Previous work on symmetry reduction ([ES96] cf. [CE+96]) defined the symmetry reduced structure, $\overline{M}$, as the quotient structure of M induced by the equivalence relation, $\equiv_G$, on states. Two states, s and s', are equivalent, $s \equiv_G s'$, iff there is an automorphism π in G which drives s to s'. We relax this relationship by defining a permutation π to be a *near automorphism* iff for every transition $s \to t$ in M either $\pi(s) \to \pi(t)$ is in M or s is invariant under the automorphisms of S, the set of states of M. The equivalence relation on states induced by the group of near automorphisms defines a quotient structure that is bisimilar, up to permutation, to M. Therefore, even asymmetric structures can be near symmetry reduced.

Near automorphisms are, however, restrictive in the sense that whenever $\pi(s) \to \pi(t)$ is not a transition then s must be a highly symmetric state. By weakening the requirements for the preservation of transitions by permutations we can apply these ideas to a wider class of problems. Specifically, we define a notion of *rough symmetry* for multi-process systems whose processes are 'almost-symmetric'. Intuitively, a system is roughly symmetric if for every pair of processes i and j, the actions of process i from local state P, in global state s, can be mimicked by process j when j is the highest priority process in local state P, in the equivalent global state s'. We then show that the rough symmetry reduced system is bisimilar to the original system M.

By further weakening the restrictions on permutations applied to structures, we define a notion of *sub-symmetry* which allows for the creation of an abstract symmetry reduced structure that is simulated by the original program. A permutation π is a *sub-automorphism* of M if π drives certain "closed" subgraphs of M back into M. This notion of sub-automorphism induces a pre-order $\leq_H$ on states such that $s \leq_H t$ iff there is a sub-automorphism π which drives a closed subgraph containing s back into M and $\pi(s) = t$. We then use $\leq_H$ to define a sub-symmetry reduced structure, $\overline{M}_{\leq_H}$, which is simulated up to permutation by M, thereby showing that $\forall$CTL* [CGL94] formulae true of $\overline{M}_{\leq_H}$ are true of M.

Finally, we show how to successfully combine symmetry with BDD-based symbolic representations of systems. For many symmetry groups, including the full symmetry group, the BDD for the orbit relation, that is for determining equivalence of two states under group action, must always be of exponential size. This orbit BDD is used to permit designation of a *specific representative* state for each equivalence class in the quotient structure. The orbit BDD must recognize as equivalent, say, the states (N_1, N_2, T_3), (N_1, T_2, N_3), and (T_1, N_2, N_3). A specific, actual state is chosen as a representative. In the case of full symmetry, we can instead use *generic representatives*, for example, $(2N, 1T)$, which obviates the need for representation of the orbit relation. This is accomplished by compiling the program text of the fully symmetric program P into the program text of the symmetry reduced program $\overline{P}$ over generic states. $\overline{P}$ defines a structure $M(\overline{P})$ isomorphic (and bisimilar up to permutation) to the symmetry reduced structure $\overline{M}$ and model checking can then be performed on $M(\overline{P})$. Assuming that M is composed of n processes, this compilation process not only obviates the need for determining the equivalence of states under permutation but also reduces the number of bits used to represent a state in the symmetry reduced program from $\mathcal{O}(n)$ in the case of $\overline{M}$ to $\mathcal{O}(\log n)$ in the case of $M(\overline{P})$. A consequence is that the BDD representing $M(\overline{P})$ is always of polynomial size in the number of processes, in contrast to the exponential size BDD based on specific representatives.

The remainder of the paper is organized as follows: Section 2 contains some preliminaries, Section 3 discusses model checking asymmetric systems, compilation of fully symmetric programs into symmetry reduced programs is outlined in Section 4 and Section 5 contains a brief conclusion.

2 Preliminaries

Our model of computation, presented formally below, can be seen to represent the interleaved computations of a program composed of n communicating processes. States are n-tuples of local states, one for each process. Transitions represent the movement of one process from one local state to another. Permutations on the set $[1..n]$ can then be interpreted as permutations of process indices.

We denote the set of natural numbers by $\mathbb{N}$. Let $\mathcal{I}$ be a finite index set $[1..n]$ for some $n \in \mathbb{N}$, $n > 0$. LP is a finite set of local states. $Sym\ \mathcal{I}$ is the set of permutations on index set $\mathcal{I}$. $M = (S, R)$ is a structure where $S \subseteq LP^n$ and $R \subseteq S \times S$ is non-empty and total. We write both $(s, t) \in R$ and $s \to t \in R$ to mean that there is a transition from state s to state t in R. For $l \in LP$, $i \in [1..n]$ and $s \in S$ we write $(l, i) \in LP \times \mathcal{I}$ as l_i and $s(i) = l$ (l_i is true at s) iff the ith element of s is l.

A permutation $\pi \in Sym\ \mathcal{I}$ acts on a state $s \in S$ in the following way: $s(i) = l$ iff the $\pi(i)$th element of $\pi(s)$ is l. π is an automorphism of $M = (S, R)$ iff $S = \{\pi(s) \mid s \in S\}$ and $R = \{(\pi(s), \pi(t)) \mid (s, t) \in R\}$. Attention is usually restricted to such permutations because they preserve both the state space and the transition relation of the structure, M. A state s is said to be fully symmetric if for all $\pi \in Sym\ \mathcal{I}$, $\pi(s) = s$. The identity permutation is denoted by id. For any M, $Aut(M)$ the set of automorphisms of M, is a group. Similarly, for state s, $Aut(s)$ is the set of permutations, π, such that $\pi(s) = s$ and $Aut(S)$ is the set of permutations, π, such that $\pi(S) = S$.

Any subgroup G of $Aut(M)$, induces the following equivalence relation, $s \equiv_G t$ iff there exists a $\pi \in G$ such that $\pi(s) = t$. M's symmetry reduced structure, with respect to G, $\overline{M} = M/_{\equiv_G} = (\overline{S}, \overline{R})$ is defined as follows: $\overline{S} = \{\overline{s} \in S \mid \overline{s}$ is the unique representative of the equivalence class $[\overline{s}]_{\equiv_G}\}$ [1] and $(\overline{s}, \overline{t}) \in \overline{R}$ iff there exists $(\overline{s}, t) \in R$ for some $t \equiv_G \hat{t}$ [ES96][CE+96] (cf. [ES96] for more details).

In the sequel we will make use of the expressive branching time temporal logic CTL* [EH86] (cf. [Em90] for more details). Let $LP \times \mathcal{I}$ be the set of atomic propositions. A path formula is formed from boolean combinations ($\wedge, \vee, \neg$) and nestings of atomic propositions, state formulae and the usual temporal operators X, G, F, U and V (the dual of U). State formulae are formed from boolean combinations of atomic propositions, state formulae and prefixing of path formulae by path quantifiers A and E. For example, the formula $AG\neg(writerC \wedge readerC)$ says that along all computations it is never the case that both the *writer* and the *reader* are accessing their critical sections. We write $M, s \models f$ to denote that state s in structure M satisfies formula f and $M \models f$ to denote that there is a state, s, in M such that $M, s \models f$. A formula is in positive normal form (PNF) if the $\neg$ operator appears only in front of atomic propositions. ECTL* is the sub-logic of CTL* in which every formula, when put in PNF, contains only E path quantifiers. Similarly, ACTL* is the sub-logic of CTL* in which every formula, when put in PNF, contains only A path quantifiers [CGL94].

---

[1] $\overline{s}$ is the distinguished element of S and $[\overline{s}]_{\equiv_G}$ is the set of $s \in S$ such that $s \equiv_G \overline{s}$.

We define symmetric versions of CTL* and its sub-logics simply for ease of exposition – all our results can be restated to handle full CTL*[2]. The syntax of Symmetric CTL* (SCTL*) is the same as for CTL* except that the atomic formulae are restricted to the following: $\forall i : l_i, \exists i : l_i, \forall i : \neg l_i, \exists i : \neg l_i$ and $\exists i \neq j : l_i \wedge l_j$. For example, $\mathsf{AG}\neg(\exists i \neq j : C_i \wedge C_j)$ is a formula of SACTL*.

2.1 Simulation up to Permutation

Let $M = (S, R)$ and $M' = (S', R')$ be structures defined over LP and $\mathcal{I}$. $B \subseteq S \times S'$ is a simulation up to permutation (cf. [Mi71] [Pa81] [HM85] [MAV96] [ES96] [CE+96]) iff for all $(s, s') \in B$

- there is a $\pi \in Sym\ \mathcal{I}$ such that $\pi(s) = s'$ and
- for all $(s, t) \in R$ there is a t' such that $(s', t') \in R'$ and $(t, t') \in B$.

$B \subseteq S \times S'$ is a bisimulation up to permutation iff for all $(s, s') \in B$ the above two conditions hold and

- for all $(s', t') \in R'$ there is a t such that $(s, t) \in R$ and $(t, t') \in B$.

Proposition 1. *([ES96] [CE+96]) Let B be a bisimulation up to permutation. For all $(s, s') \in B$ and all SCTL* formulae f, $M, s \models f$ iff $M', s' \models f$.*

Proposition 2. *([ES96] [CE+96]) Let B be a simulation up to permutation. For all $(s, s') \in B$ and all SACTL* formulae f, $M', s' \models f$ implies $M, s \models f$.*

Proposition 3. *([ES96] [CE+96]) Let B be a simulation up to permutation. For all $(s, s') \in B$ and all SECTL* formulae f, $M, s \models f$ implies $M', s' \models f$.*

3 Symmetry Reduction and Asymmetric Systems

3.1 Near Automorphism Based Reductions

Let $M = (S, R)$ be a structure. A permutation π is a *near automorphism* of M if $\pi(S) = S$ and for all $(s, t) \in R$ either $Aut(S) \subseteq Aut(s)$ or $\pi(s) \to \pi(t) \in R$. Let $NAutM = \{\pi \in Sym\ \mathcal{I} \mid \pi \text{ is a near automorphism of } M\}$.

Theorem 1. *Given $M = (S, R)$ the set $NAutM$ is a group.*

Corollary 1. *$\overline{M}_{NAut} = M/\equiv_{NAutM} = (\overline{S}, \overline{R})$ is bisimilar up to permutation to $M = (S, R)$ and for all $(s, \overline{s})$ such that $s \equiv_{NAutM} \overline{s}$, and for all SCTL* formulae f, $M, s \models f$ iff $\overline{M}_{NAut}, \overline{s} \models f$.*

---

[2] When model checking formula f over $\overline{M} = M/\equiv_G$ it is required that for every maximal propositional sub-formula p of f, and every permutation $\pi \in G$, $\pi(p) \equiv p$ [ES96] [CE+96]. SCTL*, SACTL*, and SECTL* all satisfy this requirement.

We can apply these ideas to the readers-writers problem as given in figure 1. The flip permutation (1 2) which drives index 1 to index 2 and vice versa is a near automorphism. This implies that the structure in the figure has the full symmetry group, $Sym\ \mathcal{I}$, as its group of near automorphisms and therefore the near symmetry reduced structure given in figure 2 is bisimilar up to permutation to the structure in figure 1. Model checking for safety formulae like $\mathsf{AG}\neg(\exists i \neq j : C_i \wedge C_j)$ and liveness formulae like $\mathsf{AG}[(\exists i : T_i) \Rightarrow \mathsf{AF}(\exists i : C_i)]$ – which says that along all computations it is always the case that if some process is trying to enter its critical section then it is inevitable that some process enters its critical section – can then be performed on the near symmetry reduced structure $\overline{M}_{NAut}$ instead of M.

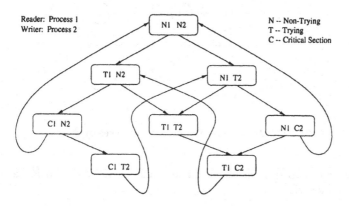

Fig. 1. Asymmetric Readers-Writers

Figure 3 contains the program skeletons which generate the structure M in figure 1. The near automorphisms for M can be generated directly from the program skeletons through the following observation. While the skeletons are not symmetric they are nearly symmetric in the following sense. Ignoring, for the moment, the transition $T_2 \rightarrow C_2$ that is enabled when T_1 is true, the two skeletons are symmetric – the flip permutations applied to the skeletons results in the same two skeletons. The asymmetry of the transition $T_2 \rightarrow C_2$ that is enabled when T_1 is true guarantees a near symmetry of the induced Kripke structure because this symmetry breaking transition is only enabled from the fully symmetric state (T_1, T_2). In the full paper we give a more detailed algorithm for determining near automorphisms from program skeletons.

Finally, we note that the near symmetry reduced quotient structure $\overline{M}_{NAut} = M/\equiv_{NAut}$ can be built directly from the program text without building M in a manner analogous to that used to build $\overline{M}$. Basically, the procedure works as follows, given a state $\overline{s} \in \overline{S}$ generate each of the states t such that $\overline{s} \rightarrow t \in R$ as described by the program text. For each t if t is equivalent to a state $\overline{t} \in \overline{S}$ then

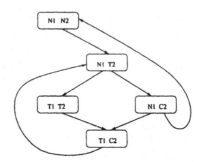

Fig. 2. Near Symmetry Reduced Asymmetric Readers-Writers

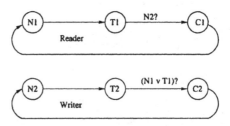

Fig. 3. Readers-Writers Program Skeletons

add an arc $\bar{s} \to \bar{t}$ to $\overline{R}$ otherwise add t to $\overline{S}$ and the arc $\bar{s} \to \bar{t}$ to $\overline{R}$ (see [ES96] for complete details).

3.2 Rough Symmetry Based Reduction

The previous section showed how the definition of near automorphism was sufficient to justify applying symmetry reduction techniques to asymmetric systems such as the reader-writer problem. That technique required a permutation to act on the structure by preserving transitions: $s \to t$ implies $\pi(s) \to \pi(t)$ or s is a symmetric state. Requiring s to be symmetric, however, implies that the technique cannot be used to handle the more general readers-writers problems. We therefore seek a relationship which will allow extensive reduction and generality. Below we formulate a notion of equivalence based on roughly symmetric programs which again leads to a bisimulation between M and the rough symmetry reduced version of M. Within this framework we can show that a more general version of the readers-writers problem may be symmetry reduced.

The intuition behind our approach is as follows. We suppose that $M = (S, R)$ is the structure corresponding to a multi-process system in which the processes have been assigned priorities. Furthermore, we assume that some portion of M is highly symmetric. For instance, in the multiple readers-writers protocol, by ignoring the extra functionality of a writer over a reader, it is possible to see this protocol as 'fully symmetric'. We view this highly symmetric M' as a 'substructure' of M from which states and actions have either been removed or added in some systematic way to form M. Taking the group, G, of automorphisms from this sub-structure we then seek to show that for every pair, (s, s'), of G equivalent states and every transition $s \to t$ of process i in M is preserved by some permutation π which drives s to s' and the transition $s \to t$ to an equivalent transition by the highest priority process, $\pi(i)$. We then say that M is roughly symmetric with respect to the symmetry group of M'.

Formally, let $M = (S, R)$ where $R = R_1 \cup \ldots \cup R_n$, each $R_i \subseteq S \times S$. R_i is the transition relation for process i. Let G be a sub-group of $Sym\ \mathcal{I}$. We say that R_i is covered (with respect to G) by R_j, iff for all transitions $s \to t \in R_i$ and all $s' \equiv_G s$, if $j = \max\{j' \mid s(i) = s'(j')\}$ then there is a permutation $\pi \in G$ such that $\pi(s) = s'$, $\pi(i) = j$, $s' \to \pi(t) \in R_j$. Then we say that M is roughly symmetric with respect to G iff for all $i, j \in \mathcal{I}$, R_i is covered by R_j.

Theorem 2. *Suppose $M = (S, R)$, $R = R_1 \cup \ldots, R_n$ is roughly symmetric with respect to G, a sub-group of $Sym\ \mathcal{I}$. Then $\overline{M} = M/\equiv_G = (\overline{S}, \overline{R})$ is bisimilar up to permutation to M.*

Proof: Let $B = \{(s, \overline{s}) \in S \times \overline{S} \mid s \equiv_G \overline{s}\}$. Let $s \equiv_G \overline{s}$ for some $s \in S, \overline{s} \in \overline{S}$. Then suppose $s \to t \in R_i$. Let $j = \max\{j' \mid \overline{s}(j') = s(i)\}$, then there is a $\pi \in G$ such that $\pi(s) = \overline{s}$, $\pi(i) = j$ and $\overline{s} \to \pi(t) \in R_j$. This implies $\overline{s} \to \overline{t} \in \overline{R}$ for some $\overline{t} \equiv_G \pi(t)$ which implies that $t \equiv_G \overline{t}$ and $(t, \overline{t}) \in B$. Suppose $\overline{s} \to \overline{t} \in \overline{R}$. Then for some i and $t' \equiv_G \overline{t}$, $\overline{s} \to t' \in R_i$. Let $j \max\{j' \mid s(j') = \overline{s}(i)\}$. Then there is some $\pi \in G$ such that $\pi(\overline{s}) = s$, $\pi(i) = j$ and $s \to \pi(t') \in R_j$. But then $\pi(t') \equiv_G t'$ which implies $\pi(t') \equiv_G \overline{t}$ and hence $(\pi(t'), \overline{t}) \in B$. $\square$

We can apply these ideas to show that a general readers-writers system can be symmetry reduced by the full symmetry group to a bisimilar rough symmetry reduced quotient structure. We have $m < n$ identical readers and $n - m$ identical writers. The writers all have priority over all the readers but no two processes may access their critical sections at the same time. The generic process skeletons for these processes are identical to the ones in figure 3 except that for reader i the arc from state T_i to state C_i is labeled by $(N_1 \vee T_1) \wedge \ldots (N_{i-1} \vee T_{i-1}) \wedge (N_{i+1} \vee T_{i+1}) \wedge \ldots \wedge (N_m \vee T_m) \wedge N_{m+1} \wedge \ldots \wedge N_n$ and for writer j the arc from state T_j to state C_j is labeled by $(N_1 \vee T_1) \wedge \ldots \wedge (N_{j-1} \vee T_{j-1}) \wedge (N_{j+1} \vee T_{j+1}) \wedge \ldots \wedge (N_n \vee T_n)$. We now show that $M = (S, R)$ for this readers writers system is roughly symmetric with respect to $Sym\ \mathcal{I}$. Consider $s \to t \in R_i$ and $s \equiv_{Sym\ \mathcal{I}} s'$. Let $j = \max\{j' \mid s'(j') = s(i)\}$. Since there is a permutation mapping s to s' there is a permutation π such that $\pi(s) = s'$ and $\pi(i) = j$. Suppose $s \to t \in R_i$ follows from $N_i \to T_i$, then $\pi(s) \to \pi(t) \in R_j$. Similarly if $s \to t \in R_i$ is due to $C_i \to N_i$ then $\pi(s) \to \pi(t) \in R_j$. Suppose $s \to t \in R_i$ follows from either a writer or a reader entering its critical section. If i is a reader then $(N_1 \vee T_1) \wedge \ldots (N_{i-1} \vee T_{i-1}) \wedge (N_{i+1} \vee T_{i+1}) \wedge \ldots \wedge (N_{m-1} \vee T_{m-1}) \wedge N_m \wedge \ldots \wedge N_n$ or if i is a writer then $(N_1 \vee T_1) \wedge \ldots \wedge (N_{j-1} \vee T_{j-1}) \wedge (N_{j+1} \vee T_{j+1}) \wedge \ldots \wedge (N_n \vee T_n)$. Then consider that j is the largest index such that $s'_j = T$. If j is a writer it cannot be blocked. If j is a reader it cannot be blocked because no writer is in its T section. Therefore process j can enter it's critical section from state s' and we have that $\pi(s) \to \pi(t) \in R_j$. Therefore, the readers-writers system is roughly symmetric with respect to $Sym\ \mathcal{I}$ and an exponential savings may be achieved through rough symmetry reduction.

3.3 Simulation Based Reductions and Asymmetry

Given $M = (S, R)$, let S' be a subset of S. S' is closed (with respect to M) iff for all $s \in S'$ and all $(s, t) \in R$, $t \in S'$. Let $\pi \in Sym\ \mathcal{I}$ and $S' \subseteq S$ be closed. π is a sub-automorphism on S' iff $\{\pi(s) \mid s \in S'\} \subseteq S$ and for all $s, t \in S'$ if $s \to t \in R$ then $\pi(s) \to \pi(t) \in R$. Let H be the subset of $Sym\ \mathcal{I} \times 2^S$ such that $(\pi, S') \in H$ iff π is a sub-automorphism on the closed subset of S, S'. $s \leq_H t$ iff there is a $(\pi, S') \in H$ such that $s \in S'$ and $\pi(s) = t$.

Proposition 4. $s \leq_H t$ and $t \leq_H u$ implies $s \leq_H u$.

Proof: $s \leq_H t$ implies there is some closed $S' \subseteq S$ and π such that $(\pi, S') \in H$ and $\pi(s) = t$. Furthermore, there is some closed $S'' \subseteq S$ and ϕ such that $(\phi, S'') \in H$ and $\phi(t) = u$. Consider $T \subseteq S'$ such that T contains s and all the states reachable from s in S'. S' closed implies such a T exists and is a closed subset of S'. $\pi(T) \subseteq S''$ is straight forward. This implies that for all $s, t \in T$, $(\phi \cdot \pi)(s) \in S$ and if $(s, t) \in R$, $(\phi \cdot \pi)(s) \to (\phi \cdot \pi)(t) \in R$ which implies that $((\phi \cdot \pi), T) \in H$. Since $(\phi \cdot \pi)(s) = u$ it is the case that $s \leq_H u$. $\square$

For, $\bar{s} \in S$, define $[\bar{s}]_{\leq_H} = \{s \in S | s \leq_H \bar{s}\}$. Then a sub-symmetry reduced version of M is $\overline{M}_{\leq_H} = M/\leq_H = (\overline{S}, \overline{R})$ where

- $\overline{S} \subseteq S$ and
- for all $s \in S$ there is an $\bar{s} \in \overline{S}$ such that $s \in [\bar{s}]_{\leq_H}$ and
- $(\bar{s}, \bar{t}) \in \overline{R}$ iff there is some $t \leq_H \bar{t}$ such that $(\bar{s}, t) \in R$.

Let $M = (S, R)$ and $\overline{M}_{\leq_H} = (\overline{S}, \overline{R})$ be structures as described above. Then let $B = \{(s, \bar{s}) \in S \times \overline{S} \mid s \in [\bar{s}]_{\leq_H}\}$.

Theorem 3. B is a simulation up to permutation.

Proof: Suppose $(s, \bar{s}) \in B$ then $s \in [\bar{s}]$ and there is a $(\pi, S') \in H$, such that $\pi(s) = \bar{s}$. Suppose $(s, t) \in R$. This implies that $(\pi(s), \pi(t)) \in R$. By the structure of $\overline{M}_{\leq_H}$ this implies that there is some $\bar{t}$ such that $(\bar{s}, \bar{t}) \in \overline{R}$ and $\pi(t) \leq_H \bar{t}$. But this implies that $t \leq_H \bar{t}$ hence $(t, \bar{t}) \in B$. $\square$

Corollary 2. For all $SACTL^*$ formulae, f, and for all $(s, \bar{s}) \in B$, $\overline{M}_{\leq_H}, s \models f$ implies $M, s \models f$.

In fact, this type of reduction is possible for any $H' \subseteq H$. $\leq_{H'}$ is defined as above and $\leq_{H'}^*$ is the reflexive, transitive closure of $\leq_{H'}$. The sub-symmetry reduced system, $\overline{M}_{\leq_{H'}^*} = M/\leq_{H'}^* = (\overline{S}, \overline{R})$ is then defined analogously.

Proposition 5. If $\leq_H$ is symmetric then $\leq_H$ is an equivalence relation.

Theorem 4. Let $M = (S, R)$ and $\overline{M}_{\leq_H} = (\overline{S}, \overline{R})$ be as above and let $\leq_H$ be symmetric, then $B = \{(s, \bar{s}) \in S \times \overline{S} | s \in [\bar{s}]_H\}$ is a bisimulation up to permutation.

Proof: Let $(s, \bar{s}) \in B$. Suppose $\bar{s} \to \bar{t} \in \overline{R}$, then there is some t such that $\bar{s} \to t \in R$ and $t \leq_H \bar{t}$. $s \leq_H \bar{s}$ implies $\bar{s} \leq_H s$ which implies there is some $(\phi, T') \in H$ such that $\bar{s} \in T'$ and $\phi(\bar{s}) = s$. Hence $s \to \phi(t)$ which implies $t \leq_H \phi(t)$ and therefore $\phi(t) \leq_H t$. This implies $\phi(t) \leq_H \bar{t}$ and therefore $(\phi(t), \bar{t}) \in B$. □

Corollary 3. *For all $SCTL^*$ formulae, f, and all $(s, \bar{s}) \in B$, $M, s \models f$ iff $\overline{M}_{\leq_H}, \bar{s} \models f$*

When $\leq_H$ (or $\leq_{H'}$) can be determined from the program text or is given *a priori* then it is possible to build the sub-symmetry reduced structure $\overline{M}_{\leq_H}$ directly from the program text without first constructing M. The procedure is analogous to building $\overline{M} = M/\equiv_{Aut(M)}$, however, it may require some backtracking as it is possible that a state s is generated in $\overline{M}_{\leq_H}$ which can then be replaced by a state $\bar{s}$ such that $s \leq_H \bar{s}$.

4 Symmetry Reduction on Fully Symmetric Programs

Representing symmetry reduced structures with BDD's is, typically, computationally intractable. The BDD representing the orbit relation of many groups, including the full symmetry group, is of size exponential in the number of processes or the number of bits in a state. In the sequel, we show that under the assumption of full symmetry, symmetry reduction can be done efficiently in the symbolic model checking paradigm without representation of the orbit relation. Let $k = |LP|$, be the number of local states of an individual process P_i. Given a program $P = //_{i \in [1..n]} P_i$, the parallel composition of n processes identical up to renaming, which defines a fully symmetric Kripke structure $M(P)$, we compile P into a program $\overline{P}$, in time linear in the size of P. $\overline{P}$ defines a symmetry reduced quotient structure $M(\overline{P})$ which is isomorphic to $\overline{M(P)}$. However, each specific representative in $\overline{M(P)}$ is replaced by its corresponding generic representative in $M(\overline{P})$. $M(\overline{P})$ can then be used to model check $M(P)$ without having to represent the orbit relation for the symmetry group on the states of $M(P)$. We have then reduced a problem of worst case size k^n which is exponential in n, to one of worst case size n^k which is polynomial for any fixed number k of local states. Furthermore, the number of bits required to symbolically represent a state has been decreased from $\mathcal{O}(n \log k)$ in $\overline{M(P)}$, the standard quotient, to $\mathcal{O}(k \log n)$ in $M(\overline{P})$ the generic quotient. We then show that in many cases the transitions in $\overline{P}$ can be represented by BDD's polynomial in the size of the text of $\overline{P}$.

The key idea is that a generic representative can be chosen for each of the equivalences classes of states under the assumption of full symmetry [ES96] [CE+96]. Equivalence under full symmetry means that two states $s, t \in LP^n$ are equivalent iff they have exactly the same number of processes in local state l for each state $l \in LP$. Hence the generic representative needs only track the number of processes in each local state and not any information regarding which processes are in a particular local state.

Let a program $P = //_{i \in [1..n]} P_i$ be the parallel composition of processes $P_1, \ldots, P_n$ which are identical up to renaming. Each process is specified by a program skeleton similar to the ones in figure 3. The skeletons give rise to generic transitions of the processes which are specified by $l : g \to l'$ where $l, l' \in LP$ are local states and g is a guard. Guards are positive boolean combinations of the following elements: $\forall j : l_j$, $\forall j : \neg l_j$, $\exists j : l_j$, $\exists j : \neg l_j$ and $\exists j \neq j' : l_j \wedge l_{j'}$. Since the processes are identical up to renaming this syntax gives rise to fully symmetric structures.

The intended meaning of $l_i : g \to l'_i$ is that if P is in state s, where process i is in local state l_i and guard g is true of s then P may transit to the state t, everywhere the same as s, except that process i is in state l'_i. P executes the enabled transitions – there may be multiple enabled transitions for a single process – non-deterministically. We further stipulate that P defines an initial state s_0 of the form $l^n = (l_1, \ldots, l_n)$ for some $l \in LP$.

Given $P = //_{i \in [1..n]} P_i$ with initial state l^n, as above, P defines a Kripke structure $M(P) = (S, R, s_0)$ as follows: $s_0 = l^n$ is the initial state, $S = LP^n$ and $s \to t \in R$ iff there exists a generic transition statement $l : g \to l'$ such that $s(i) = l$, $t(i) = l'$, g is true at s and for all $i' \neq i$, $s(i') = t(i')$. For a Kripke structure M with an initial state s_0, we say that $M \models f$ iff $M, s_0 \models f$. $\overline{M(P)} = M(P)/\equiv_{Sym} \mathcal{I} = (\overline{S}, \overline{R}, s_0)$ is the symmetry reduced quotient structure.

Theorem 5. *[ES96] For any SCTL* formula f, $M(P), s_0 \models f$ iff $\overline{M(P)}, s_0 \models f$.*

We define the symmetry reduced program $\overline{P}$ as follows: $\overline{P}$ has variables $x_1, \ldots, x_k$ each of type $[0..n]$ and we assume the existence of a bijective function $\iota : LP \to [1..k]$. Suppose each process P_i has c different transitions of the form $l_i : g \to l'_i$ each generated by the generic transition $l : g \to l'$. Then $\overline{P}$ has c transitions of the form $x_{\iota(l)} > 0 \wedge \mathcal{T}(g) \to x_{\iota(l)}, x_{\iota(l')} := x_{\iota(l)} - 1, x_{\iota(l')} + 1$. The intended meaning being that if $\overline{P}$ is in a state $s \in [0..n]^k$ where the variable $x_{\iota(l)} \geq 0$ and the guard $\mathcal{T}(g)$ is true, then $\overline{P}$ may non-deterministically transit to a state $t \in [0..n]^k$ such that $x_{\iota(l)}$ has decreased by 1, $x_{\iota(l')}$ has increased by 1 and all other variables are unchanged.

The symmetry reduced guard $\mathcal{T}(g)$ is derived from g as follows: $\mathcal{T}(\forall j : l_j) = {}'x_{\iota(l)} = n'$, $\mathcal{T}(\forall j : \neg l_j) = {}'x_{\iota(l)} = 0'$, $\mathcal{T}(\exists j : l_j) = {}'x_{\iota(l)} > 0'$, $\mathcal{T}(\exists j : \neg l_j) = {}'x_{\iota(l)} < n'$, $\mathcal{T}(\exists j \neq j' : l_j \wedge l_{j'}) = {}'x_{\iota(l)} \geq 2'$, $\mathcal{T}(g_1 \vee g_2) = \mathcal{T}(g_1) \vee \mathcal{T}(g_2)$ and $\mathcal{T}(g_1 \wedge g_2) = \mathcal{T}(g_1) \wedge \mathcal{T}(g_2)$. Finally, if the initial state of P is l^n then the initial state of $\overline{P}$ is $x_{\iota(l)} = n$ and $x_{\iota(l')} = 0$ for all $l' \neq l$.

$\overline{P}$ defines a Kripke structure $M(\overline{P}) = (S', R', s'_0)$ as follows: if $x_i = n$ and for all $i' \neq i$, $x_{i'} = 0$ is the initial state of $\overline{P}$ then $s'_0(i) = n$ and for all $i' \neq i$, $s'_0(i') = 0$, $S' = [0..n]^k$ and $R' \subseteq S' \times S'$ where $s \to t \in R'$ iff there is a transition in $\overline{P}$, of the form $x_{\iota(l)} \geq 0 \wedge \mathcal{T}(g) \to x_{\iota(l)}, x_{\iota(l')} := x_{\iota(l)} - 1, x_{\iota(l')} + 1$ where the $\iota(l)$th element of s is greater than 0, $\mathcal{T}(g)$ is true at s and for all $j \in [1..k]$, $j = \iota(l)$ implies $t(j) = s(j) - 1$, $j = \iota(l')$ implies $t(j) = s(j) + 1$ and otherwise $s(j) = t(j)$.

Theorem 6. *$M(\overline{P})$ is isomorphic to $\overline{M(P)}$.*

We can also show that $M(\overline{P})$ is bisimilar to $\overline{M(P)}$ by translating the labels of the states in $\overline{M(P)}$ into the generic state format, that is by representing only the number of processes in a particular local state. Similarly, by translating the formulae of SCTL* into this generic format we have the following result.

Corollary 4. *For all SCTL* formulae f, $M(\overline{P}), s_0' \models f$ iff $M(P), s_0 \models f$*

In the sequel we describe how S' and R' can be succinctly represented by BDD's. States in S' are represented by tuples in $[0..n]^k$. Such a state space can be represented by $k \cdot (\log(n) + 1)$ boolean variables (for ease of explanation we assume that n is a power of two). Bits $b_0 \ldots b_{\log n}$ represent x_1, bits $b_{\log(n)+1} \ldots b_{2 \log n}$ represent the variable x_2, etc. Assuming that k is fixed, then generic states of S' can be represented in $\mathcal{O}(\log n)$ bits. It follows that, for any type of transition relation R' over S', the BDD representing R' is of size at most $poly(n)$. This should be contrasted with the size of the BDD representing the orbit relation in the conventional symmetry reduced quotient which has a lower bound $exp(\min(n, k))$ [CE+96]. But for this model of computation we can obtain better bounds as described below.

We now show that transitions of the form $x_{\iota(l)} \geq 0 \wedge \mathcal{T}(g) \rightarrow x_{\iota(l)}, x_{\iota(l')} := x_{\iota(l)} - 1, x_{\iota(l')} + 1$ can be represented succinctly when $\mathcal{T}(g)$ is of a particular form. Firstly, $x_{\iota(l)} \geq 0$ can be checked with a BDD of size $\mathcal{O}(log(n) + 1)$ since the BDD need only check that the bits $(\iota(l) - 1) \cdot \log n \ldots [\iota(l) \cdot \log n] - 1$ are not all 0 (false). Consider the set of atomic boolean guards $\{x_j = n, x_j = 0, x_j > 0, x_j < n, x_j \geq 2\}$, for $j \in [1..k]$ and assume that $\mathcal{T}(g)$ is either a conjunction of atomic boolean guards or a disjunction of atomic boolean guards. For the case where $\mathcal{T}(g)$ is conjunctive, extend the set of atomic boolean guards to include $x_j > 0 \wedge x_j < n$ and $x_j < n \wedge x_j \geq 2$.

In a manner similar to the above it is possible to show that each of the extended atomic boolean guards is representable by a BDD polynomial in the number of bits used to represent the value of the variable which the guard restricts. Conjunctive guard $\mathcal{T}(g)$ can be rewritten so that it first mentions only those atomic boolean guards which mention variable x_1 then x_2 and so on. Consider the conjunctive portion of $\mathcal{T}(g)$ in which x_j occurs, $j \in [1..k]$. Under the assumption that $n \geq 1$, it is not hard to prove that any conjunctive combination of boolean atomic guards reduces to the constant 0 or a single instance of one of the extended set of conjunctive boolean guards. Since the BDD's for the separate variables in $\mathcal{T}(g)$ are independent, they can be put together to form the BDD for $\mathcal{T}(g)$ which is of size additive in the sizes of the BDD's for each of the separate variables and hence polynomial in the length of $\mathcal{T}(g)$.

A similar argument can be made for the case when $\mathcal{T}(g)$ is disjunctive. However, in that instance the set of atomic boolean guards is extended by $x_j = n \vee x_j = 0$ and $x_j = 0 \vee x_j \geq 2$. Furthermore, arbitrary disjunctions of the atomic boolean guards never result in the constant 0 (false) but they do result either in a single instance of the extended set of atomic boolean guards or the constant 1 (true). Finally, it is not hard to see that a BDD can be built to check whether two states are related by the assignments of the form

$x_{\iota(l)}, x_{\iota(l')} := x_{\iota(l)} - 1, x_{\iota(l')} + 1$ which is of size polynomial in $k \cdot (\log(n) + 1)$. The bits representing the variable $x_{\iota(l)}$ ($x_{\iota(l')}$) increase (decrease) by 1 and all other variables remain unchanged. Finally, by combining all three sections of the BDD representing a transition we see that the BDD is at most cubic in $\mathcal{O}(k \cdot (\log(n) + 1))$ and hence polynomial in the size of the transition. These BDD's for individual program statements can be combined to get a BDD for R' of size $poly(n)$. However, they combine disjunctively which can be advantageous in terms of possible disjunctive partitioning.

When $P = ||_{i \in [1..n]} P_i$ is the synchronous composition of processes $P_1, \ldots, P_n$ a similar but slightly more complex translation is required. $\overline{P}_{||}$, the symmetry reduced program, contains two variables $x_{\iota(l)}$ and $x'_{\iota(l)}$ for each local state l. The generic transitions of the synchronous program P are translated in the same manner as the generic transitions in the asynchronous case except for the following: guards refer to unprimed variables while the assignments are made to the primed variables. Computation then proceeds in rounds. For each local state l, if the unprimed variable $x_{\iota(l)}$ has value b then up to b enabled transitions from place l – compiled transitions with $x_{\iota(l)} > 0$ in their guard – are executed. At the end of the round, each unprimed variable x_j is set to the value of the primed variable x'_j.

5 Conclusion

Many researchers have investigated the exploitation of symmetry in order to expedite verification but 'almost' symmetric designs have received little attention. A different type of partial symmetry has been explored in [HI+95], without precise formalization and only in relation to preservation of reachability properties of petri nets. Our formalizations of near and rough symmetry are new and our use of near and rough symmetries of M in the reduction of M to an abstract quotient structure is new. The term partial symmetry has been used for quite some time (cf. [Ko78]) in switching theory. There, however, a system is partially symmetric if its group of symmetries over index set $\mathcal{I}$ is isomorphic to the full symmetry group of an index set $\mathcal{I}' \subseteq \mathcal{I}$. This type of partial symmetry has been handled explicitly by [ES96] and [CE+96]. [AHI98] considers partial symmetry in a manner more analogous to our definition of sub-symmetry. However, they deal only with partial symmetries of the formula (or its automaton representation) to be model checked, rather than reduction of the structure itself. Abstraction of M, on the other hand, has the potential to be of greater benefit in ameliorating the state explosion problem [LP85]. We have shown that near automorphisms are sufficient for the preservation of temporal properties, a generalization of the results of [ES96] [CE+96], we have extended these ideas to rough symmetries, and we have shown how to obtain simulated symmetry reduced quotient structures from asymmetric systems via sub-symmetries.

With respect to full symmetry, we have shown how to exploit the symmetry of program text without the need to represent the symmetry reduced Kripke structure or the orbit relation induced by the symmetry group. [ID96] deals with similar symmetry groups but they do so explicitly. That is, the state spaces are not represented by BDD's and they therefore a priori do not have to cope with the problem of representing the symmetry induced equivalence classes by a BDD. [CE+96] shows BDD's representing the orbit relation of the full symmetry group are of exponential size. They suggested a heuristic to mitigate this problem using multiple representatives, but did not prove it to yield a tractable representation in general. Our technique consists in compiling the symmetric program P with Kripke structure $M(P)$ to a symmetry reduced program $\overline{P}$ over generic states whose structure $M(\overline{P})$ is bisimilar up to permutation to $M(P)/\equiv_{Sym}\, \mathcal{I}$. This can be seen to be an example of the utility of compiling programs into Petri Nets to achieve an exponential reduction. We believe we are the first to show that it is just such a reduction strategy which can usefully combine symmetry reduction with BDD based state representation. Previous work on Petri Nets, BDD's and symmetry reduction has not dealt explicitly with the fact that BDD representation of the symmetry induced equivalence classes is a self-defeating proposition for many symmetry groups.

For the future, we are implementing a preprocessor front end to a symbolic model checking tool to take advantage of our results on full symmetry. We are also investigating extending our technique to a larger class of groups [CE+98] for which symmetry reduction can be applied directly to program text. With respect to near symmetry and full symmetry we are interested in exploring the applicability of our work here to symmetry reduction techniques which use the annotated symmetry reduced structure which preserves the truth of all CTL* (and μ-calculus) properties [ES96][ES97][ET98].

Acknowledgment: The authors would like to thank Bob Kurshan for many stimulating comments and discussions.

References

[AHI98]   Ajami, K., Haddad, S. and Ilie, J.-M., Exploiting Symmetry in Linear Time Temporal Logic Model Checking: One Step Beyond. In *Tools and Algorithms for the Construction and Analysis of Systems, 4th Interntational Conference, ETAPS98* LNCS 1384, Springer Verlag, 1998.

[BC+92]   Burch, J. R., Clarke, E. M., McMillan, K. L., Dill, D. L. and Hwang, L. J., Symbolic Model Checking: 10^{20} states and beyond. In *Information and Computation*, 98(2):142-170, June, 1992.

[Br86]   Bryant, R. E., Graph-Based Algorithms for Boolean Function Manipulation. In *IEEE Transactions on Computers*, Vol. C-35, No. 8, Aug. 86, pp. 677-691.

[CE81]   Clarke, E. M., and Emerson, E. A., Design and Verification of Synchronization Skeletons using Branching Time Temporal Logic. In *Logics of Programs Workshop*, Springer, LNCS no. 131., pp. 52-71, May 1981.

[CE+98]   Clarke, E. M., Emerson, E. A., Jha, S. and Sistla A. P., Symmetry Reductions in Model Checking. In *Computer Aided Verification, 10th International Conference* LNCS 1427, Springer- Verlag, 1998.

[CES86]   Clarke, E. M., Emerson, E. A., and Sistla, A. P., Automatic Verification of Finite State Concurrent System Using Temporal Logic. In *ACM Trans. on Prog. Lang. and Sys.*, vol. 8, no. 2, pp. 244-263, April 1986.

[CE+96] Clarke, E. M., Enders, R., Filkorn, T., and Jha, S., Exploiting Symmetry in Temporal Logic Model Checking. In *Formal Methods in System Design*, Kluwer, vol. 9, no. 1/2, August 1996.

[CGL94] Clarke, E. M., Grumberg, O. and Long, D. E., Model Checking and Abstraction. In *Transactions on Programming Languages and Systems* ACM, vol 16, no. 5, 1994.

[Em90] E. Allen Emerson, Temporal and Modal Logic. In J. van Leeuwen editor *Handbook of Theoretical Computer Science* vol. B, Elsevier Science Publishing, 1990.

[EH86] Emerson, E. A., and Halpern, J. Y., 'Sometimes' and 'Not Never' Revisited: On Branching versus Linear Time Temporal Logic, *JACM*, vol. 33, no. 1, pp. 151-178, Jan. 86.

[ES96] Emerson, E. A. and Sistla, A. P., Symmetry and Model Checking. In *Formal Methods in System Design*, Kluwer, vol. 9, no. 1/2, August 1996.

[ES97] Emerson, E. A. and Sistla, A. P., Utilizing Symmetry when Model Checking under Fairness Assumptions. In *TOPLAS* 19(4): 617-638 (1997).

[ET98] Emerson, E. A. and Trefler, R. J., Model Checking Real-Time Properties of Symmetric Systems. In *Mathematical Foundations of Computer Science, 23rd International Symposium* LNCS 1450, Springer-Verlag, 1998.

[GS97] Gyuris, V. and Sistla, A. P., On-the-Fly Model checking under Fairness that Exploits Symmetry. In *Proceedings of the 9th International Conference on Computer Aided Verification, Haifa, Israel*, 1997.

[HI+95] Haddad, S., Ilie, J.M., Taghelit, M. and Zouari, B., Symbolic Reachability Graph and Partial Symmetries. In *Application and Theory of Petri Nets 1995*, Springer-Verlag, LNCS 935, 1995.

[HM85] Hennessy, M., Milner, R., Algebraic Laws for Nondeterminism and Concurrency. In *Journal of the ACM*, Vol 32, no. 1, January, 1985, pp 137-161.

[ID96] Ip, C-W. N., Dill, D. L., Better Verification through Symmetry. In *Formal Methods in System Design*, Kluwer, vol. 9, no. 1/2, August 1996.

[JR91] Jensen, K. and Rozenberg, G. (eds.), High-Level Petri Nets: Theory and Application, Springer- Verlag, 1991.

[Ko78] Kohavi, Zvi, *Switching and Finite Automata Theory*, second edition, McGraw-Hill Book Company, New York, 1978.

[LP85] Lichtenstein, O., and Pnueli, A., Checking That Finite State Concurrent Programs Satisfy Their Linear Specifications, POPL85, pp. 97-107, Jan. 85.

[MAV96] Michel, F., Azema, P. and Vernadat, F., Permutable Agents in Process Algebra. In *Tools and Algorithms for the Construction and Analysis of Systems, 96*, Springer Verlag, LNCS 1055, 1996.

[Mi71] Milner, R., An Algebraic Definition of Simulations Between Programs. In *Proceedings of the Second International Joint Conference on Artificial Intelligence*, British Computer Society, 1971, pp 481-489.

[Mc92] McMillan, K. L., *Symbolic Model Checking: An Approach to the State Explosion Problem*, Ph.D. Thesis, Carnegie Mellon University, 1992.

[Pa81] Park, D., Concurrency and Automata on Infinite Sequences. In *Theoretical Computer Science: 5th GI-Conference, Karlsruhe*, Springer-Verlag, LNCS 104, pp 167-183, 1981.

[QS82] Queille, J. P., and Sifakis, J., Specification and verification of concurrent programs in CESAR, Proc. 5th Int. Symp. Prog., Springer LNCS no. 137, pp. 195-220, 1982.

Automatic Error Correction of Large Circuits Using Boolean Decomposition and Abstraction*

Dirk W. Hoffmann and Thomas Kropf

Institut für Technische Informatik,
Universität Tübingen, D-72076 Tübingen, Germany
{hoff,kropf}@informatik.uni-tuebingen.de

Abstract. Boolean equivalence checking has turned out to be a powerful method for verifying combinatorial circuits and has been widely accepted both in academia and industry.

In this paper, we present a method for localizing and correcting errors in combinatorial circuits for which equivalence checking has failed. Our approach is general and does not assume any error model. Thus, it allows the detection of arbitrary design errors. Since our method is *not* structure-based, the produced results are independent of any structural similarities between the implementation circuit and its specification. It can even be applied if the specification is given, e.g., as a propositional formula, a BDD, or in form of a truth table.

Furthermore, we discuss two kinds of circuit abstractions and prove compatibility with our rectification method. In combination with abstractions, we show that our method can be used to rectify large circuits.

We have implemented our approach in the **AC/3** equivalence checker and circuit rectifier and evaluated it with the Berkeley benchmark circuits [6] and the ISCAS85 benchmarks [7] to show its practical strength.

keywords: Automatic error correction, design error diagnosis, equivalence checking, formal methods

1 Introduction

In recent years, formal verification techniques [11] have become more and more sophisticated and for several application domains they have already found their way into industrial environments. Boolean equivalence checking [13, 4, 16], mostly based on BDDs [8, 9], is unquestionably one of these techniques and is usually applied during the optimization process to ensure that an optimized circuit still exhibits the same behavior as the original "golden" design. When using BDDs for representing Boolean functions, the verification task mainly consists of creating a BDD for the Boolean function of each output signal. Then, due to the normal form property of BDDs, both signals implement the same function if and only if they have the same BDD representation. Hence, equivalence can be decided by simply comparing both BDDs.

A major requirement for successful application of formal methods in industrial environments is the ability of a verification tool to provide useful information even when

---

* This work is supported by the ESPRIT LTR Project 26241 (PROSPER)

the verification attempt fails. Then, the application domain of formal verification is no longer restricted to proving correctness of a specific design, but it can also be used as a debugging tool and therefore helps speeding up the whole design cycle.

If equivalence checking fails, most verification tools only allow the computation of counterexamples in the form of combinations of input values for which the output of the optimized circuit differ from its specification. Therefore, in many cases it remains extremely hard to detect the error causing components. Counterexamples as produced by most equivalence checkers can only serve as hints for debugging a circuit, while a deeper understanding of the design is still needed.

In recent years, several approaches have been presented for extending equivalence checkers with capabilities not only to compute counterexamples, but to locate and rectify errors in the provided designs. The applicability of such a method is strongly influenced by the following aspects:

- Which types of errors can be found ?
- Does the method scale to large circuits ?
- How many modifications in the original circuit are needed for correction ?
- Does the method perform well if both circuits are structurally different ?

Most earlier research [20, 10, 17, 18, 21, 19] in the area of automatic error correction assumes a concrete error model based on a classification of typical design errors going back to Abadir et. al. [1]. Errors are divided into *gate errors* (*missing gate, extra gate, wrong logical connective*) and *line errors* (*missing line, extra line*). Each gate is basically checked against these error classes and most approaches can only handle circuits with exactly one error (*single error assumption*).

In [15] and [14], no error model is assumed. The method presented in [15] propagates meta-variables through the circuit. Erroneous single gates are determined by solving formulas in quantified propositional logic. However, the method is very time consuming and needs to invoke a propositional prover.

In [14], the implementation circuit and the specification circuit are searched for equivalent signal pairs and a back substitution algorithm is used for rectifying the circuit. The success of this method highly depends on structural similarities between the implementation and the specification.

Incremental synthesis [3, 5] is a field closely related to automatic error correction. An *old implementation*, an *old specification*, and a *new specification* are given. The goal is to create a *new implementation* fulfilling the new specification while reusing as much of the old implementation as possible. In [5], structural similarities between the new specification and the old specification are exploited to figure out subparts in the old implementation that can be reused. The method is based on the structural analysis technique in [2] and the method presented in [4] which uses a test generation strategy to determine equivalent parts in two designs.

In this paper, we present a method for localizing and correcting errors in combinatorial circuits based on Boolean decomposition and abstraction. The main contributions of our approach can be summarized as follows:

- Unlike [20, 10, 17, 18, 21], our approach does not assume any error model. Thus, arbitrary design errors can be detected.

- Our method is *not* structure based. Only the abstract BDD representation of the specification is considered. Thus, the success of our algorithm does not depend on any structural similarity between the implementation and the specification. Our technique can even be applied in scenarios where the specification is given as a Boolean formula, a truth table, or directly in form of a BDD. This is in contrast to structure based methods such as [14, 3, 5] that can only be applied if both circuits are structurally similar.
- Circuit rectifications are computed in form of a BDD and then converted back to a net-list description. This is in contrast to techniques such as [14, 3, 5] which basically modify a given design by putting the implementation and specification together and "rewiring" erroneous parts.
- Computed solutions are weighted by a cost function in order to find a minimal solution – a solution that requires minimal number of modifications in the implementation.
- Our rectification procedure can be combined with circuit abstractions. In this paper, we discuss two kinds of circuit abstractions which are compatible with our method. We prove a correctness theorem showing how rectifications computed for abstract circuits can be lifted to circuit corrections for the original unabstracted circuits.
- We have implemented the rectification and abstraction algorithms in the AC/3 [12] equivalence checker and circuit rectifier and evaluated the method with the Berkeley benchmark circuits [6] and the ISCAS85 benchmarks [7]. Our experimental results show that in combination with abstraction, our rectification method can be applied to large designs.

This paper is organized as follows: In Section 2, we give a brief introduction to the theoretical background and Section 3 defines the formalism how combinatorial circuits are represented. Section 4 describes the rectification algorithm and Section 5 introduces circuit abstractions. Section 6 addresses the problem of rectifying multiple output circuits. We close our paper with experimental results in Section 7 and a conclusion in Section 8.

2 Preliminaries

In the following, $f, g, h, \ldots$ denote propositional formulas and $X, Y, Z, \ldots$ represent propositional variables. We use the symbol $\equiv$ to denote logical equivalence between propositional formulas while $=$ is used for expressing syntactical similarity.

The *positive* and *negative cofactor* of f, written as $f|_X$ and $f|_{\neg X}$, represent the functions obtained from f where X is instantiated by the truth values 1 and 0, respectively. A formula f is said to be *independent* of X, if $f|_X \equiv f|_{\neg X}$.

$f_{\downarrow g}$ represents some Boolean function that agrees with f for all valuations which satisfy g. For all other valuations, $f_{\downarrow g}$ is not defined and can be chosen freely.

Assume we are given three propositional formulas f, g, and h. The pair (g, h) is called a *decomposition* of f, if there exists a variable X in g with

$$f \equiv g[X \leftarrow h] \tag{1}$$

If formulas f, g, and variable X are given, the decomposition problem is to compute a formula h satisfying (1).

Example 1. Consider $f = (A \wedge B) \vee A$ and $g = A \vee X$. $(g, A \wedge B)$ and (g, A) are both decompositions of f since $f \equiv g[X \leftarrow (A \wedge B)]$ and $f \equiv g[X \leftarrow A]$. Assuming $g = C \wedge X$, there exists no decomposition for f since there is no term h such that $f \equiv g[X \leftarrow h]$.

3 Representing Combinatorial Circuits with Circuit Graphs

For the rest of this paper, we will use *circuit graphs* for representing combinatorial circuits. For a given circuit C, the corresponding circuit graph is constructed by introducing a new node v_c for each logical gate c in C. Wires are translated into transitions such that there is a transition from v_{c_1} to v_{c_2} iff some input of gate c_1 is connected with the output of gate c_2. Formally, we define circuit graphs as follows:

Definition 1. *A circuit graph $\mathcal{F}$ is a rooted, directed acyclic graph (V, l, e). V is a set of nodes with $|V| < \infty$. The function l labels every inner node of V with a logical connective and every leaf of V with a propositional variable. The edge-function e maps every node of $v \in V$ to an element $(v_1, \ldots, v_n) \in V^n$ where n is the arity of $l(v)$. The root-node of $\mathcal{F}$ is denoted by root($\mathcal{F}$).*

To simplify notation, we define the following abbreviations:

$$l(\mathcal{F}) := l(\text{root}(\mathcal{F}))$$
$$e_i(v) := i\text{-th element of } e(v), i \geq 1$$
$$\mathcal{F}_i(v) := \text{sub-graph of } \mathcal{F} \text{ with root-node } e_i(v)$$
$$\mathcal{F}_i := \mathcal{F}_i(\text{root}(\mathcal{F}))$$

Definition 1 is slightly stronger than the usual definition of labeled graphs where edges are represented by a relation $E \subset V \times V$. Using an edge-function e as defined above, multiple edges to the same successor node (Fig. 1 (a)) are possible. Furthermore, the successor nodes are implicitly ordered (Fig. 1 (b,c)). These properties cannot be expressed by using a relational edge-representation.

In the following, we restrict the set of logical connectives to $\neg$ (negation), $\wedge$ (conjunction), $\vee$ (disjunction), $\rightarrow$ (logical implication), $\leftrightarrow$ (logical equivalence), and $\oplus$ (exclusive-or). Every node v of a circuit-graph $\mathcal{F}$ induces a Boolean function f^v by

$$f^v = \begin{cases} l(v) & \text{if } v \text{ is a leaf} \\ \neg f^{e_1(v)} & \text{if } l(v) = \neg \\ f^{e_1(v)} \ l(v) \ f^{e_2(v)} & \text{if } l(v) \in \{\wedge, \vee, \rightarrow, \leftrightarrow, \oplus\} \end{cases}$$

To simplify notation we often identify a circuit graph $\mathcal{F}$ with its corresponding Boolean function and write $\mathcal{F}$ instead of $f^{\text{root}(\mathcal{F})}$ if it is clear from context.

We define two substitutions on circuit-graphs. *Node substitutions* replace a single node v in a graph $\mathcal{F}$ by some other graph $\mathcal{G}$ (denoted by $\mathcal{F}[v \leftarrow \mathcal{G}]$) while *variable*

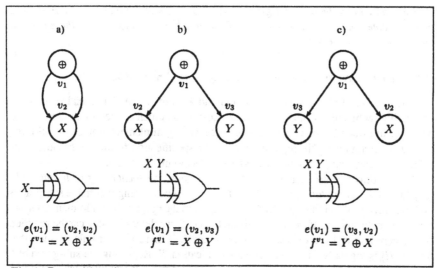

Fig. 1: Graph a) is a circuit-graph where an edge appears twice. Graph b) and c) demonstrate that an ordering on successors is necessary to distinguish the corresponding circuits.

substitutions replace every occurrence of some variable X by another graph $\mathcal{G}$ (denoted by $\mathcal{F}[X \leftarrow \mathcal{G}]$). If σ is the name of the substitution, we also write $\sigma\mathcal{F}$ instead of $\mathcal{F}[v \leftarrow \mathcal{G}]$ or $\mathcal{F}[X \leftarrow \mathcal{G}]$, respectively.

Note that node substitutions as defined here only replace a single node in the circuit-graph whereas variable substitutions replace *all* nodes that are labeled with the specified variable.

Lemma 1. *Let $\mathcal{F}, \mathcal{G}$ be two circuit-graphs. v is a node in $\mathcal{F}$. Then, for all variable substitutions σ,*

$$\sigma\left(\mathcal{F}[v \leftarrow \mathcal{G}]\right) = (\sigma\mathcal{F})\left[v \leftarrow \sigma\mathcal{G}\right] \qquad (2)$$

4 The Rectification Method

Let $\mathcal{G}$ be a combinatorial circuit represented as circuit graph. Further assume that $\mathcal{G}$ is a single output circuit. $\mathcal{F}$ denotes some Boolean function represented either as another combinatorial circuit (e.g., $\mathcal{G}$ could be the result of an optimization step applied to $\mathcal{F}$), a propositional formula, a truth table, or directly as a BDD. Since we only use the abstract BDD representation of $\mathcal{F}$ in our algorithm, the computed solutions are totally independent of the structure of $\mathcal{F}$. For the rest of this paper, we treat $\mathcal{F}$ as the specification and $\mathcal{G}$ as the implementation.

Our goal is to modify the circuit-graph of $\mathcal{G}$ with a minimal number of changes such that $\mathcal{F} \equiv \mathcal{G}$ holds. Each such modification is called a rectification of $\mathcal{G}$:

Definition 2. *Let $\mathcal{G}$ be a circuit-graph and $\mathcal{F}$ some Boolean function with $\mathcal{F} \not\equiv \mathcal{G}$. v denotes some node in the node-set of $\mathcal{G}$. $\mathcal{G}$ is called $\mathcal{F}$-rectifiable at v if there exists a circuit-graph $\mathcal{H}$ such that*

$$\mathcal{F} \equiv \mathcal{G}[v \leftarrow \mathcal{H}] \tag{3}$$

If $\mathcal{F}$ and v are clear from the context, we simply call $\mathcal{G}$ rectifiable.

The number of changes we have to apply to a given circuit is a crucial issue when computing rectifications since we want to preserve as much of the circuit structure as possible. In principle, we can always correct a wrong implementation by substituting the whole circuit by a DNF-representation of the specification-formula. Obviously, this is far away from what a designer would accept as circuit correction.

Our rectification procedure consists of two steps: the *location of rectifiable sub-graphs* and the *computation of circuit rectifications*. For locating rectifiable sub-graphs in $\mathcal{G}$, we traverse the circuit-graph of $\mathcal{G}$ starting from the outputs. In our implementation, we use a depth-first search strategy. For each node ξ, we determine if $\mathcal{G}$ can be rectified at ξ. According to Definition 2, we have to check if there is a formula $\mathcal{H}$ such that $\mathcal{G}[\xi \leftarrow \mathcal{H}]$ is logically equivalent to the specification $\mathcal{F}$. Replacing the sub-graph at ξ by a newly introduced variable X, we can easily perform this test by checking if there exists a term $\mathcal{H}$ such that $(\mathcal{G}[\xi \leftarrow X], \mathcal{H})$ is a decomposition of $\mathcal{F}$. For doing this, we first create a BDD-representation for $\mathcal{F}$ and $\mathcal{G}[\xi \leftarrow X]$. Then, decomposability can be decided with standard BDD operations according to the following lemma which is a direct result from the theory of Boolean equations:

Lemma 2. *Let f and g be two propositional formulas. X is a variable occurring in g. Then, there exists a formula h with $f \equiv g[X \leftarrow h]$ if and only if*

$$f \wedge (g|_{\neg x} \leftrightarrow g|x) \equiv g \wedge (g|_{\neg x} \leftrightarrow g|x) \tag{4}$$

Basically, Lemma 2 reflects the idea that we can find some h with $f \equiv g[X \leftarrow h]$ iff f and g agree on all valuations that are independent of X (expressed by $g|_{\neg x} \leftrightarrow g|x$).

For computing circuit corrections, we first compute a formula $\mathcal{H}$ such that $\mathcal{G}[\xi \leftarrow \mathcal{H}] \equiv \mathcal{F}$. Again, this can be done by applying elementary BDD operations as the following lemma states:

Lemma 3. *Assume f and g are decomposable in respect to variable X. Then,*

$$f \equiv g[X \leftarrow [(g|x \leftrightarrow f) \wedge (g|x \oplus g|_{\neg x})]] \tag{5}$$

In Lemma (5), the solution formula is obviously not unique. Here, the solution with the smallest 1-set is being computed.

Using BDDs for representing Boolean functions, the application of Lemma 3 returns a formula h which is also represented in form of a BDD. Thus, the BDD for h first has to be converted back into a circuit-graph $\mathcal{H}$ before the rectification can be performed. This conversion, however, directly influences the resulting graph structure. To maximize the syntactical similarities between $\mathcal{G}$ and $\mathcal{G}[\xi \leftarrow \mathcal{H}]$, we try to reuse as many sub-graphs of $\mathcal{G}$ as possible. The heuristic implemented in **AC/3** is to reuse the current gate inputs or the set of inputs of the component containing ξ (when dealing with hierarchical circuits). Assume $\mathcal{G}_1, \ldots, \mathcal{G}_n$ are the sub-graphs of $\mathcal{G}$ we want to reuse. Hence,

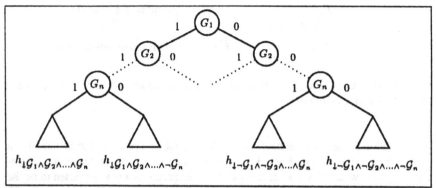

Fig. 2: Reuse of sub-graphs. Variables $G_1, \ldots, G_n$ denote newly introduced meta-variables representing sub-graphs $\mathcal{G}_1, \ldots, \mathcal{G}_n$, respectively.

our goal is to create a syntax-graph $\mathcal{H}$ for h containing $\mathcal{G}_1, \ldots, \mathcal{G}_n$. To achieve this, we construct a second BDD h' as shown in Fig. 2. $G_1, \ldots, G_n$ are newly introduced BDD variables. Since for all m,

$$h_{\downarrow \mathcal{G}_1 \wedge \ldots \mathcal{G}_{m-1}} \equiv \mathcal{G}_m h_{\downarrow \mathcal{G}_1 \wedge \ldots \mathcal{G}_{m-1} \wedge \mathcal{G}_m} \vee \neg \mathcal{G}_m h_{\downarrow \mathcal{G}_1 \wedge \ldots \mathcal{G}_{m-1} \wedge \neg \mathcal{G}_m}$$
$$\equiv (G_m h_{\downarrow \mathcal{G}_1 \wedge \ldots \mathcal{G}_{m-1} \wedge \mathcal{G}_m} \vee \neg G_m h_{\downarrow \mathcal{G}_1 \wedge \ldots \mathcal{G}_{m-1} \wedge \neg \mathcal{G}_m})[G_m \leftarrow \mathcal{G}_m]$$

the newly constructed BDD h' in Fig. 2 is logical equivalent to h if we substitute $G_1, \ldots, G_n$ by the BDDs for $\mathcal{G}_1, \ldots, \mathcal{G}_n$, respectively.

A crucial issue in the construction process is to check the possibility if h can be exclusively constructed out of $\mathcal{G}_1, \ldots, \mathcal{G}_n$ and the logical connectives $\wedge$, $\vee$, and $\neg$. If h has this property, the sub-BDDs $h_{\downarrow f}$ in Fig. 2 can always be simplified to 0 or 1. Then, h' only contains the meta-variables $G_1, \ldots, G_n$. This property becomes important when dealing with hierarchical circuit descriptions. If we have located an erroneous sub-component in a circuit, we first try to replace it by another component that does not require changes in the component-interfaces. Thus, after computing a circuit correction h, we first try to convert h to a formula only involving the current component inputs as sub-terms. Every solution that keeps the component-interfaces unchanged is called *hierarchy preserving*.

5 Circuit Abstractions

Assume we are given some combinatorial circuit $\mathcal{F}$ represented as circuit graph. Any circuit $\mathcal{F}'$ obtained from $\mathcal{F}$ by replacing one or more inner nodes by leaves labeled with newly introduced, so called abstraction variables, is called an abstraction of $\mathcal{F}$. More precisely, we define circuit abstractions as follows:

Definition 3. *Let $\mathcal{F}, \mathcal{F}'$ be two circuit-graphs. $\mathcal{F}'$ is called an abstraction of $\mathcal{F}$ if there exists a variable substitution σ with*

$$\mathcal{F} = \sigma \mathcal{F}' \tag{6}$$

$$\begin{array}{lll} \mathcal{F}, \mathcal{G} & & \text{circuit modification for } \mathcal{G} \\ \Downarrow \text{ abstraction} & & \Uparrow \text{ lift} \\ \mathcal{F}', \mathcal{G}' & \overset{\text{rectification}}{\Longrightarrow} & \text{circuit modification for } \mathcal{G}' \end{array}$$

Fig. 3: Circuit-rectifications for abstracted circuits can be lifted to rectifications of the original unabstracted circuits.

We now address the question how the rectification method presented in Section 4 can benefit from using circuit abstractions. Assume we are given two combinatorial circuits $\mathcal{F}$ and $\mathcal{G}$ where $\mathcal{F}$ serves as the specification circuit, and $\mathcal{G}$ is supposed to be the implementation. Both circuits are represented with circuit graphs. Further assume that $\mathcal{F} = \sigma \mathcal{F}'$ and $\mathcal{G} = \sigma \mathcal{G}'$ for some variable substitution σ, i.e., $\mathcal{F}'$ and $\mathcal{G}'$ are abstractions of $\mathcal{F}$ and $\mathcal{G}$, respectively. Then, Theorem 1 guarantees that every $\mathcal{F}'$-rectification of $\mathcal{G}'$ can be lifted to an $\mathcal{F}$-rectification of $\mathcal{G}$ (Fig. 3). As the experimental results in Section 7 will show, this can dramatically reduce rectification time. Moreover, it becomes possible to rectify much larger circuits. However, as every abstraction technique, our approach has some drawbacks which will be discussed in detail in Section 5.2.

Theorem 1. *Let σ be a variable substitution and $\mathcal{F}, \mathcal{G}, \mathcal{F}', \mathcal{G}'$ be circuit-graphs with $\mathcal{F} = \sigma \mathcal{F}'$ and $\mathcal{G} = \sigma \mathcal{G}'$ Further assume that $\mathcal{G}'$ is $\mathcal{F}'$-rectifiable at v, i.e., $\mathcal{F}' \equiv \mathcal{G}'[v \leftarrow \mathcal{H}]$ for some circuit-graph $\mathcal{H}$. Then, $\mathcal{G}$ is $\mathcal{F}$-rectifiable at v with circuit-graph $\sigma \mathcal{H}$, i.e.,*

$$\mathcal{F} \equiv \mathcal{G}[v \leftarrow \sigma \mathcal{H}] \tag{7}$$

Proof. By the assumptions of Theorem 1, we get $\mathcal{F} \equiv \sigma \mathcal{F}' \equiv \sigma \left(\mathcal{G}' [v \leftarrow \mathcal{H}] \right)$. Applying Lemma 1, we get $\sigma \left(\mathcal{G}' [v \leftarrow \mathcal{H}] \right) \equiv \left(\sigma \mathcal{G}' \right) [v \leftarrow \sigma \mathcal{H}]$ and rewriting $\sigma \mathcal{G}'$ by $\mathcal{G}$ finally proves $\mathcal{F} \equiv \mathcal{G} [v \leftarrow \sigma \mathcal{H}]$.

5.1 Computing Abstractions

In this section, we examine two specific kinds of abstractions, i.e., *structural abstractions (S-abstractions)* and *logical abstractions (L-abstractions)*. For computing S-abstractions, we replace structurally identical sub-graphs in $\mathcal{F}$ and $\mathcal{G}$ by a newly introduced abstraction variable. Similarly, L-abstractions prune away logically equivalent circuit parts. Thus, L-abstractions are stronger than S-abstractions. However, S-abstractions are often sufficient, especially when using the method after an optimization step or within an incremental synthesis environment. Then, both circuits usually differ in a very small part of the circuit, only.

Note that – strictly speaking – L-abstractions are not abstractions in the sense of Definition 5. When we construct $\mathcal{F}'$ by substituting logically equivalent sub-graphs in $\mathcal{F}$ by a common abstraction variable, we cannot always guarantee the existence of a variable substitution σ with $\mathcal{F} = \sigma \mathcal{F}'$ since $\mathcal{F}$ and $\sigma \mathcal{F}'$ are usually not structurally identical. However, $\mathcal{F}$ and $\sigma \mathcal{F}'$ are logically equivalent. Thus, there always exists some circuit graph $\mathcal{F}''$ with $\sigma \mathcal{F}' = \mathcal{F}''$ and $\mathcal{F}'' \equiv \mathcal{F}$. Since $\mathcal{F}$ serves as the specification and is logically equivalent to $\mathcal{F}''$, we can also use $\mathcal{F}''$ as specification instead. Hence, soundness of the abstraction method is not affected.

For computing abstractions for some specification circuit $\mathcal{F}$ and some implementation circuit $\mathcal{G}$, we proceed as follows: In the first step, we determine nodes in $\mathcal{F}$ and $\mathcal{G}$ that are going to be substituted by a common abstraction variable. In case of S-abstractions, we use a hash-table for storing circuit-graphs. For each node, a hash table index is created such that two nodes are mapped to the same index iff their subgraphs are structurally identical. The hash-table index can be computed in constant time. Thus, index computation of the whole graph can be done linear in the number of graph nodes. In case of L-abstractions, a BDD is computed for each node. Using the BDD reference pointer as index, nodes have the same index if and only if their associated sub-graphs are logically equivalent. For a circuit graph with n nodes, n BDDs have to be computed. Since BDDs can grow exponentially, index computation may also take exponential time.

After computing indices for all graph-nodes, correlated nodes in $\mathcal{F}$ and $\mathcal{G}$ can now be determined for both abstraction types by simply comparing their indices.

In the second step, all nodes with the same index are replaced by a common abstraction variable. For both abstraction types, this can be done in linear time.

When dealing with large circuits, we embed the procedure above in an iterated abstraction algorithm. Using a threshold value τ, we only compute indices for graph nodes having less than τ successor nodes or a BDD representation with less than τ BDD nodes (depending on the computed abstraction type). After abstracting away correlated sub-graphs, the complete abstraction process is repeated until a fix-point is reached. Obviously, the computed results may differ depending on the threshold τ. The bigger the threshold, the more equivalences are usually detected, but the more computation time will be spent within the abstraction algorithm.

5.2 Drawbacks of the Abstraction Method

Performing automatic error correction on abstracted circuits can dramatically decrease computation time and broaden the range of rectifiable circuits. However, it has some drawbacks that are going to be discussed in this Section.

The first noteworthy property is that two equivalent circuits can become inequivalent after abstractions have been computed (see Fig. 4 for an example). This may cause the rectification algorithm to compute unnecessary changes for the implementation circuit. However, the correctness theorem proven in Section 1 guarantees that both circuits remain equivalent after the modifications have been applied. To avoid unnecessary changes to the implementation circuit, equivalence of both designs should be decided beforehand (e.g. by the method presented in [4]).

Another aspect is that the erroneous part of the implementation circuit may be pruned away when computing abstractions. Fig. 5 shows two circuits where the specification is shown on the left and the implementation on the right. Both circuits are not equivalent due to a missing NOT gate at the upper input of component C_1. Both abstraction types result in a circuit where the erroneous position has been abstracted away. Therefore, in this example, all solutions computed for the abstract circuits will rectify the implementation by modifying component C_2. Hence, it is most likely that

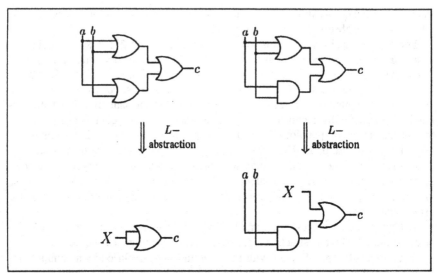

Fig. 4: The two originally equivalent circuits become inequivalent after abstraction. Both unabstracted circuits implement the function $c = a \vee b$.

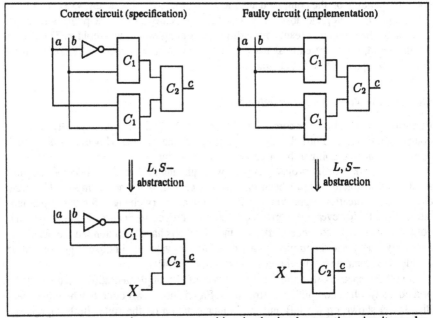

Fig. 5: In some cases, the erroneous position in the implementation circuit can be pruned away which may lead to unnecessary complex solutions.

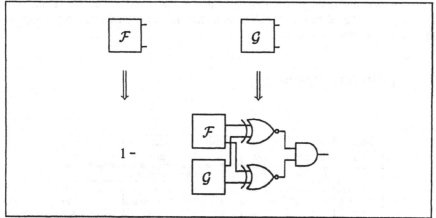

Fig. 6: Transforming $\mathcal{G}$ from a multiple output circuit to a single output circuit. (here, $\mathcal{F}$ and $\mathcal{G}$ have two output signals.)

more modifications are needed than necessary when rectifying the unabstracted circuits directly. However, the experimental results in Section 7 show that especially S-abstractions are very robust in respect to this problem.

A third aspect worth mentioning is that the abstraction algorithm adds new inputs (variables) to the circuit. As BDDs can grow exponentially in the number of variables, this can in principle cause runtime to increase exponentially. Applied to a circuit with n input variables, it is in theory possible that the abstraction algorithm creates a circuit with up to $2^{(2^n)}$ input variables.

However, such an example had to be constructed explicitly and we have observed this phenomenon for none of our example circuits in practice.

6 Rectifying Multiple-Output Circuits

For rectifying multiple output-circuits, there are two possibilities: The first possibility is to rectify every output-signal separately and intersect the solution sets. Then, every solution has to be tested if it also corrects the other signals. If yes, the solution fixes the whole circuit, otherwise it has to be discarded. In practice, this approach finds correct solutions for most multiple output circuits. However, this method is not complete in the sense that it is possible that no rectification is computed that corrects all output signals, even if such a solution exists.

In these cases, we first transform the implementation circuit to a single output circuit as shown in Fig. 6. This transformation basically puts the (abstracted) specification $\mathcal{F}$ and the (abstracted) implementation $\mathcal{G}$ together in one circuit and replaces the old specification by logical true. If $\mathcal{F}$ is not given as a circuit net-list, it has to be synthesized to some net-list equivalent to $\mathcal{F}$.

The transformation assures that whenever the newly created output is logically equivalent to true, all output signals of $\mathcal{F}$ are equivalent to the outputs of $\mathcal{G}$. Obviously, when applying the rectification algorithm to the newly constructed circuit, we have to restrict the solution set to solutions modifying the circuit at nodes belonging to the old implementation circuit.

7 Experimental Results

circuit name	no. of inputs	no. of gates	signal name	rectification time	no. of solutions
The Berkeley Benchmark circuits (unabstracted)					
x1dn	27	108	32_out	0.14 sec	10
x9dn	27	89	31_out	0.20 sec	8
x6dn	38	285	42_out	0.89 sec	51
jbp	36	397	87_out	0.58 sec	16
chkn	29	511	539_out	0.83 sec	4
signet	39	240	40_out	2.97 sec	3
in6	33	188	41_out	0.12 sec	9
in7	26	143	31_out	0.21 sec	26
in3	34	302	49_out	0.43 sec	7
in5	24	213	25_out	0.72 sec	47
in4	32	568	33_out	5.10 sec	90
cps	24	936	942_out	4.86 sec	27
bc0	21	952	927_out	8.21 sec	7
The ISCAS85 Benchmark circuits (unabstracted)					
C432	36	160	1355_out	16.08 sec	4
C499	41	202	23_out	25.26 sec	12
C880	60	383	2899_out	0.01 sec	4
C1355	41	546	3882_out	215.78 sec	7
C1908	33	880	5361_out	509.44 sec	48
C2670	233	1193	432_out	> 20 min	—
C3540	50	1669	747_out	> 20 min	—
C5315	178	2307	7754_out	> 20 min	—
C6288	32	2406	6288_out	> 20 min	—
C7552	207	3512	420_out	> 20 min	—

Table 1. Rectification of the Berkeley benchmark circuits and the ISCAS85 benchmark circuits without abstraction.

We have implemented our rectification and abstraction method presented in Section 4 and Section 5 in an equivalence checker and circuit rectifier called AC/3 [12].

Using AC/3, we have evaluated our abstraction method with various benchmark examples, i.e., the Berkeley benchmark circuits [6], and the ISCAS85 benchmark circuits [7]. In each implementation circuit, we have induced a single error ranging from a missing line, a wrong logical connective, to a missing or double inverter. Computation time has been measured on a Sun Ultra 10 with 128 MB main memory and 300 MHz.

circuit name	Structural abstraction			Logical abstraction		
	abstraction time	rectification time	no. of solutions	abstraction time	rectification time	no. of solutions
The Berkeley Benchmark Circuits:						
	Threshold $\tau = \infty$			Threshold $\tau = \infty$		
x1dn	0.23 sec	0.01 sec	4	0.26 sec	0.01 sec	4
x9dn	0.25 sec	< 0.01 sec	7	0.24 sec	0.01 sec	4
x6dn	0.84 sec	0.03 sec	4	0.82 sec	0.03 sec	4
jbp	1.09 sec	0.01 sec	4	1.09 sec	0.02 sec	4
chkn	1.55 sec	0.04 sec	4	1.63 sec	0.03 sec	4
signet	0.73 sec	0.02 sec	3	0.78 sec	0.02 sec	2
in6	0.48 sec	0.02 sec	2	0.5 sec	0.01 sec	2
in7	0.34 sec	0.02 sec	3	0.36 sec	0.02 sec	3
in3	0.79 sec	< 0.01 sec	3	0.79 sec	< 0.01 sec	3
in5	0.60 sec	0.02 sec	5	0.63 sec	0.02 sec	5
in4	1.72 sec	0.16 sec	16	1.79 sec	0.04 sec	8
cps	6.19 sec	0.89 sec	6	10.84 sec	0.02 sec	7
bc0	3.92 sec	0.02 sec	6	3.71 sec	0.02 sec	6
The ISCAS85 Benchmark Circuits:						
	Threshold $\tau = \infty$			Threshold $\tau = 2000$ (200 for C7552)		
C432	0.44 sec	0.01 sec	4	0.82 sec	0.01 sec	4
C499	0.57 sec	0.01 sec	4	3.01 sec	0.02 sec	4
C880	6.09 sec	0.01 sec	4	8.11 sec	0.01 sec	4
C1355	1.84 sec	0.09 sec	7	8.15 sec	0.03 sec	5
C1908	2.35 sec	0.81 sec	13	9.22 sec	0.71 sec	10
C2670	4.78 sec	0.01 sec	10	14.57 sec	0.01 sec	9
C3540	5.51 sec	0.47 sec	12	18.13 sec	0.45 sec	10
C5315	11.53 sec	0.02 sec	7	22.93 sec	0.01 sec	6
C6288	15.74 sec	0.09 sec	30	74.28 sec	0.07 sec	28
C7552	20.74 sec	0.07 sec	11	24.02 sec	0.1 sec	5

Table 2. Rectification of the Berkeley benchmark circuits and the ISCAS85 benchmark circuits using abstraction. All circuits can be rectified.

The experimental results for the Berkeley benchmark circuits are shown in the upper half of Table 1 and 2. The first column contains the benchmark's name and the second column is the output signal to which the algorithm has been applied. Without abstraction, all Berkeley benchmark circuits can be rectified. Using structural abstraction, computation time can be reduced drastically for all examples. In 13 out of 15 examples the original error has been found so that the best solution (the solution which requires the minimal number of changes in the implementation circuit) was identical to the best solution computed on the unabstracted circuits. Only for 2 circuits, S-abstraction produced solutions that require more modifications in the implementation circuit. A drawback of logical abstraction seems to be that they are less robust with respect to pruning away the original error position which often results in unnecessary complex solutions (see Section 5.2). However, for the Berkeley Benchmark circuits, in 11 out of 16 examples, the optimal solution has been found.

The experimental results for the ISCAS85 benchmark circuits are shown in the lower half of Table 1 and 2. Without abstraction, 5 out of 10 circuits can be processed successfully. Using abstraction, we have been able to rectify all ISCAS85 circuits. Again, structural abstraction turned out to be more robust in respect to finding the optimal solution. While L-abstraction didn't find the optimal solution in 8 cases, S-abstraction produced the optimal solution in 8 out of 10 examples.

8 Summary

We have presented a method for localizing and correcting errors in combinatorial circuits. Unlike most other approaches, our method does not assume any error model. Thus, arbitrary design errors can be found.

Our method is split into two parts: the *location of erroneous sub-components* and the *computation of circuit corrections*. For both tasks, we have presented efficient solutions based on Boolean decomposition.

Since our method is not structure based, our technique can even be applied in scenarios where the specification is given as a Boolean formula, a truth table, or directly in form of a BDD. When computing circuit corrections, our approach tries to reuse parts of the old circuit in order to minimize the number of modifications and therefore to increase the quality of the computed solutions. Our method is powerful if the error causing elements are concentrated in a comparably small sub-part of the circuit since our algorithm tries to locate the smallest sub-component containing the erroneous parts. This is obviously true, e.g., for all circuits fulfilling the single error assumption.

To be able to handle large circuits, we have combined the rectification method with circuit abstractions. Two classes of abstractions have been examined: *structural abstractions* and *logical abstractions*. Whereas structural abstractions prune away structurally identical regions of a circuit, logical abstractions remove logically equivalent parts. We have shown correctness of our method by proving that circuit rectifications computed for abstract circuits can be lifted to circuit corrections for the original, unabstracted circuits.

We have implemented the presented methods in the **AC/3** verification tool and evaluated it with the Berkeley benchmark circuits [6] and the ISCAS85 benchmarks [7]. In combination with circuit abstractions, we have been able to rectify all ISCAS85 benchmarks. The experimental results show that together with the abstraction techniques discussed in this paper, our rectification approach is a powerful method for performing automatic error correction of large circuits.

In future, we will extend the rectification method with capabilities to simultaneously rectify circuits at multiple positions. Furthermore, we are currently extending our method to circuits containing tri-states and bus architectures.

References

1. M.S. Abadir, J. Ferguson, and T.E. Kirkland. Logic design verification via test generation. *IEEE Transactions on CAD*, 7(1):138–148, January 1988.

2. D. Brand. The taming of synthesis. In *International Workshop on Logic Synthesis*, May 1991.

3. D. Brand. Incremental synthesis. In *Proceedings of the International Conference on Computer Aided Design*, pages 126–129, 1992.

4. D. Brand. Verification of Large Synthesized Designs. In *IEEE/ACM International Conference on Computer Aided Design (ICCAD)*, pages 534–537, Santa Clara, California, November 1993. ACM/IEEE, IEEE Computer Society Press.

5. D. Brand, A. Drumm, S. Kundu, and P. Narain. Incremental synthesis. In *Proceedings of the International Conference on Computer Aided Design*, pages 14–18, 1994.

6. R.K. Brayton, G.D. Hachtel, C.T. McMullen, and A.L. Sangiovanni-Vincentelli. *Logic Minimization Algorithms for VLSI Synthesis*. The Kluwer International Series in Engineering and Computer Science. Kluwer Academic Publishers, 1986.

7. F. Brglez and H. Fujiwara. A neutral netlist of 10 combinatorial benchmark circuits and a target translator in FORTRAN. In *Int. Symposium on Circuits and Systems, Special Session on ATPG and Fault Simulation*, 1985.

8. R.E. Bryant. Graph-Based Algorithms for Boolean Function Manipulation. *IEEE Transactions on Computers*, C-35(8):677–691, August 1986.

9. R.E. Bryant. Symbolic boolean manipulation with ordered binary decision diagrams. *ACM Computing Surveys*, 24(3):293–318, September 1992.

10. P.Y. Chung, Y.M. Wang, and I.N. Hajj. Diagnosis and correction of logic design errors in digital circuits. In *Proceedings of the 30th Design Automation Conference (DAC)*, 1993.

11. A. Gupta. Formal Hardware Verification Methods: A Survey. *Journal of Formal Methods in System Design*, 1:151–238, 1992.

12. D. W. Hoffmann and T. Kropf. AC/3 V1.00 - A Tool for Automatic Error Correction of Combinatorial Circuits. Technical Report 5/99, University of Karlsruhe, 1999. available at http://goethe.ira.uka.de/~hoff.

13. Alan J. Hu. Formal hardware verification with BDDs: An introduction. In *IEEE Pacific Rim Conference on Communications, Computers, and Signal Processing (PACRIM)*, pages 677–682, October 1997.

14. S.Y. Huang, K.C. Chen, and K.T. Cheng. Error correction based on verification techniques. In *Proceedings of the 33rd Design Automation Conference (DAC)*, 1996.

15. J.C. Madre, O. Coudert, and J.P. Billon. Automating the diagnosis and the rectification of design errors with PRIAM. In *Proceedings of ICCAD*, pages 30–33, 1989.

16. S.M. Reddy, W. Kunz, and D.K. Pradhan. Novel Verification Framework Combining Structural and OBDD Methods in a Synthesis Environment. In *ACM/IEEE Design Automation Conference*, pages 414–419, 1995.

17. M. Tomita and H.H. Jiang. An algorithm for locating logic design errors. In *IEEE International Conference of Computer Aided Design (ICCAD)*, 1990.

18. M. Tomita, T. Yamamoto, F. Sumikawa, and K. Hirano. Rectification of multiple logic design errors in multiple output circuits. In *Proceedings of the 31st Design Automation Conference (DAC)*, 1994.

19. A. Veneris and I. N. Hajj. Correcting multiple design errors in digital VLSI circuits. In *IEEE International Symposium on Circuits and Systems (ISCAS)*, Orlando, Florida, USA, May 1999.

20. A. Wahba and D. Borrione. A method for automatic design error location and correction in combinational logic circuits. *Journal of Electronic Testing: Theory and Applications*, 8(2):113–127, April 1996.

21. A. Wahba and D. Borrione. Connection errors location and correction in combinational circuits. In *European Design and Test Conference ED&TC-97*, Paris, France, March 1997.

Abstract BDDs: A Technique for Using Abstraction in Model Checking *

Edmund Clarke, Somesh Jha, Yuan Lu, and Dong Wang

Carnegie Mellon University, Pittsburgh, PA 15213, USA
{emc,sjha,yuanlu,dongw}@cs.cmu.edu

Abstract. We propose a new methodology for exploiting abstraction in the context of model-checking. Our new technique uses abstract BDDs as its underlying data structure. We show that this technique builds a more refined model than traditional compiler-based methods proposed by Clarke, Grumberg and Long. We also provide experimental results to demonstrate the usefulness of our method. We have verified a pipelined carry-save multiplier and a simple version of the PCI local bus protocol. Our verification of the PCI bus revealed a subtle inconsistency in the PCI standard. We believe this is an interesting result by itself.

Keywords: Abstract BDDs, Model checking, and abstraction.

1 Introduction

Model-checking has attracted considerable attention because of existence of tools that can automatically prove temporal properties about designs. However, the state-explosion problem still remains a major hurdle in dealing with large systems. Most approaches for solving the state-explosion problem fall into two major categories: *efficient data-structures* and *state reduction techniques*. Symbolic model checking, which uses Binary Decision Diagrams(BDDs) [1, 2, 14] is an example of the first approach. State reduction methods apply transformations to the system and model-check the transformed system instead of the original system. Examples of such techniques are abstraction [6], symmetry reduction [4, 7, 10], and partial order reduction [8, 15].

Among the state-reduction approaches mentioned above, manual abstraction is the most widely used technique. However, this method is ad hoc and error-prone. Furthermore, since different properties of large systems usually require different abstraction techniques, manual abstraction is often difficult to use. For this reason, property driven automatic abstraction techniques are desirable for verifying actual hardware designs. Clarke, Grumberg and Long [6] have proposed

* This research is sponsored by the Semiconductor Research Corporation (SRC) under Contract No. 97-DJ-294 and the National Science Foundation (NSF) under Grant No. CCR-9505472. Any opinions, findings and conclusions or recommendations expressed in this material are those of the authors and do not necessarily reflect the views of SRC, NSF, or the United States Government.

a compiler-based abstraction technique. Their technique applies abstraction directly to variables during the compilation phase. It is automatic and efficient.

However, there are true ACTL properties which cannot be proved using their techniques since the compiler-based abstraction introduces many spurious behaviors. In this paper, we propose an automatic post-compilation abstraction technique using *abstract BDDs* (aBDDs) to avoid such problems. Our approach is closely related to those discussed in [6] and [13]. The difference is that we apply abstraction after the partitioned transition relation for the system has been constructed. The advantage of our approach is that we produce a *more refined* model of the system than theirs. Therefore, we can prove more properties about the system. Note that extraction of a partitioned transition relation is usually possible even though model checking may not be feasible.

Intuitively, an abstract BDD collapses paths in a BDD that have the same abstract value with respect to some abstraction function. They were originally used to find errors in combinational circuits. In this paper, we show how they can be used to provide a general framework for generating abstract models for sequential circuit designs. Our methodology consists of the following steps:

- The user provides the abstraction function along with the partitioned transition relation of the system represented in terms of BDDs. The user generally indicates how a particular variable should be abstracted.
- Next, the *abstract BDDs* corresponding to the transition relation are built. The procedure for building the abstract BDDs will be described in detail later in this paper.
- ACTL properties of the system are checked using the *abstract BDDs* determined by the transition relation.

We have modified SMV [14] in order to support our methodology. Notice that once the user provides the abstraction function, the entire methodology is automatic. We have selected two different designs to demonstrate the power of our approach. The first is a pipelined carry-save multiplier similar to the one described in Hennessy and Patterson [9]. The second is the PCI Local Bus Protocol [16]. We show that our technique can be used to establish correctness of certain key properties of these designs.

The paper is organized as follows. In Section 2, we provide a brief overview of how abstraction is used in model checking. We also explain how abstract BDDs are constructed. Section 3 provides a modified definition for abstract BDDs that is more suitable for the purposes of this paper. Section 4 discusses how abstract BDDs can be used to build an abstract model of a sequential circuit. Experimental results are provided in Section 5 which demonstrate that our method of using abstraction with model checking is superior to the one described in [6]. Section 6 concludes with some directions for future research.

2 Background

Throughout this paper, we assume that there are n state variables $x_1, \cdots, x_n$ with a domain $D = \{0,1\}^k$. The abstract state variables $\hat{x_1}, \hat{x_2}, \cdots, \hat{x_n}$ take values in an arbitrary domain A. For each x_i there is a *surjection* $h_i : D \rightarrow A$, which is the *abstraction function* for that variable. When it is clear from the context, we will suppress the index and write the abstraction function simply as $h : D \rightarrow A$. The abstraction function h induces an equivalence relation $\equiv_h$ on D as follows:

$$(d_1 \equiv_h d_2) \leftrightarrow h(d_1) = h(d_2).$$

The set of all possible equivalence classes of D under the equivalence relation $\equiv_h$ is denoted by $[D]_h$ and defined as: $\{[d] | d \in D\}$. Assume that we have a function $rep : [D]_h \rightarrow D$ that selects a unique representative from each equivalence class $[d]$. In other words, for a 0-1 vector $d \in D$, $rep([d])$ is the unique representative in the equivalence class of d. Moreover, the abstraction function h generates an abstraction function $\mathcal{H} : D \rightarrow D$ as follows:

$$\mathcal{H}(d) = rep([d]).$$

We call $\mathcal{H}$ the *generated abstraction function*. From the definition of $\mathcal{H}$ it is easy to see that $\mathcal{H}(rep([d])) = rep([d])$. Notice that the image of D under the function $\mathcal{H}$ is simply the set of representatives. The set of representatives will be denoted by $Img(\mathcal{H})$.

2.1 Abstraction for ACTL

Given a structure $M = (S, S_0, R)$, where S is the set of states, $S_0 \subseteq S$ is a set of initial states, and $R \subseteq S \times S$ is the transition relation, we define the *existential abstraction* $M_h = (S_h, S_{0,h}, R_h)$ as follows:

$$S_{0,h} = \exists x_1 \cdots x_n [h(x_1) = \hat{x}_1 \wedge \cdots \wedge h(x_n) = \hat{x}_n \wedge S_0(x_1, \cdots, x_n)]$$

$$R_h = \exists x_1 \cdots x_n \exists x'_1 \cdots x'_n [h(x_1) = \hat{x}_1 \wedge \cdots \wedge h(x'_1) = \hat{x'}_1 \wedge \cdots \wedge R(x_1, \cdots, x'_1, \cdots)]$$

In [6], the authors define a relation $\sqsubseteq_h$ between structures. For a structure $\tilde{M} = (\tilde{S}, \tilde{S}_0, \tilde{R})$, if

1. $S_{0,h}$ implies $\tilde{S}_0$ and
2. R_h implies $\tilde{R}$

then we say that $\tilde{M}$ *approximates* M (denoted by $M \sqsubseteq_h \tilde{M}$). Intuitively, if $M \sqsubseteq_h \tilde{M}$, then $\tilde{M}$ is *more* abstract than M_h, i.e., has more behaviors than M_h.

Since the number of states in the abstract structure M_h is usually much smaller than the number of states in M, it is usually desirable to prove a property on M_h instead of M. However, building M_h is often computationally expensive. In [6], Clarke, Grumberg and Long define a practical transformation $\mathcal{T}$ which applies the existential abstraction operation directly to variables at the innermost level of the formula. This transformation generates a new structure $M_{app} = (\mathcal{T}(S), \mathcal{T}(S_0), \mathcal{T}(R))$ and $M \sqsubseteq_h M_{app}$. As a simple example consider a system M which is a composition of two systems M_1 and M_2, or in other words $M = M_1 \| M_2$. In this case M_{app} is equal to $M_{1,h} \| M_{2,h}$. Note that the existential abstraction operation is applied to each process individually. Since $\mathcal{T}$ is applied at the innermost level, abstraction can be performed before building the BDDs for the transition relation. This abstraction technique is usually fast and easy to implement. However, it has potential limitations in checking certain properties. Since M_{app} is a coarse abstraction, there exist many properties which cannot be checked on M_{app} but can still be verified using a finer approximation. The following small example will highlight some of these problems.

A sensor-based traffic light example is shown in Figure 1. It includes two finite state machines (FSMs), one for a traffic light and one for an automobile. The traffic light M_t has four states {red, $green1$, $green2$, $yellow$}, and the automobile M_a also has four states {$stop1$, $stop2$, $drive$, $slow$}. M_t starts in the state red, when it senses that the automobile has waited for some time (in state $stop2$), it triggers a transition to state $green1$ which allows the automobiles to move. M_a starts from state $stop1$ and transitions according to the states of M_t. The safety property we want to prove is that when traffic light is red, the automobile should either slow down or stop. The property given above can be written in ACTL as follows:

$$\phi \equiv \mathbf{AG}[\neg(State_t = red \wedge State_a = drive)]$$

The composed machine is shown in Figure 1(c). It is easy to see that the property ϕ is true. Let us assume that we want to collapse states {$green1$, $green2$, $yellow$} into one state go. If we use the transformation $\mathcal{T}$, which applies abstraction before we compose M_t and M_a, property ϕ does not hold (the shaded state in Figure 1(d)). On the other hand, if we apply this abstraction after composing M_t and M_a, states $(green2, drive)$ and $(yellow, drive)$ are collapsed into one state(Figure 1(c)), and the property ϕ still holds. Basically, by abstracting the individual components and then composing we introduce too many spurious behaviors. Our methodology remedies this disadvantage by abstracting the transition relation of the composed structure $M_t \| M_a$.

The methodology presented in this paper constructs an approximate structure $\tilde{M}$ which is more precise than the structure M_{app} obtained by the technique proposed in [6]. All the transitions in the abstract structure M_h are included in both $\tilde{M}$ and M_{app}. Note that the state sets of M_h, $\tilde{M}$ and M_{app} are the same. The relationship between M, M_h, $\tilde{M}$ and M_{app} is shown in Figure 2. Roughly speaking, $\tilde{M}$ is a more refined approximation of M_h than M_{app}, or $\tilde{M}$ has less extra behaviors than M_{app}.

2.2 Abstract BDDs

In this subsection, we briefly review abstract BDDs. Additional information about this data structure can be found in [11]. Intuitively, an abstract BDD collapses paths in a BDD that have the same abstract value with respect to some abstraction function. The concepts underlying abstract BDDs are most easily explained using Binary Decision Trees (BDTs) but apply to BDDs as well. A binary decision tree (BDT) is simply a BDD in which there is no graph sharing.

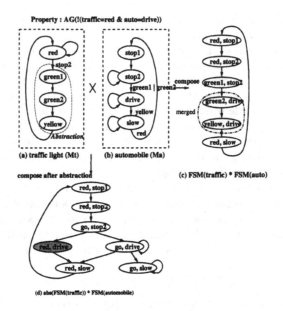

Fig. 1. Traffic light example

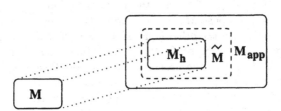

Fig. 2. Relationship between different structures

Given a boolean function $f : \{0,1\}^n \to \{0,1\}$ and its corresponding BDT T_f, let v denote the path from root to the node v at level $k+1$. It is easy to see that the path is a 0-1 vector in the domain $D = \{0,1\}^k$, i.e. $v \in D$. As we described before, an abstraction function $h : D \to A$ induces a generated abstraction function $\mathcal{H} : D \to D$. Assume that $w = rep([v])$, then the path $\mathcal{H}(v) = w$ ends at a node w, which is at the same level as v. Intuitively, in the abstraction procedure, The BDT rooted at v is replaced by the BDT rooted at w. We call node w the *representative* of node v. More formally, the abstract BDT $\mathcal{H}(f)$ of T_f rooted at v is defined as

$$\mathcal{H}(f)(v) = f(\mathcal{H}(v)).$$

In [11], the authors show a procedure which can build the abstract BDD of a function directly from the BDD of f instead of building the BDT of f. We also prove the following formula

$$f = p \circ q \rightarrow \mathcal{H}(f) = \mathcal{H}(p) \circ \mathcal{H}(q)$$

where $\circ$ is any logic operation. Notice that this means that the abstract BDD for f can be built incrementally, i.e., we do not have to build the BDD for f and then apply the abstraction function. Details can be found in [11].

The construction of an abstract BDD is illustrated by the following example. Assume that we have a boolean function $f(x_1, x_2, x_3) = (x_1 \wedge \neg x_3) \vee (x_2 \wedge x_3)$ (see Figure 3(a)). Consider the abstraction function $h(x_1, x_2) = x_1 + x_2$ (where "+" is ordinary addition). The abstraction function h induces an equivalence relation $\equiv_h$ on 0-1 vectors of length 2. Note that vectors of length 2 terminate at nodes of level 3. Therefore, we have $B \equiv_h C$ since $h(B) = h(C) = 1$. Assume that B is chosen as a representative, i.e. $\mathcal{H}(B) = \mathcal{H}(C) = B$. In other words, B is a representative node. Then the directed graph after abstraction is shown in Figure 3(b) and the final reduced BDD in Figure 3(c). Intuitively, the construction maintains some "useful" *minterms* and ignores other "uninteresting" *minterms*.

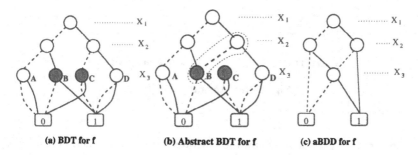

(a) BDT for f   (b) Abstract BDT for f   (c) aBDD for f

Fig. 3. Abstract BDD for f

Notice that the BDD of a representative node remains unaltered under the present definition. The non-representative nodes, however, are replaced by the corresponding representative nodes. In this paper, we extend aBDDs by allowing other operations on the nodes in an equivalence class. Details are described in next section.

3 New definition of abstract BDDs

Next, we define a new kind of abstract BDD which is more suitable for the purposes of this paper. The BDDs rooted at node v defines a boolean function $f(v)$. Recall that v is the path from the root to node v. The abstract BDD $\mathcal{H}(f)$

corresponding to the boolean function $f(v)$ is defined by the following equation:

$$\mathcal{H}(f)(v) = \begin{cases} \bigvee f(v') & v = \mathcal{H}(v') \\ 0 & \text{otherwise} \end{cases}$$

Notice that if v is a representative node, then $\mathcal{H}(v) = v$. Basically, the boolean function corresponding to a representative node is the *or* of all the boolean functions corresponding to the nodes in the same equivalence class. For non-representative nodes, the boolean function is defined to be **false** or 0. We use the example discussed previously to illustrate the new definition. Consider the boolean function $f(x_1, x_2, x_3) = (x_1 \wedge \neg x_3) \vee (x_2 \wedge x_3)$, $B \equiv_h C$ and B is the representative vector, so $\mathcal{H}(f)(B) = f(B) \vee f(C) = 1$ and $\mathcal{H}(f)(C) = 0$ (Figure 4).

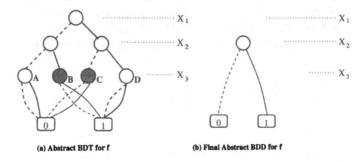

(a) Abstract BDT for f   (b) Final Abstract BDD for f

Fig. 4. New Abstract BDD for f

The next lemma describes how the generated abstraction function $\mathcal{H}$ interacts with conjunction and disjunction.

Lemma 1. *Let* $f, p, q : \{0,1\}^n \to \{0,1\}$ *be boolean functions, and let* $\mathcal{H} : D \to D$ *be the generated abstraction function corresponding to the abstraction function* $h : D \to A$. *The following equations hold:*

$$(f = p \vee q) \to (\mathcal{H}(f) = \mathcal{H}(p) \vee \mathcal{H}(q))$$

$$(f = p \wedge q) \to (\mathcal{H}(f) \to \mathcal{H}(p) \wedge \mathcal{H}(q))$$

The proof of this lemma is similar to the proof of Lemma 1 in [11]. A formal proof is provided in an appendix to this paper. The new definition of abstract BDDs can easily be extended to deal with multiple abstraction functions. For example, assume that we have m generated abstraction functions $\mathcal{H}_i : D \to D$ and a boolean function $f : D^m \to \{0,1\}$, the new abstract BDD of f can be defined as

$$\mathcal{H}(f)(x_1, \cdots, x_n) = \begin{cases} \bigvee f(y_1, \cdots, y_n) & x_1 = \mathcal{H}(y_1), \cdots, x_n = \mathcal{H}(y_n) \\ 0 & \text{otherwise} \end{cases}$$

Vectors x_i and y_i belong to the domain D, and $\mathcal{H}(y_i) = x_i$ implies that x_i is the representative in the equivalence class of y_i. It is easy to prove that Lemma 1 holds for multiple abstraction functions.

Lemma 2. *Given a boolean function* $f(x_1, \cdots x_n)$ *and an abstraction function* $\mathcal{H} : D \to D$, *the following formula holds*

$$\mathcal{H}(f) = \exists x_1, \cdots, x_n[\mathcal{H}(x_1) = y_1 \wedge \cdots \mathcal{H}(x_n) = y_n \wedge f(x)]$$

Lemma 2 states that the aBDD $\mathcal{H}(f)$ of the boolean function f corresponds to applying the existential abstraction operation to f (see the first paragraph in Section 2.1).

4 Methodology using abstract BDDs

In this section, we will discuss how to use abstract BDDs to construct an abstract Kripke structure. As we discussed in Section 2.1, $M_h = (S_h, S_{0,h}, R_h)$ is the abstract Kripke structure corresponding to $M = (S, S_0, R)$ using the abstraction function $h : D \to A$. Next we define an abstract structure $M_H = (S_H, S_{0,H}, R_H)$, which we can construct using aBDDs. The structure M_H is defined as follows:

- *State set* S_H is the image of S under the generated abstraction function $\mathcal{H}$.
- *Initial set of states* $S_{0,H}$ is the image of S_0 under the function $\mathcal{H}$. Notice that if S_0 is represented as a boolean function, then $S_{0,H}$ corresponds to the aBDD $\mathcal{H}(S_0)$ (see Lemma 2).
- *Transition Relation* R_H is the image of R under the function $\mathcal{H}$. Notice that if R is represented as a boolean function, then R_H corresponds to the aBDD $\mathcal{H}(R)$ (see Lemma 2).

Lemma 3. $M_h \cong M_H$

Lemma 3 states that M_h and M_H are *isomorphic structures*, i.e., there is a bijection $u : S_h \to S_H$ such that $u(S_0) = S_{0,H}$ and $u(R_h) = R_H$. Lemma 3 is proved in the appendix.

In general, it is intractable to build directly the BDD for transition relation R. Instead, the transition relation is usually partitioned [2]. Suppose that the transition relation R is partitioned into m clusters $R_1, \cdots, R_m$. Each cluster R_i is the transition relation of the composition of some set of components of the entire system. The transition relation R has one of the following forms depending on the nature of the composition (synchronous or asynchronous):

$$R = \begin{cases} R_1 \vee R_2 \vee \cdots \vee R_m & \text{asynchronous} \\ R_1 \wedge R_2 \wedge \cdots \wedge R_m & \text{synchronous} \end{cases}$$

The obvious way of applying abstraction is to distribute $\mathcal{H}$ over the cluster R_i. Using Lemma 1, we have

$$\mathcal{H}(R) = \mathcal{H}(R_1) \vee \mathcal{H}(R_2) \vee \cdots \vee \mathcal{H}(R_m) \quad \text{asynchronous}$$
$$\mathcal{H}(R) \to \mathcal{H}(R_1) \wedge \mathcal{H}(R_2) \wedge \cdots \wedge \mathcal{H}(R_m) \quad \text{synchronous}$$

If we use partitioned transition relation, then the abstract structure $\tilde{M}$ which is constructed is not isomorphic to M_h. But, because of the equations given above, we have $M \sqsubseteq \tilde{M}$.

If we apply $\mathcal{H}$ to the innermost components of the system, we build an abstract structure $M_{app,H}$. It is easy to prove that $M_{app,H}$ is isomorphic to the structure M_{app}. Recall that M_{app} is the abstract structure built using the techniques given in [6]. The technique presented in this section builds a more refined model than $M_{app,H}$ which is isomorphic to M_{app}. We emphasize this point once again using a small example. Assume that we decide to abstract the domain of state variable x to a single value $\perp$. Intuitively, the actual value of x is irrelevant for the property we are interested in. Suppose there are two state variables y and z whose values in the next state depend on x in the following way: $y' = x$ and $z' = \neg x$. In the structure M_{app_H}, y' and z' will both become free variables. If we combine the two transitions together as $y' = x \wedge z' = \neg x$ and then abstract x, the result will be $y' \neq z'$. Clearly the structure produced by the second technique is more precise than M_{app_H}. Intuitively, in [6] the abstraction is applied to every transition of each component and therefore can produce very coarse models.

4.1 Building the abstract structure M_h

Recall that the transition relation of the abstract structure M_h is given by the following formula:

$$R_h = \exists x_1 \cdots x_n \exists x_1' \cdots x_n' [h(x_1) = \hat{x}_1 \wedge \cdots \wedge h(x_1') = \hat{x}'_1 \wedge \cdots \wedge R(x_1, \cdots, x_1', \cdots)]$$

If we have BDDs encoding the transition relation R and the abstraction functions h_i, we could use standard traditional *relational product* technique to build the abstract transition relation R_h. We call this straightforward approach the traditional approach or method. Our new methodology has advantages over the traditional approach. First, in the traditional method the BDD for the abstraction functions has to be constructed before applying the method. For many abstraction functions, these BDDs are very hard to build. Second, in our experience a good variable ordering for an abstraction function might be different from a good variable ordering for the transition relation of the system. Our approach using abstract BDDs does not suffer from these problems since we never explicitly build the BDDs for the abstraction functions. Abstraction functions are employed while building the abstract BDD corresponding to the transition relation.

5 Experimental results

In order to test our ideas we modified the model-checker SMV. In our implementation, the user gives an abstraction function for each variable of interest. Once the user provides a system model and the abstraction functions, our method is completely automatic. We consider two examples in this paper: a pipelined multiplier design and the PCI local bus protocol.

5.1 Verification of a pipelined multiplier

In [6], Clarke, Grumberg, and Long propose an approach based on the *Chinese Remainder Theorem* for verifying *sequential* multipliers. The statement of the *Chinese Remainder Theorem* can be found in most texts on elementary number theory and will not be repeated here. Clarke, Grumberg, and Long use the modulus function $h(i) = i \bmod m$ for abstraction. They exploit the distributive property of the modulus function over addition, subtraction, and multiplication.

$$((i \bmod m) + (j \bmod m)) \bmod m \equiv (i + j) \bmod m$$

$$((i \bmod m) \times (j \bmod m)) \bmod m \equiv (i \times j) \bmod m$$

Let $\bullet$ represent the operation corresponding to the *implementation*. The goal is to prove that $\bullet$ is actually multiplication $\times$, or, in other words, for all x and y (within some finite range) $x \bullet y$ is equal to $x \times y$. If the actual implementation of the multiplier is composed of *shift-add* components, then the modulus function will distribute over the $\bullet$ operation. Therefore, we have the following equation:

$$(x \bullet y) \bmod m = [(x \bmod m) \bullet (y \bmod m)] \bmod m$$

Using this property and the Chinese Remainder Theorem, Clarke, Grumberg, and Long verify a sequential multiplier.

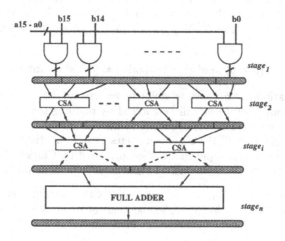

Fig. 5. Carry-save-adder pipeline multiplier

Unfortunately, this approach may not work if the multiplier is not composed of shift-add components. Suppose there is a mistake in the design of the multiplier, then there is no guarantee that the modulus operator will distribute over the operation $\bullet$ (corresponding to the actual implementation). For example, the mistake might scramble the inputs in some arbitrary way which breaks the distributive property of the • operation. In this case, the method proposed by Clarke, Grumberg and Long is not complete and may miss some errors. Therefore, before we apply the methodology in [6] it is necessary to check the distributive property of the modulus function with respect to the • operator. In other words, we must show that the following equation holds:

$$(x \bullet y) \bmod m = [(x \bmod m) \bullet (y \bmod m)] \bmod m$$

We illustrate our ideas by verifying a 16 × 16 pipelined multiplier which uses carry-save adders (see Figure 5). Notice that the first stage consists of *shift* operations and the last stage corresponds to the *add* operation. It easy to show that the first and the last stages satisfy the distributive property. In fact, this can be determined using classical equivalence checking methods. We will focus our attention on the intermediate stages.

Notice that the Chinese Remainder Theorem implies that it is enough to verify the multiplier by choosing $m = 5, 7, 9, 11, 13, 16, 17, 19, 23$ because of the following equation:

$$5 * 7 * 9 * 11 * 13 * 16 * 17 * 19 * 23 = 5354228880 > 2^{32} = 4294967296.$$

Our technique works as follows:

- First verify that each pipelined stage satisfies the distributive property using numbers in the set $\{5, 7, 9, 11, 13, 16, 17, 19, 23\}$). Formally, let $\bullet_i$ correspond to the operation of the i-th stage in the pipeline. We want to verify the following equation for all m in the set $\{5, 7, 9, 11, 13, 16, 17, 19, 23\}$ and $1 \leq i \leq 6$:

$$(x \bullet_i y) \bmod m = (x \bmod m \bullet_i y \bmod m) \bmod m$$

  If the equation given above is violated, we have found a error. Notice that the equation given above can be checked by building the abstract BDD for the transition relation corresponding to the i-th stage.
- Next, assume that all the pipelined stages satisfy the distributive property. In this case, we can apply the method proposed by Clarke, Grumberg, and Long because the entire design will also satisfy the distributive property.

In Figure 6 we give our experimental results for the first step. The row for space usage corresponds to the largest amount of memory that is used during verification.

modulus	5	7	9	11	13	16	17	19	23
time(s)	99	137	199	372	636	130	1497	2648	6977
space(MB)	7.7	12.8	21.5	51.7	92.5	9.2	210	231	430

Fig. 6. Experimental Results for various modulus

5.2 The PCI local bus

This subsection briefly describes our results for the PCI local bus protocol. During verification, we found a potential error in the PCI bus protocol specification. In particular, we discovered an inconsistency between the textual specification and one of the state machines given in the PCI standard [16]. The precise nature of the error will be explained later.

During model-checking, we used following abstraction functions on various state variables:

- $h(x) = \bot$, where $\bot$ means constant;
- $h(x) = $ if $x \neq 0$ then 1 else 0;
- $h(x) = $ if $x > 1$ then 1 else 0;

Incidentally, the bug we discovered was *not* found when we applied the techniques proposed in [6].

We briefly describe the PCI local bus protocol. There are three types of devices that can be connected to the PCI local bus: masters, targets, and bridges. Masters can start transactions. Targets respond to transactions and bridges connect buses. Masters and targets are controlled by a finite-state machine. We considered a simple model which consists of one master, one target, and one bus arbiter. The model includes different timers to meet the timing specification. The master and target both include a *lock* machine to support exclusive read/write. The master also has a data counter to remember the number of data phases.

In the verification, we applied different abstractions to some of the timers, the lock machine and the data counter in the master. We also clustered the transition relations of the major state controllers in both master and the target. We checked various properties dealing with handshaking, read/write transactions, and timing in this simplified model. Next, we describe in detail the property which demonstrates the inconsistency in the design. Description of all the properties that we checked is not given here because of space restrictions.

One of the textual requirements is that *the target responds to every read/write transaction issued by the master*. This important property turns out to be false for the state machine given in the standard when the master reads or writes a single data value. The negation of this property can be expressed in ACTL as follows:

$$\mathbf{AG}(\text{m.req} \wedge \text{m.data\_cnt=1}) \rightarrow \mathbf{A}[(\text{m.req} \wedge \neg\text{t.ack})\mathbf{U}(\text{m.timeout})]) \quad (*)$$

where *m.req* corresponds to the master issuing a transaction; *m.data_cnt=1* means that the master requests one data value; *t.ack* means that the target acknowledges the master's request; and *m.timeout* means that the time the master has allowed for the transaction has expired. If this ACTL formula is true in the abstract model, it is also true in the concrete model. We verified that this formula is true, so there must be an inconsistency in the standard.

The experimental results are shown in Figure 7. the first row in Figure 7 (*Error*) corresponds to the inconsistency we discovered. The remaining properties are not described here. The second and third columns show the running time and maximum BDD nodes for the original version of SMV. The fourth and fifth columns show the results obtained using our methodology. For some cases our approach reduces the space needed for verification by a factor of 20.

Properties	SMV		SMV_ABS	
	Time(s)	# nodes	Time(s)	# nodes
Error	278	727K	65	33K
Property 1	20	164K	18	14K
Property 2	137	353K	30	66K
Property 3	99	436K	138	54K
Property 4	185	870K	40	36K
Property 5	67	352K	42	57K

Fig. 7. Experimental Results for Verifying PCI using Abstraction

6 Related work and directions for future research

In this paper, we propose a new technique for exploiting abstraction using abstract BDDs. The work of Clarke, Grumberg, and Long [6] is closest to that described here. The main advantage of our method is that we generate more accurate abstractions than existing methods. Moreover, we do not need to build BDDs for the abstraction functions.

A technique for verifying *combinational* multipliers is described in [12, 17]. Their methods use *residue BDDs* which are a special case of abstract BDDs (see [11]). However, the method proposed in [17] is not general and does not readily extend to the problem of model-checking arbitrary systems. Errors have been found in many other bus protocols by using model checking techniques [5]. By using our methodology, it should be possible to handle larger systems consisting of multiple devices connected by buses and bus bridges.

We are pursuing several directions for future research. First, we intend to try our techniques on larger designs. In fact, we are currently in the process of verifying the entire PCI local bus protocol. We also want to find ways of generating abstraction functions automatically. A methodology for refining the abstraction functions automatically would also be extremely useful.

References

1. R. E. Bryant, "Graph-Based Algorithms for Boolean Function Manipulation", *IEEE Trans. on Comput., Vol. C-35, No.8, pp.677-691, Aug. 1986.*
2. J. R. Burch, E. M. Clarke, D. E. Long, K. L. McMillan, and D. L. Dill, "Symbolic Model Checking for Sequential Circuit Verification", *IEEE Trans. on CAD of Integrated Circuits and System, Vol.13, No.4, pp.401-424, 1994.*

3. E. M. Clarke, E. A. Emerson, and A. P. Sistla, "Automatic verification of finite-state concurrent system using temporal logic", *Proceedings of the Tenth Annual ACM Symposium on Principles of Programming Languages (POPL)*, January, 1983.

4. E.M. Clarke, R. Enders, T. Filkorn and S. Jha, "Exploiting Symmetry in Temporal Logic Model Checking", *Formal Methods in System Design 9(1/2):77-104*, 1996.

5. E. M. Clarke and O. Grumberg and H. Hiraishi and S. Jha and D. E. Long and K. L. McMillan and L. A. Ness, "Verification of the Futurebus+ Cache Coherence Protocol", *Formal Methods in System Design 6(2):217-232*, 1995.

6. E. M. Clarke, O. Grumberg, D. E. Long, "Model Checking and Abstraction", *ACM Transactions on Programming Languages and System (TOPLAS), Vol.16, No.5, pp.1512-1542*, Sept. 1994.

7. E.A. Emerson and A.P. Sistla, "Symmetry and Model Checking", *Formal Methods in System Design 9(1/2):105-130*, 1996.

8. P. Godefroid, D. Peled, and M. Staskauskas, "Using Partial Order Methods in the Formal Verification of Industrial Concurrent Programs", *ISSTA '96 International Symposium on Software Testing and Analysis, pp.261-269*, San Diego, California, USA, 1996. ACM Press.

9. J. L. Hennessy, D. A. Patterson, *Computer Architecture: A Quantitative Approach*, second edition, 1996. Morgan Kaufman Press.

10. C.N. Ip and D.L. Dill, "Better Verification Through Symmetry", *Formal Methods in System Design 9(1/2):41-76*, 1996.

11. S. Jha, Y. Lu, M. Minea, E. M. Clarke, "Equivalence Checking using Abstract BDDs", *Intl. Conf. on Computer Design (ICCD)*, 1997.

12. Shinji Kimura, "Residue BDD and Its Application to the Verification of Arithmetic Circuits", *32nd Design Automation Conference (DAC)*, 1995.

13. D. E. Long, "Model Checking, Abstraction and Compositional Verification", School of Computer Science, Carnegie Mellon University publication CMU-CS-93-178, July 1993.

14. K. L. McMillan, "Symbolic Model Checking: An Approach to the State Explosion Problem". Kluwer Academic Publishers, 1993.

15. D. Peled, "Combining Partial Order Reduction with on-the-fly model-checking", *Journal of Formal Methods in System Design, 8(1):39-64*.

16. PCI SGI, "PCI Local Bus Specification", Production Version Revision 2.1, June 1, 1995.

17. Kavita Ravi, Abelardo Pardo, Gary D. Hachtel, Fabio Somenzi, "Modular Verification of Multipliers", *Formal Methods in Computer-Aided Design, pp.49-63*, Nov. 1996.

A Proofs of the Lemmas

Proof of Lemma 1: Consider the binary decision trees of the functions f, p, and q. Note that these trees are exactly the same for f, p and q except the terminal nodes since all of the functions depend on the same set of variables. So if we can prove that for any node v in the trees, $f(v) = p(v) \vee q(v) \rightarrow \mathcal{H}(f)(v) = \mathcal{H}(p)(v) \vee \mathcal{H}(q)(v)$ and $f(v) = p(v) \wedge q(v)) \rightarrow \mathcal{H}(f)((v) = \mathcal{H}(p)(v) \wedge \mathcal{H}(q)(v)$, then the formula holds for the special case of the root. This implies that the formula holds for the original function. If v is a representative, from the definition, $\mathcal{H}(f)(v) = \bigvee_{\mathcal{H}(v')=v} f(v')$; Otherwise, $\mathcal{H}(f)(v) = 0$. The same formula holds when f is replaced by p and by q. If $f = p \vee q$, when v is non-representative, $\mathcal{H}(p)(v) \vee \mathcal{H}(q)(v) = 0 \vee 0 = \mathcal{H}(f)(v)$; otherwise, when v is a representative,

$$\mathcal{H}(f)(v) = \bigvee_{\mathcal{H}(v')=v} (p(v') \vee q(v')) = (\bigvee_{\mathcal{H}(v')=v} p(v')) \vee (\bigvee_{\mathcal{H}(v')=v} q(v'))$$

In general, we have $\mathcal{H}(f) = \mathcal{H}(p) \vee \mathcal{H}(q)$. On the other hand, If $f = p \wedge q$, when v is a non-representative, it is easy to see that $\mathcal{H}(p)(v) \wedge \mathcal{H}(q)(v) = 0 \wedge 0 = \mathcal{H}(f)(v)$; when v is a representative, we have

$$\mathcal{H}(f) = \bigvee_{\mathcal{H}(v')=v} f(v') = \bigvee_{\mathcal{H}(v')=v} (p(v') \wedge q(v'))$$

Likewise,

$$\mathcal{H}(p)(v) \wedge \mathcal{H}(q)(v) = \bigvee_{\mathcal{H}(v')=v} p(v') \wedge \bigvee_{\mathcal{H}(v')=v} q(v')$$

It is easy to see that $\bigvee_{\mathcal{H}(v')=v}(p(v') \wedge q(v'))$ implies $\bigvee_{\mathcal{H}(v')=v} p(v') \wedge \bigvee_{\mathcal{H}(v')=v} q(v')$. Consequently, $\mathcal{H}(f)(v) \rightarrow \mathcal{H}(p)(v) \wedge \mathcal{H}(q)(v)$. In general, $\mathcal{H}(f) \rightarrow \mathcal{H}(p) \wedge \mathcal{H}(q)$.

Proof of Lemma 3. Assume that $I : A \rightarrow img(\mathcal{H})$ is a function which is defined as $I(h(d)) = rep([d])$. First, we will show that I is well-defined and that I is a bijection. Second, using I we will build a bijection between the states of M_h and M_H.

From the definition, $h(d_1) = h(d_2)$ implies that $rep([d_1]) = rep([d_2])$ which in turn implies that $I(h(d_1)) = I(h(d_2))$. Therefore, I is a well defined function. If $d_1 \in img(\mathcal{H})$, then there exists a $d_2 \in D$, where $d_1 = rep([d_2])$. Moreover, $I(h(d_2)) = rep(d_2) = d_1$, so I is a surjection. On the other hand, if $I(h(d_1)) = I(h(d_2))$, then $rep([d_1]) = rep([d_2])$ which implies that $h(d_1) = h(d_2)$. Hence I is an injection. Since I is injective and surjective, I is a bijection.

As defined before, $S \subseteq D^n$ is the set of states of M; $S_h \subseteq A^n$ is the set of states of M_h; and $S_H \subseteq img(\mathcal{H})^n$ is the set of states of M_H. We define a mapping $u : S_h \rightarrow S_H$ as follows:

$$u(< \hat{x}_1, \cdots, \hat{x}_n >) = < I(\hat{x}_1), \cdots, I(\hat{x}_n) >$$

where $< \hat{x}_1, \cdots, \hat{x}_n > \in S_h$ and $< I(\hat{x}_1), \cdots, I(\hat{x}_n) > \in S_H$. Next we will show that $u(S_{0,h}) = S_{0,H}$ and $u(R_h) = R_H$, i.e., the bijection u preserves the initial states and the transitions. Consider an arbitrary state $< \hat{x}_1, \cdots, \hat{x}_n > \in S_{0,h}$ and an arbitrary transition $(< \hat{x}_1, \cdots, \hat{x}_n >, < \hat{x}'_1, \cdots, \hat{x}'_n >) \in R_h$ Since $< \hat{x}_1, \cdots, \hat{x}_n > \in S_{0,h}$, there exists a state $< x_1, \cdots, x_n > \in S$ such that $h(x_i) = \hat{x}_i$ and $< x_1, \cdots, x_n > \in S_0$. Since $\mathcal{H}(x_i) = rep([x_i]) = I(h(x_i)) = I(\hat{x}_i)$, and $S_{0,H}$ is the existential abstraction of S_0, it follows that $< I(\hat{x}_1), \cdots, I(\hat{x}_n) > S_{0,H}$.

The proof for the transition relation, is very similar. Therefore, $u(S_{0,h}) \subseteq S_{0,H}$ and $u(R_h) \subseteq R_H$. Since I is a bijection, the argument given above holds in the reverse direction. Thus, $u(S_{0,h}) = S_{0,H}$ and $u(R_h) = R_H$ This proves that $M_h \cong M_H$.

Formal Synthesis at the Algorithmic Level *

Christian Blumenröhr and Viktor Sabelfeld

Institute for Circuit Design and Fault Tolerance (Prof. Dr.-Ing. D. Schmid)
University of Karlsruhe, Germany
{blumen,sabelfel}@ira.uka.de   http://goethe.ira.uka.de/fsynth

Abstract. In our terminology, the term "formal synthesis" stands for a synthesis process where the implementation is derived from the specification by applying elementary mathematical rules within a theorem prover. As a result the implementation is guaranteed to be correct. In this paper we introduce a new methodology to formally derive register-transfer structures from descriptions at the algorithmic level via program transformations. Some experimental results at the end of the paper show how the run-time complexity of the synthesis process in our approach could be.

1 Introduction

The synthesis of hardware systems is heading toward more and more abstract design levels. This is due to the fact that the systems are becoming more complex and so does the synthesis process for deriving them. Therefore, the correctness of hardware components has become an important matter — especially in safety-critical domains. By correctness we mean that the synthesis result (implementation) satisfies the synthesis input (specification), in a formal mathematical sense. It is assumed that the specifications are correct, which has to be examined separately, e.g., by model-checking certain properties or by simulation. For proving the correctness of implementations, simulation is no longer suitable, since it is normally (i.e. for large designs) not exhaustive in reasonable time. Formal post-synthesis verification [1] on the other hand needs manual interactions at higher abstraction levels; it can be automated at the gate level, but is extremely costly — and can only be applied, if some very simple synthesis steps have been performed. Therefore, it is our objective to perform synthesis via logical transformations and thus to guarantee "correctness by construction".

There are many approaches, that claim to fulfill this paradigm. When regarding the state of the art in this area, one can distinguish two concepts: *transformational design* and *formal synthesis*. In transformational design [2], the synthesis process is based on correctness-preserving transformations. However, in most cases a lot of intuition is used during the proofs [3]. Furthermore the proofs are often based on non-mathematical formalizations [4] and are performed in a

---

* This work has been partly financed by the Deutsche Forschungsgemeinschaft, Project SCHM 623/6-2.

paper&pencil style [5], which means that they have to be examined by others to verify them. However, the most restrictive fact in transformational design is that the implementations of the transformations are not proven to be correct. The transformations are realized by complex software programs, that might be error-prone. Therefore these approaches do not fulfill the above mentioned paradigm.

In formal synthesis approaches the synthesis process is performed within some logical calculus. The circuit descriptions are formalized in a mathematical manner and the transformations are based on some logical rules. The DDD system [6], e.g., starts from a specification in a Lisp-like syntax. The behavior is specified as an iterative system of tail-recursive functions. This is translated into a sequential description which can be regarded as a network of simultaneous signal definitions comprising variables, constants, delays and expressions involving operations. Then a series of transformations are applied to refine the description into an implementation. The disadvantage of this method is that the description language is not strongly typed. Therefore the consistency of an expression has to be checked separately. Furthermore, although all the transformations are based on functional algebra, their implementations have not been formally verified, nor are they based on a small core of elementary rules. Finally, the derivation process needs manual interactions. An automatic design space exploration method is not provided.

Our work is based on a functional hardware description language named Gropius, which ranges from the gate level to the system level. Gropius is strongly-typed, polymorphic and higher-order. Each construct of Gropius is defined within the higher-order logic theorem prover HOL [7] and since it is a subset of higher-order logic, Gropius has a mathematically exact semantics. This is the precondition for proving correctness. The implementation of HOL is not formally verified. However, since the implementation of the correctness-critical part of HOL — i.e. deriving new theorems — is very small and is independent of the size of our formal synthesis system, our approach can be considered to be extremely safe as to correctness. In the next section, we briefly introduce the way we represent circuit descriptions at the algorithmic level and give a small program as a running example.

Existing approaches in the area of formal synthesis deal with lower levels of abstraction (register-transfer (RT) level, gate level) [8, 9, 6, 10, 11] or with pure dataflow graphs at the algorithmic level [12]. This paper addresses formal synthesis at the algorithmic level. The approach goes beyond pure basic blocks and allows synthesizing arbitrary computable, i.e. μ-recursive programs.

The starting point for high-level synthesis (HLS) is an algorithmic description. The result is a structure at the RT level. Usually, hardware at the RT-level consists of a data-path and a controller. In conventional approaches [13], first, all loops in a given control/dataflow graph (CDFG) are cut, thus introducing several acyclic program pieces each corresponding to one clock tick. The number of these cycle-free pieces hereby grows exponentially with the size of the CDFG. Afterwards scheduling, allocation and binding are performed separately on these parts leading to a data-path and a state transition table. Finally, the controller and the communication part are generated.

We have developed a methodology that absolutely differs from this standard. In our approach, the synthesis process is not reduced to the synthesis of pure dataflow graphs, but the circuit description always remains compact and the RT-level structure is derived via program transformations. Besides an RT-level structure, our approach additionally delivers an accompanying proof in terms of a theorem telling that this implementation is correct. High-level synthesis is performed in four steps. In the first two steps which are explained in Section 3, scheduling and register allocation/binding are performed. Based on pre-proven program equations which can be steered by external design exploration techniques, the program is first transformed into an equivalent but optimized program, and then this program is transformed into an equivalent program with a single while-loop. The third step (Section 4) performs interface synthesis. An interface behavior can be selected and the program is mapped by means of a pre-proven implementation theorem to a RT-level structure, that realizes the interface behavior with respect to the program. In the last step, which is not addressed explicitly here, functional units are allocated and bound. Section 5 will give some experimental results.

2 Formal representation of programs

At the algorithmic level, behavioral descriptions are represented as pure software programs. The concrete timing of the circuit, that has to be synthesized, is not yet considered. In Gropius, we distinguish between two different algorithmic descriptions. DFG-terms represent non-recursive programs that always terminate (<u>D</u>ata <u>F</u>low <u>G</u>raphs). They have some type $\alpha \to \beta$. P-terms are means for representing arbitrary computable functions (<u>P</u>rograms). Since P-terms may not terminate, we have added an explicit value to represent nontermination: a P-term either has the value Defined (x) indicating that the function application terminates with result x, or in case of nontermination the value is Undefined. The type of P-terms is expressed by $\alpha \to (\beta)$partial.

In our approach, P-terms are used for representing entire programs as well as blocks. Blocks are used for representing inner pieces of programs. In contrast to programs, the input type equals the output type. This is necessary for loops which apply some function iteratively. In Gropius, there is a small core of 8 basic control structures for building arbitrary computable blocks and programs based on basic blocks and conditions. Basic blocks (type $\alpha \to \alpha$) and conditions (type $\alpha \to$ bool) itself are represented by DFG-terms. In Table 1, only those control structures are explained that are used in this paper. Based on this core of control structures further control structures like for- and repeat-loops can be derived by the designer.

In the rest of the paper a specific pattern called single-loop form (SLF) plays an important role. Programs in SLF have the following shape:

$$\text{PROGRAM } \textit{out\_init} \text{ (LOCVAR } \textit{var\_init} \text{ (WHILE } c \text{ (PARTIALIZE } a\text{)))} \qquad (1)$$

PARTIALIZE	Type: $(\alpha \to \alpha) \to \alpha \to (\alpha)$partial. Turns a basic block a to a block (PARTIALIZE a). Since basic blocks always terminate, (PARTIALIZE a) maps some x to Defined$(a(x))$. Undefined is never reached.
WHILE	Type: $(\alpha \to$ bool$) \to (\alpha \to (\alpha)$partial$) \to \alpha \to (\alpha)$partial. Basis for formalizing true μ–recursion. Given a block A and a condition c, (WHILE $c\,A$) maps some parameter x to some value Defined(y) by iterating A until the value y is reached with $\neg(c\,y)$. In case of nontermination the result becomes Undefined.
THEN	Type: $(\alpha \to (\alpha)$partial$) \to (\alpha \to (\alpha)$partial$) \to \alpha \to (\alpha)$partial. Binary function used in infix notation. Two blocks A and B are executed consecutively. The result of $(A$ THEN $B)$ becomes Undefined, iff one of the two blocks does not terminate.
LOCVAR	Type: $\beta \to ((\alpha \times \beta) \to (\alpha \times \beta)$partial$) \to$ $\alpha \to (\alpha)$partial. Introduce a local variable. Given an arbitrary initial value $init$ for the local variable, the function (LOCVAR $init\,A$) first maps some input x to $A(x, init)$. If the result becomes Defined$(x', init)$, then Defined(x') is returned. In case of nontermination Undefined is returned.
PROGRAM	Type: $\beta \to ((\alpha \times \beta) \to (\alpha \times \beta)$partial$) \to$ $\alpha \to (\beta)$partial. Turns a block into a program. Given an arbitrary initial value $init$ for the output variable, the function (PROGRAM $init\,A$) first maps some input x to $A(x, init)$. If $A(x, init)$ terminates with Defined$(x', init)$, then Defined$(init')$ is returned. In case of nontermination Undefined is returned.

Table 1. Core control structures

The expressions *out_init* and *var_init* denote arbitrary constants, c is an arbitrary condition and a an arbitrary basic block.

Basically, no front-end is required, since Gropius is both the input language and the intermediate format for transforming the circuit description. However, since the "average designer" may not be willing to specify in a mathematical notation, it would also be possible to automatically translate Gropius-descriptions from other languages like Pascal. But on the other hand this adds an error-prone part into the synthesis process you can abandon, since Gropius is easy to learn — there are only few syntax rules.

Fig. 1 shows a program in an imperative representation style (Pascal) and a corresponding description in Gropius. The program computes the nth Fibonacci number and uses a fast algorithm which has a complexity of $O(\log_2 n)$.

The program fib has type num $\to$ (num)partial (num is the type of natural numbers). The input variable is n. The construct (PROGRAM1) introduces a variable (here called $y1$) which indicates the output of the program and hence the nth Fibonacci number. A special "return"-statement is not necessary since the result is stored in $y1$. The initial value of the output is the same as for

Imperative Program	Representation in Gropius
FUNCTION FIB	val fib $=$
(var n : int) : int;	PROGRAM 1
VAR $a1, a2, y1, y2, m$: int;	LOCVAR $(1, 1, 0, 0)$
VAR r, s : int;	PARTIALIZE $(\lambda((n, y1), a1, a2, y2, m).$
BEGIN	$((n, y1), a1, a2, y2, (n$ DIV $2) + 1))$
$a1 := 1;\ a2 := 1;$	THEN
$y1 := 1;$	WHILE $(\lambda((n, y1), a1, a2, y2, m).\neg(m = 0))$
$y2 := 0;$	PARTIALIZE $(\lambda((n, y1), a1, a2, y2, m).$
$m := n$ div $2 + 1;$	let $c =$ ODD m in
WHILE $m <> 0$ DO	let $m1 = m - 1$ in
IF odd m	let $m2 = m$ DIV 2 in
THEN BEGIN	let $m3 =$ MUX$(c, m1, m2)$ in
$r := y1;$	let $x = a1 + a2$ in
$y1 := y1 * a1 + y2 * a2;$	let $x1 =$ MUX$(c, y1, a1)$ in
$y2 := r * a2 + y2 * (a1 + a2);$	let $x2 =$ MUX$(c, y2, a2)$ in
$m := m - 1$	let $x3 = x1 * a1$ in
END	let $x4 = x2 * a2$ in
ELSE BEGIN	let $x5 = x3 + x4$ in
$s := a1;$	let $x6 = x1 * a2$ in
$a1 := a1 * a1 + a2 * a2;$	let $x7 = x2 * x$ in
$a2 := s * a2 + a2 * (s + a2);$	let $x8 = x6 + x7$ in
$m := m$ div 2	let $y1' =$ MUX$(c, x5, y1)$ in
END;	let $a1' =$ MUX$(c, a1, x5)$ in
IF odd n	let $a2' =$ MUX$(c, a2, x8)$ in
THEN	let $y2' =$ MUX$(c, x8, y2)$ in
RETURN $y2$	$((n, y1'), a1', a2', y2', m3))$
ELSE	THEN
RETURN $y1$	PARTIALIZE $(\lambda((n, y1), a1, a2, y2, m).$
END;	$(n, MUX($ODD $n, y2, y1)), a1, a2, y2, m))$

Fig. 1. Program for calculating the nth Fibonacci number

the local variable $y1$ in the imperative program. (LOCVAR$(1, 1, 0, 0)$) is used to introduce local variables with initial values 1, 1, 0 and 0, respectively. These local variables correspond to the local variables $a1, a2, y2$ and m in the imperative program. DFG-terms are formalized using λ-expressions [14]. An expression $(\lambda x.a[x])$ denotes a function, which maps the value of a variable x to the expression $a[x]$, which has some free occurrences of x. Two basic laws of the λ-calculus are β-conversion and η-conversion:

$$(\lambda x.a[x])\ b \xrightarrow{\beta} a[b/x] \qquad (\lambda x.a\ x) \xrightarrow{\eta} a$$

let-terms (let $x = y$ in z) are a syntactic variant of β-redices $(\lambda x.z)\ y$. By applying β-conversion some or all of the let-terms can be eliminated. In contrast to LOCVAR, let-terms introduce local variables only within basic blocks. The expression MUX is an abbreviation for a conditional expression within basic blocks. $\text{MUX}(c, a, b)$ returns a, if the condition c is true, otherwise it returns b.

The correspondence between the two descriptions in Fig. 1 is not one-to-one. To yield a more efficient description with less addition and multiplication operations, the following two theorems for conditionals have been applied:

$$\vdash f\,\text{MUX}(c, a, b) = \text{MUX}(c, f\,a, f\,b) \quad \vdash \text{MUX}(c, a, b)\,g = \text{MUX}(c, a\,g, b\,g) \quad (2)$$

The imperative program can first be translated into a Gropius description containing the same eight multiplications and six additions in the loop-body. Then the theorems (2) can be applied to generate the description shown in Fig. 1, which needs only four multiplications and three additions in the loop-body.

3 Program transformations

The basic idea of our formal synthesis concept is to transform a given program into an equivalent one, which is given in SLF. This is motivated by the fact that hardware implementations are nothing but a single while-loop, always executing the same basic block. Every program can be transformed into an equivalent SLF-program (KLEENE's normal form of μ-recursive functions). However for a given program there might not be an unique SLF, but there are infinitely many equivalent SLF-programs. In the loop-body of a SLF-program, all operations of the originally given program are scheduled. The loop-body behaves like a case-statement, in which within a single execution certain operations are performed according to the control state indicated by the local variables (LOCVAR var_init). After mapping the SLF-program to a RT-level structure (see Section 4), every execution of the loop-body corresponds to a single control step. The cost of the RT-implementation therefore depends on which operations are performed in which control step. Thus every SLF corresponds to a RT-implementation with certain costs. Performing high-level synthesis therefore requires to transform the program into a SLF-program, that corresponds to a cost-minimal implementation.

In the HOL theorem prover we proved several program transformation theorems which can be subdivided into two groups. The first group consists of 27 theorems. One can prove that these theorems are sufficient to transform every program into an equivalent program in SLF. The application of these theorems is called the standard-program-transformation (SPT). During the SPT, control structures are removed and instead auxiliary variables are introduced holding the control information. Theorem (3) is an example:

$$\vdash \text{WHILE}\,c_1\,(\text{LOCVAR}\,v\,(\text{WHILE}\,c_2\,(\text{PARTIALIZE}\,a))) =$$
$$\text{LOCVAR}\,(v, \text{F})$$
$$\text{WHILE}\,(\lambda(x, h_1, h_2).\,c_1\,x\,\vee\,h_2)$$
$$\text{PARTIALIZE}\,(\lambda(x, h_1, h_2).\text{MUX}\,(c_2\,(x, h_1), (a\,(x, h_1), \text{T}), (x, v, \text{F}))) \quad (3)$$

Two nested while-loops with a local variable at the beginning of the outer loop-body are transformed to a single while-loop. The local variable is now outside the loop and there is an additional local variable with initial value F. This variable holds the control information, whether the inner while-loop is performed (value is T) or not (value is F).

Although the SLF representation is not unique, the SPT always leads to the same SLF for a given program by scheduling the operations in a fixed way. Therefore, the SPT unambiguously assigns costs to every program. To produce other, equivalent SLF representations, which result in another scheduling and thus in other costs for the implementation, the theorems of the second group have to be applied before performing the SPT. Currently, we proved 19 optimization-program-transformation (OPT) theorems. These OPT-theorems can be selected manually, but it is also possible to integrate existing design space exploration techniques which steer the application of the OPT-theorems. The OPT-theorems realize transformations which are known from the optimization of compilers in the software domain [15]. Two of these transformations are loop-unrolling and loop-cutting.

Loop unrolling reduces the execution time since several operations are performed in the same control step. On the other hand, it increases the combinatorial depth and therefore the amount of hardware. Theorem (4) shows the loop unrolling theorem. It describes the equivalence between a while-loop and an n-fold unrolled while-loop with several loop-bodies which are executed successively. Between two loop-bodies, the loop-condition is checked to guarantee that the second body is only executed if the value of the condition is still true. FOR_N is a function derived from the core of Gropius. (FOR_N n A) realizes an n-fold application of the same block A. Theorem (5) can be used to remove FOR_N after having instantiated n (SUC is the successor function).

$$\vdash \text{WHILE } c \text{ (PARTIALIZE } a) =$$
$$\text{WHILE } c \left((\text{PARTIALIZE } a) \text{ THEN}\right.$$
$$\left.(\text{FOR\_N } n \text{ (PARTIALIZE } (\lambda x. \text{ MUX } (c\, x, a\, x, x)))))\right) \tag{4}$$

$$\vdash (\text{FOR\_N } 1\, A = A) \; \wedge \; (\text{FOR\_N } (\text{SUC } n)\, A = A \text{ THEN } (\text{FOR\_N } n\, A)) \tag{5}$$

The counterpart to loop unrolling is the loop cutting: the loop is cut into several smaller parts. Each part then corresponds to a separate control step. This results in a longer execution time; however, the hardware consumption might be reduced, if the parts can share function units.

$$\vdash \text{WHILE } c \text{ (PARTIALIZE } (\text{list\_o } (k :: r))) =$$
$$\text{LOCVAR } (\text{enum } (\text{SUC } (\text{LENGTH } r))\, 0)$$
$$\text{WHILE } (\lambda(x,h). c\, x \; \vee \; \neg(h = 0)) \tag{6}$$
$$\text{PARTIALIZE } (\lambda(x,h). ((\text{CASE } (k :: r)\, h)\, x, \text{ next } (\text{SUC } (\text{LENGTH } r))\, h))$$

In (6) the loop cutting theorem is shown. It assumes that the body of a while-loop has been scheduled into a composition of functions. The term (list_o L)

denotes a composition of functions given by the list L. By restricting L to be of the form $(k :: r)$ it is guaranteed that the list is not empty. Each function of the list L must have the same type $\alpha \to \alpha$. The input and output types must be equal, since loop-bodies are executed iteratively. Given a while-loop with such a scheduled loop-body (list_o $(k :: r)$), theorem (6) turns it into an equivalent while-loop, which executes its body (LENGTH r)[1] times more often than the original while-loop. Within one execution of the loop-body exactly one of the functions of the list $(k :: r)$ is applied. The control information, which function is to be performed is stored in a local variable that has been introduced. This variable has an enumeration datatype. Its initial value is 0 and its value ranges from 0 to (LENGTH r). The semantics of enum is shown in (7). If the local variable has value 0, the loop-condition c is checked, whether to perform the loop-body or not. If the local variable's value differs from 0, the loop-body will be executed independent of c. (CASE L i) picks the ith function of the list L. Therefore, within one execution of the loop-body, a function of the list $(k :: r)$ is selected according to the value h of the local variable and then this function is applied to the value x of the global input variable. Furthermore, the new value of the local variable is determined. The semantics of next is shown in (7).

$$\vdash \text{enum} \, n \, m \; = \; \text{MUX}(m < n, m, 0) \quad \vdash \text{next} \, n \, x \; = \; \text{MUX}(\text{SUC} \, x \; < \; n, \text{SUC} \, x, 0) \quad (7)$$

Returning to our example program fib, the body of the while-loop can be scheduled in many ways. The decision on how the body should be scheduled can be made outside the logic by incorporating existing scheduling techniques for data-paths. Table 2 shows the results of applying the ASAP (as-soon-as-possible), force-directed [16] and list-based scheduling techniques to our example program. The ASAP algorithm delivers the minimal number of control steps. In addition to this, the force-directed-algorithm tries to minimize the amount of hardware. The list-based scheduling on the other hand restricts the amount of hardware components and tries to minimize the execution time. For each control step, we list the local variables of the loop from Fig. 1 that hold the result of the corresponding operations. Note that no chaining was allowed in the implementation of these scheduling programs. However, this is not a general restriction. For performing the list-based scheduling, the number of multiplications and additions was each restricted to two.

In the next step, the number and types of the registers will be determined that have to be allocated. This is also listed in Table 2. Before actually scheduling the loop-body by a logical transformation, additional input variables have to be introduced, since the number of input and output variables directly corresponds to the number of registers necessary at the RT-level. The number of additional variables is (#$regalloc$ − #$invars$) with #$invars$ being the number of input variables of the loop-body and #$regalloc$ being the number of allocated registers. For our example fib in the case of allocation after the ASAP scheduling, this value is $11 - 6 = 5$. Therefore 5 additional input variables for the loop have to

---

[1] LENGTH L returns the length of a list L. Instantiating the theorem with a concrete list, yields a concrete value for this expression. Similarly, (list_o L) then corresponds to a concrete term.

Control step	ASAP	Force-directed	List-based
1	$c, m1, m2, x$	c	$c, m1, m2, x$
2	$m3, x1, x2$	$m1, m2, x, x1, x2$	$m3, x1, x2$
3	$x3, x4, x6, x7$	$x3, x4, x6, x7$	$x3, x4$
4	$x5, x8$	$m3, x5, x8$	$x5, x6, x7$
5	$y1', a1', a2', y2'$	$y1', a1', a2', y2'$	$x8, y1', a1'$
6	$-\ -\ -$	$-\ -\ -$	$a2', y2'$
Allocated registers	Type bool : 1 Type num : 10	Type bool : 1 Type num : 10	Type bool : 1 Type num : 11

Table 2. Control information extracted by different scheduling algorithms

be introduced. This is done by theorem (8). Applying it to the loop in Fig. 1 with appropriately instantiating i gives program (9).

$$\vdash \forall i.\ \mathsf{WHILE}\ c\ (\mathsf{PARTIALIZE}\ a) = \qquad\qquad\qquad\qquad\qquad (8)$$
$$\mathsf{LOCVAR}\ i\ (\mathsf{WHILE}\ (\lambda(x,h).cx)\ (\mathsf{PARTIALIZE}(\lambda(x,h).(ax,i))))$$

$$\mathsf{LOCVAR}(F,0,0,0,0)$$
$$\mathsf{WHILE}(\lambda(((n,y1),a1,a2,y2,m),h1,h2,h3,h4,h5).\neg(m=0))$$
$$\mathsf{PARTIALIZE}(\lambda(((n,y1),a1,a2,y2,m),h1,h2,h3,h4,h5). \qquad\qquad (9)$$
$$\mathsf{let}\ c = \mathsf{ODD}\ m\ \mathsf{in}\ \ \ \ldots\ \ \mathsf{in}\ (((n,y1'),a1',a2',y2',m3),F,0,0,0,0))$$

The additional variables are only dummies for the following scheduling and register allocation/binding. They must not be used within the loop-body. Since the original input variables are all of type num, one variable of type bool ($h1$) and four variables of type num ($h2,\ldots,h5$) are introduced. Some default initial values are used for each type. Since the output type must equal the input type, additional outputs have to be introduced as well.

Now the DFG-term representing the loop-body can be scheduled and register binding can be performed, both by logical conversions within HOL. Fig. 2 shows the resulting theorem after this conversion. The equivalence between the original and the scheduled DFG-term is proven by normalizing the terms, i.e. performing β-conversions on all β-redices [19]. The register binding was performed based on the result of a heuristic that tries to keep a variable in the same register as long as possible to avoid unnecessary register transfer. The right hand side of the theorem in Fig. 2 is actually an expression of the form (list_o L) with L being a list of five DFG-terms. The theorem in Fig. 2 can be used to transform the circuit description fib by rewriting and afterwards the loop-cutting theorem (6) can be applied.

Besides loop-unrolling and loop-cutting, several other OPT-theorems can be applied. After that the SPT is performed generating a specific program in SLF. Fig. 3 shows the theorem after performing the SPT without applying any OPT-theorem before. When the program in SLF is generated without any OPT, then the operations in the three blocks before, within and after the while-loop in the original program fib, will be performed in separate executions of the resulting loop-body. Therefore, function units can be shared among these three blocks.

$\vdash \lambda(((n, y1), a1, a2, y2, m), h1, h2, h3, h4, h5).$
  let $c = $ ODD m in $\ \ \ldots \ \ $ in $\ \ \ (((n, y1'), a1', a2', y2', m3), \mathsf{F}, 0, 0, 0, 0)$
$=$
$\lambda(((r1, r2), r3, r4, r5, r6), r7, r8, r9, r10, r11).$
  let $y1' = $ MUX$(r7, r2, r6)$ in let $a1' = $ MUX$(r7, r1, r2)$ in
  let $a2' = $ MUX$(r7, r8, r3)$ in let $y2' = $ MUX$(r7, r3, r9)$ in
  $(((r5, y1'), a1', a2', y2', r4), \mathsf{F}, 0, 0, 0, 0)$
$\circ$
$\lambda(((r1, r2), r3, r4, r5, r6), r7, r8, r9, r10, r11).$
  let $r2' = r11 + r10$ in let $r3' = r3 + r2$ in
  $(((r1, r2)', r3', r4, r5, r6), r7, r8, r9, r10, r11)$
$\circ$
$\lambda(((r1, r2), r3, r4, r5, r6), r7, r8, r9, r10, r11).$
  let $r11' = r2 * r1$ in let $r10' = r10 * r8$ in let $r3' = r2 * r8$ in
  let $r2' = r10 * r3$ in $(((r1, r2'), r3', r4, r5, r6), r7, r8, r9, r10', r11')$
$\circ$
$\lambda(((r1, r2), r3, r4, r5, r6), r7, r8, r9, r10, r11).$
  let $r4' = $ MUX$(r7, r2, r4)$ in let $r2' = $ MUX$(r7, r6, r1)$ in
  let $r10' = $ MUX$(r7, r9, r8)$ in $(((r1, r2'), r3, r4', r5, r6), r7, r8, r9, r10', r11)$
$\circ$
$\lambda(((n, y1), a1, a2, y2, m), h1, h2, h3, h4, h5).$
  let $r7' = $ ODD m in let $r2' = m - 1$ in let $r4' = m$ DIV 2 in
  let $r3' = a1 + a2$ in $(((a1, r2'), r3', r4', n, y1), r7', a2, y2, r10, r11)$

Fig. 2. Theorem after scheduling and register allocation/binding

Although ,e.g., two DIV-operations appear in Fig. 3, only one divider is necessary. One division comes from the block before the loop and the other results from the loop-body. These two division-operations are therefore needed in different executions of the new loop-body. They can be shifted behind the multiplexer by using one of the theorems (2). The allocation and binding of functional units is the last step in our high-level synthesis scenario. Before this, interface synthesis must be applied.

4 Interface synthesis

At the algorithmic level, circuit representations consist of two parts. An algorithmic part describes the functional relationship between input and output. Time is not yet considered. During high-level synthesis the algorithmic description is mapped to a RT-level structure. To bridge the gap between these two different abstraction levels one has to determine how the circuit communicates with its environment. Therefore, as second component of the circuit representation, an interface description is required.

   In contrast to most existing approaches, we strictly separate between the algorithmic and the interface description. We provide a set of at the moment

```
⊢ fib =
  PROGRAM1
    LOCVAR((1, 1, 0, 0), F, F)
      WHILE(λ(((n, y1), a1, a2, y2, m), h1, h2).¬h1 ∨ ¬h2)
        PARTIALIZE(λ(((n, y1), a1, a2, y2, m), h1, h2).
          let x1' = ¬(m = 0) in
          let c = ODD m in
          let x1'' = MUX(c, y1, a1) in
          let x2 = MUX(c, y2, a2) in
          let x5 = x1'' * a1 + x2 * a2 in
          let x8 = x1'' * a2 + x2 * (a1 + a2) in
          ( ((n, MUX(h2, MUX(x1', MUX(c, x5, y1), MUX(ODD n, y2, y1)), y1)),
            MUX(h2 ∧ x1', MUX(c, a1, x5), a1),
            MUX(h2 ∧ x1', MUX(c, a2, x8), a2),
            MUX(h2 ∧ x1' ∧ c, x8, y2),
            MUX(h2, MUX(x1', MUX(c, m − 1, m DIV 2), m), (n DIV 2) + 1)),
            MUX(h2, ¬x1', h1), T ) )
```

Fig. 3. Single-loop-form of fib

nine interface patterns, of which the designer can select one. Some of these patterns are used for synthesis of P-terms and others for synthesis of DFG-terms. The orthogonal treatment of functional and temporal aspects supports reuse of designs in a systematic manner, since the designer can use the same algorithmic description in combination with different interface patterns.

Remark: At the algorithmic level we only consider single processes that do not communicate with other processes. In addition to this, we have developed an approach for formal synthesis at the system level, where several processes interact with each other [17].

Fig. 4 shows the formal definition of two of those interface patterns. Beside the data signals of an algorithmic description P, the interface descriptions contain additional control signals which are used to steer the communication, to stop or to start the execution of the algorithm.

The two patterns are functions which map an arbitrary program P and the signals $(in, start, out, ready)$ and $(in, reset, out, ready)$, respectively, to a relation between these signals with respect to the program P. The pattern P_IFC_START states that at the beginning the process is idle, if the $start$-signal is not active. As long as the process is idle and no calculation is started, the process will be idle in the next clock tick and out holds its last value. If $start$ is active, i.e. a new calculation begins, two alternatives may occur:

– the calculation $P(in\ t)$ will terminate after some time steps m. As long as the calculation is performed, the process is active, i.e. $ready$ is F. When the calculation is finished at time step $t + m$, out holds the result y and $ready$ is T, indicating that the process is idle again. However, the calculation can only be finished, if $start$ is not set to T while the calculation is performed.

```
P_IFC_START =                                    P_IFC_CYCLE =
λP. λ(in, start, out, ready).                    λP. λ(in, reset, out, ready).
¬(start 0) ⇒ ready 0                             ∀t.
∧                                                  ((t = 0) ∨ ready (t − 1) ∨ reset t) ⇒
∀t.                                                case (P (in t)) of
  (ready t ∧ ¬(start (t + 1))) ⇒                    Defined y :
  (ready (t + 1) ∧ (out (t + 1) = out t))            ∃m.
  ∧                                                    ∀n. n < m ⇒
  start t ⇒                                              (∀p. p < n ⇒ ¬(reset (t + p + 1))) ⇒
  case (P (in t)) of                                     ¬(ready(t + n))
    Defined y :                                        ∧
      ∃m.                                              (∀p. p < m ⇒ ¬(reset (t + p + 1))) ⇒
        ∀n. n < m ⇒                                    ((out(t + m) = y) ∧ ready(t + m))
          (∀p. p < n ⇒ ¬(start(t + p + 1))) ⇒    Undefined:
          ¬(ready (t + n))                           ∀m.
        ∧                                              (∀n. n < m ⇒ ¬(reset(t + n + 1))) ⇒
        (∀p. p < m ⇒ ¬(start (t + p + 1))) ⇒           ¬(ready (t + m))
        ((out (t + m) = y) ∧ ready (t + m))
    Undefined:
      ∀m.
        (∀n. n < m ⇒ ¬(start(t + n + 1))) ⇒
        ¬(ready(t + m))
```

Fig. 4. Formal definition of two interface patterns

– the calculation $P (in\ t)$ will not terminate. Then the process will be active producing no result until a new calculation is started by setting *start* to T.

The pattern P_IFC_CYCLE describes a process, which always performs a calculation and starts a new one as soon as the old one has been finished. The *reset*-signal can be used here to stop a (non-terminating) calculation and to start a new one.

For each interface pattern that we provide, we also have proven a correct implementation theorem. All the implementation theorems corresponding to patterns for P-terms expect the programs to be in SLF. The formal Gropius-descriptions of implementations at the RT-level can be found in [18]. In (10), an implementation theorem is shown, stating that an implementation pattern called IMP_START fulfills the interface pattern P_IFC_START for each program being in SLF (see also the pattern of the SLF in (1)).

$$
\begin{aligned}
&\vdash \forall a\ c\ out\_init\ var\_init. \\
&\quad \text{IMP\_START}\ (a, c, out\_init, var\_init)\ (in, start, out, ready) \\
&\quad \Rightarrow \\
&\quad \text{P\_IFC\_START} \\
&\qquad (\text{PROGRAM}\ out\_init\ (\text{LOCVAR}\ var\_init\ (\text{WHILE}\ c\ (\text{PARTIALIZE}\ a)))) \\
&\qquad (in, start, out, ready)
\end{aligned}
\tag{10}
$$

$$
\begin{aligned}
&\vdash \text{IMP\_START}\ (a_{SLF}, c_{SLF}, out\_init_{SLF}, var\_init_{SLF})\ (in, start, out, ready) \\
&\quad \Rightarrow \\
&\quad \text{P\_IFC\_START}\ \text{fib}\ (in, start, out, ready)
\end{aligned}
\tag{11}
$$

The final theorem (11) is achieved by first instantiating the universal quantified variables in theorem (10) with the components $a_{SLF}, c_{SLF}, out\_init_{SLF}$ and var_init_{SLF} of the SLF in Fig. 3. Afterwards, the SLF-theorem $\vdash$ fib = fib$_{SLF}$ in

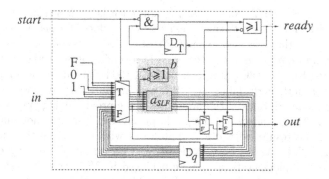

Fig. 5. RT implementation IMP_START for the program fib

Fig. 3 is turned to ⊢ fib$_{SLF}$ = fib (symmetry of equivalence) and rewriting is performed with this theorem. Theorem (11) says that the implementation sketched in Fig. 5[2] satisfies the program fib from Fig. 1 with respect to the interface pattern P_IFC_START.

The last step in our high-level synthesis scenario is allocation and binding of functional units (FU). This is done within the DFG-term b (grey area in Fig. 5). When synthesizing fib without any OPT, four multipliers, three adders and one divider have to be allocated besides some boolean function units. For sake of space, the logical transformation for this task is not shown here. When OPT-theorems like loop-unrolling or loop-cutting are applied, the amount of allocated hardware may be different. For more information on FU allocation/binding see also [19], where the synthesis of pure DFG-terms is described. Since the allocation and binding of functional units is performed within the DFG-term b, the method can be applied which is described there.

5 Experimental results

Our formal synthesis approach consists of four steps. OPT and SPT for scheduling and register allocation/binding, applying an implementation theorem for interface synthesis and allocation/binding of functional units within the resulting basic block of the RT-implementation. SPT and interface synthesis consist of rewriting and β-conversions, which can be done fully automatically within the HOL theorem prover. For the OPT and the FU-allocation/binding, however, heuristics are needed to explore the design space. Those non-formal methods can be integrated in the formal synthesis process, since the design space exploration part is separated from the transformation within the theorem prover. After establishing the formal basis for our synthesis approach, it is our objective in the future to develop further heuristics and to integrate more existing

---

[2] Due to lack of space the loop-body of the SLF (component a_{SLF}) is not explicitly shown in Fig. 5. q denotes an arbitrary initial value in the eight allocated registers.

techniques for the OPT. An interesting approach is proposed in [20] where a method is described in which the design space is explored for performing similar transformations that are used in the OPT of our synthesis process.

To give an impression about the costs in formal synthesis, Fig. 6 shows the run-times (on SUN UltraCreator, Solaris 5.5.1, 196 MB main memory) of several programs for both performing the SPT and instantiating an implementation theorem. The descriptions of the programs in the programming language C can be found in [21]. Since we did not perform any OPT, it does not make sense to compare the results with other approaches with respect to the number of registers and FUs. As we have demonstrated on our small running example, very different implementations can be achieved if OPT-theorems are applied. Since only few heuristics for automatic invoking the OPT-theorems have been implemented, we have considered only the SPT and the interface synthesis. The cost for the SPT mainly increases with the number of the control structures but also with the number of operations in the program. The cost for the interface synthesis mainly increases with the size of the loop-body of the SLF-program.

The experiments have been run using a slight variant of the HOL theorem prover. As compared to the original HOL system, it has been made more efficient by changing the term representation and adding two core functions. See [22] for a detailed description and discussion about this. As we have demonstrated in the paper, the result of our synthesis process is a guaranteed correct implementation. A proof is given together with the implementation, stating that the implementation fulfills the specification. Therefore, one should be aware that the run-times must be compared with conventional synthesis plus exhaustive simulation. Furthermore, we believe that due to the complexity it is very hard (or even impossible) to develop automatic post-synthesis verification methods at this abstraction level which could prove the correctness of the synthesis process.

Program	SPT	Interface synthesis
fibonacci	2.1	0.1
gcd	1.5	0.2
bubble	4.4	0.5
fuzzy	19.0	1.2
kalman	103.9	2.8
diffeq	3.4	0.3
fidelity	5.1	0.6
dct	50.0	5.2
atoi	1.9	0.2

Fig. 6. Time [s] for synthesis experiments

6 Conclusion

In this paper, we presented a formal way for performing high-level synthesis. The main contribution is that we perform the whole synthesis process within a theorem prover. The result is therefore not only an implementation but also an accompanying proof that this implementation is correct. Furthermore, we developed a new synthesis method, where the implementation is derived by applying program transformations instead of generating and analyzing a number of control paths that grows exponentially with the CDFG size. Last but not least we orthogonalize the treatment of algorithmic and temporal aspects and therefore support a systematic reuse of designs.

References

1. A. Gupta. Formal hardware verification methods: A survey. *Formal Methods in System Design*, 1(2/3):151–238, 1992.
2. P. Middelhoek. *Transformational Design*. PhD thesis, Univ. Twente, NL, 1997.
3. M. McFarland. Formal analysis of correctness of behavioral transformations. *Formal Methods in System Design*, 2(3), 1993.
4. Z. Peng, K. Kuchcinski. Automated transformation of algorithms into register-transfer implementations. *IEEE Transactions on CAD*, 13(2):150–166, 1994.
5. R. Camposano. Behavior-preserving transformations for high-level synthesis. In *Hardware Specification, Verification and Synthesis: Mathematical Aspects*, number 408 in LNCS, pp. 106–128, Ithaca, New York, 1989. Springer.
6. S.D. Johnson, B. Bose. DDD: A system for mechanized digital design derivation. In *Int. Workshop on Formal Methods in VLSI Design*, Miami, Florida, 1991. Available via "ftp://ftp.cs.indiana.edu/pub/techreports/TR323.ps.Z" (rev. 1997).
7. M.J.C. Gordon, T.F. Melham. *Introduction to HOL: A Theorem Proving Environment for Higher Order Logic*. Cambridge University Press, 1993.
8. R. Sharp, O. Rasmussen. The T-Ruby design system. In *IFIP Conference on Hardware Description Languages and their Applications*, pp. 587–596, 1995.
9. E.M. Mayger, M.P. Fourman. Integration of formal methods with system design. In *Int. Conf. on VLSI*, pp. 59–70, Edinburgh, Scotland, 1991. North-Holland.
10. R. Kumar et al. Formal synthesis in circuit design-A classification and survey. In *FMCAD'96*, number 1166 in LNCS, pp. 294–309, Palo Alto, CA, 1996. Springer.
11. F.K. Hanna et al. Formal synthesis of digital systems. In *Applied Formal Methods For Correct VLSI Design*, volume 2, pp. 532–548. Elsevier, 1989.
12. M. Larsson. An engineering approach to formal digital system design. *The Computer Journal*, 38(2):101–110, 1995.
13. D. Gajski et al. *High-Level Synthesis, Introduction to Chip and System Design*. Kluwer, 1992.
14. H.P . Barendregt. *Handbook of Theoretical Computer Science, Volume B: Formal Models and Semantics*, chapter 7: Functional Programming and Lambda Calculus, pp. 321–364. Elsevier, 1992.
15. A. Aho et al. *Compilers: Principles, Techniques and Tools*. Addison Wesley, 1986.
16. P.G. Paulin, J. P. Knight. Force-directed scheduling for the behavioral synthesis of ASIC's. *IEEE Transactions on CAD*, 8(6):661–679, 1989.
17. C. Blumenröhr. A formal approach to specify and synthesize at the system level. In *GI Workshop Modellierung und Verifikation von Systemen*, pp. 11–20, Braunschweig, Germany, 1999. Shaker-Verlag.
18. D. Eisenbiegler, R. Kumar. An automata theory dedicated towards formal circuit synthesis. In *TPHOL'95*, number 971 in LNCS, pp. 154–169, Aspen Grove, Utah, 1995. Springer.
19. D. Eisenbiegler et al. Implementation issues about the embedding of existing high level synthesis algorithms in HOL. In *TPHOLs'96*, number 1125 in LNCS, pp. 157–172, Turku, Finland, 1996. Springer.
20. J. Gerlach, W. Rosenstiel. A Scalable Methodology for Cost Estimation in a Transformational High-Level Design Space Exploration Environment. In *DATE'98*, pp. 226–231, Paris, France, 1998. IEEE Computer Society.
21. *http: //goethe. ira.uka.de/fsynth/Charme/<name>.c.*
22. C. Blumenröhr et al. On the efficiency of formal synthesis — experimental results. *IEEE Transactions on CAD*, 18(1):25–32, 1999.

Xs Are for Trajectory Evaluation, Booleans Are for Theorem Proving

Mark D. Aagaard[1], Thomas F. Melham[2], John W. O'Leary[1]

[1] {*maagaard,joleary*} *@ichips.intel.com*
Strategic CAD Labs
Intel Corporation, JFT-102
5200 NE Elam Young Parkway
Hillsboro, OR 97124, USA

[2] *tfm@dcs.gla.ac.uk*
Department of Computing Science
University of Glasgow
Glasgow, Scotland, G12 8QQ

Abstract. This paper describes a semantic connection between the symbolic trajectory evaluation model-checking algorithm and relational verification in higher-order logic. We prove a theorem that translates correctness results from trajectory evaluation over a four-valued lattice into a shallow embedding of temporal operators over Boolean streams. This translation connects the specialized world of trajectory evaluation to a general-purpose logic and provides the semantic basis for connecting additional decision procedures and model checkers.

1 Introduction

The well-known limits to BDD-based model-checking techniques have motivated a great deal of interest in combining model-checking with theorem proving [3, 11, 9, 6]. The foundation of any such hybrid verification approach is a semantic connection between the logic of properties in the model checker and the logic of the theorem prover. Symbolic trajectory evaluation [16] is a highly effective model checker for datapath verification. It has been combined with theorem proving and the combination has been used effectively on complex industrial circuits [1, 12]. However, two of the features that make trajectory evaluation so effective as a model checker create difficulties or limitations in the theorem proving domain. Trajectory evaluation's temporal logic has limited expressability and operates over a lattice of values containing notions of contradiction ($\top$) and unknown (X).

In this paper, we formally verify a semantic link from symbolic trajectory evaluation to higher-order logic. This link allows trajectory evaluation to be used as a decision procedure without encumbering the theorem proving world with the complications and limitations of trajectory evaluation. The trajectory evaluation temporal operators are defined in a shallow embedding of predicates over streams and the lattice domain is converted to simple Booleans. This translates trajectory evaluation results into the conventional "relations-over-Boolean-streams" approach to hardware modeling in higher-order logic [7].[1]

---

[1] In the rest of the paper we will take the phrase "relational" style to mean "relations over Boolean streams" with a shallow embedding of temporal operators as predicates.

We believe that the relational world is the right target domain for connecting model checking engines. Each model checking algorithm typically has its own temporal logic. By translating results into higher-order logic, the vestiges of the individual model checkers are removed, allowing the full power of general-purpose theorem proving to be brought to bear.

To give some intuition about the two worlds we reason about and the connection between them, consider the NAND-DELAY circuit in Figure 1. Figure 2 presents simple correctness statements for both the trajectory evaluation and relational styles. We will use this circuit as a running example throughout this paper. In this example, for simplicity, we consider the NAND gate to be zero delay. Trajectory evaluation typically uses a more detailed timing model of circuits.

Fig. 1. Simple example circuit NAND_DELAY

$$ckt_{\text{HOL}} \; i \; o \; \stackrel{def}{=}$$
$$\exists m.$$
$$(\forall t. \; m \; t \quad = \text{NAND}(i \; t)(o \; t)) \; \wedge$$
$$(\forall t. \; o \; (t+1) = m \; t)$$

$$i \stackrel{def}{=} "i"$$
$$o \stackrel{def}{=} "o"$$

$$\models_{ckt} \left[\begin{array}{l} (i \text{ is } b_1) \text{and} \\ (o \text{ is } b_2) \end{array} \implies (\text{N} \; (o \text{ is } \neg(b_1 \wedge b_2))) \right]$$

$$spec_{\text{HOL}} \; i \; o \; \stackrel{def}{=}$$
$$\forall t.o \; (t+1) = \text{NAND}(i \; t)(o \; t)$$

$$\forall i, o. \; ckt_{\text{HOL}} \; i \; o \implies spec_{\text{HOL}} i \; o$$

Trajectory-evaluation verification   Relational style verification

Fig. 2. Example stream and trajectory evaluation verifications

Trajectory evaluation is based on symbolic simulation. Correctness statements are of the form $\models_{ckt} [ant \implies cons]$, where $\implies$ is similar, but not identical to, implication (details are given in Section 2). The *antecedent ant* gives an initial state and input stimuli to the circuit *ckt*, while the *consequent cons* specifies the desired response of the circuit. Circuits are black boxes—their implementations are not user visible. Circuit *nodes* are named by strings. In the example, the antecedent drives the nodes "i" and "o" with the Boolean variables b_1 and b_2 at the initial step of the verification. The consequent says that at the next time step the node "o" has the value $\neg(b_1 \wedge b_2)$.

In relational verification, correctness statements are of the form $ckt \; i \; o \implies spec \; i \; o$, where $\implies$ is true implication. Signals (e.g., i, m, and o) are modelled by *streams*, which are functions from time to values. Both the circuit and the specification are relations over these infinite streams. A stream satisfies a circuit if it is a series of values that could be observed on the corresponding signals in the circuit. The correctness criterion for the example says that if the streams i and o satisfy the circuit model, then they must conform to the specification.

At a very cursory level, a mapping from the trajectory evaluation result in Figure 2 to the relational world would result in:

$$\forall i, o. \; \forall t.$$
$$((i \; t) = b_1) \wedge ((o \; t) = b_2) \wedge (ckt \; i \; o)$$
$$\Longrightarrow$$
$$(o \; (t+1)) = \neg(b_1 \wedge b_2)$$

Substituting for b_1 and b_2 throughout the expression gives:

$$\forall i, o. \; \forall t. \; ckt \; i \; o \implies (o \; (t+1)) = \neg((i \; t) \wedge (o \; t))$$

Technically, this description is not quite correct, but it gives the intuition behind our result that correctness statements in trajectory evaluation imply relational correctness statements. Section 5.2 shows the actual process and results for the NAND-DELAY circuit. The difficulties arise in details such as translation between different semantic domains (e.g., Boolean and lattice valued streams) and the treatment of free variables in trajectory formulas.

Our main result is a formal translation from trajectory evaluation's temporal operators over lattices to a shallow embedding of the temporal operators over Boolean streams. We prove that any result verified by the trajectory evaluation algorithm will hold in the relational world. This allows trajectory evaluation to be used as a decision procedure in a theorem prover without changing the relational style of verification used in the theorem prover.

It is interesting to note that our result is an implication, not an "if-and-only-if"; that is, we do not guarantee that every statement provable in the relational world will also hold in trajectory evaluation. The problem stems from the differences in how the relational world and the trajectory evaluation world handle contradictions in the circuit and antecedent. Joyce and Seger gave an extra constraint on trajectory evaluation that can be used to prove an if-and-only-if relationship [10]. The constraint requires reasoning about contradictions and top values. Because we use trajectory evaluation as a decision procedure, we are able to avoid the burden of reasoning about contradictions.

1.1 Organization of the Paper

Figure 3 is a roadmap of the paper. We begin with a presentation of trajectory assertions (the specifications for symbolic trajectory evaluation) over the standard four-valued lattice (Section 2). Our verification relies on two major steps: from the four-valued lattice to Booleans and from a deep embedding of the temporal operators to a shallow embedding. In Section 3 we introduce our definition of trajectory assertions over Booleans and prove that a trajectory assertion over lattice-valued streams implies the same result over Boolean-valued streams. We then prove that a shallow embedding of trajectory formulas as relations over streams is equivalent to a deep embedding (Section 4).

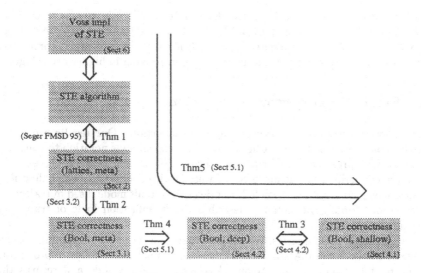

Fig. 3. Roadmap of the paper

Our proof relating trajectory assertions over lattices to trajectory assertions over Booleans is a meta-logical proof about the semantics of the two languages and links the free (Boolean) variables that appear in these assertions. The conversion from the deep to shallow embedding is done in higher-order logic. In Section 5 we connect connect all of the pieces together to prove our final result.

1.2 Related Work

An early experiment in combining model checking and theorem proving was the HOL-Voss System [10]. Within HOL [7], Joyce and Seger proved the correspondence between a deep embedding of the Voss [8] "5-tuple" implementation of trajectory assertions (see Section 6) and a simple specification language that was deeply embedded in HOL. This allowed them to use Voss as an external decision procedure for HOL. Our work focuses on the logical content of trajectory assertions, independent of any particular implementation (e.g., Voss), and connects this to a shallow embedding of temporal operators in higher-order logic.

The modal mu-calculus has been embedded in HOL [2] and in PVS [15]. In the PVS work, Rajan *et al* implemented a shallow embedding upon which they defined the ∀CTL* temporal logic. They connected a mu-calculus model checker and verified a number of abstraction opertations. It should be possible to translate a subset of the ∀CTL* formulas into predicates over Boolean streams, but given the complexity of ∀CTL*, it is difficult to estimate the feasibility of this approach.

Chou has given a set-theoretic semantics of trajectory evaluation, focusing on extensions to the basic algorithm [5]. Our work translates trajectory evaluation results into a form that can be seamlessly integrated with current practice in higher-order-logic theorem proving. It would be interesting to explore connections between the work presented here and Chou's, to find a seamless connection from extended trajectory evaluation to theorem-proving in higher-order logic.

2 Symbolic Trajectory Evaluation

This section presents the logic of trajectory assertions. Our results are meta-logical, but for convenience we use a **bold** face logical-style notation to state our results. Our presentation of trajectory evaluation is comprised of three parts. After a few preliminary definitions; we proceed with Section 2.1, which describes the four-valued lattice. Section 2.2 overviews the circuit model used in trajectory evaluation. Finally, Section 2.3 describes the specification logic of trajectory evaluation.

First, some preliminaries. We suppose there is a set of *nodes*, naming observable points in circuits. A *stream* is a function from natural numbers representing time to data values in the stream. A *sequence* takes a node and returns the stream for that node. A *state* is a mapping from *nodes* to values. We typically use σ for sequences and s for states. Two convenient sequence operations are taking the suffix and transposing a sequence so that it is a stream of states.

$$\alpha \text{ stream } \stackrel{type}{=} \mathcal{N} \to \alpha$$
$$\alpha \text{ sequence } \stackrel{type}{=} \text{node} \to \alpha \text{ stream}$$
$$\alpha \text{ state } \stackrel{type}{=} \text{node} \to \alpha$$
$$\text{suffix:} \quad \sigma_i \stackrel{def}{=} \lambda n. \ \lambda t. \ (\sigma \ n \ (t+i))$$
$$\text{transpose:} \ \sigma^\mathsf{T} \stackrel{def}{=} \lambda t. \ \lambda n. \ (\sigma \ n \ t)$$
$$\sigma^\mathsf{T} \ :: \ (\alpha \text{ state}) \text{ stream}$$

2.1 The Four Valued Lattice

In this paper, the only lattice that we use is the four-valued lattice shown below. The theory of trajectory evaluation works equally well over all complete lattices. However this lattice simplifies the presentation and is the lattice used by the Voss implementation of trajectory evaluation [8]. Our mathematical development also is based on this lattice—generalizing our results would be of theoretical interest, but not of immediate practical benefit to us.

The ordering over lattice values shown above defines the ordering relation $\sqsubseteq$, which we lift pointwise and overload over streams and states. We inject the set of Boolean values to lattice values with the postfix operator $\downarrow$ (read "drop", Definition 1), which maps the Boolean values T and F to their counterparts in the lattice. Drop is lifted pointwise to states, to sequences, and to state streams.

Definition 1. *Dropping from Boolean to lattice values*

$$F{\downarrow} \overset{def}{=} 0$$
$$T{\downarrow} \overset{def}{=} 1$$

2.2 Circuit Models

In our description of circuit models, we will refer to Table 1, which gives the lattice transition function for the example NAND-DELAY circuit. In trajectory evaluation, the circuit model is given by a next state function **Y** that takes a circuit and maps states to states:

$$\mathbf{Y} :: \texttt{ckt} \to \texttt{lattice state} \to \texttt{lattice state}$$

A lattice state is an assignment to circuit nodes of values drawn from the four valued lattice. The first argument to **Y** identifies the particular circuit of interest, and for the present purposes may be regarded as an uninterpreted constant. Intuitively, the next state function expresses a constraint on the set of possible states into which the circuit may go for any given state. Suppose the circuit is in state s, then $\mathbf{Y}(s)$ will give the least specified state the system can transition to. Here, "least specified" means that if a node can take on both 1 and 0 values in the next state, then $\mathbf{Y}(s)$ will assign the value X to that node.

Table 1. Lattice transition function for NAND-DELAY circuit

$\langle\, i\ o\,\rangle$	$\langle\, i'\ o'\,\rangle$	
0 0	X 1	
0 1	X 1	
0 X	X 1	← o can initially be X, and o' is still defined
1 0	X 1	
1 1	X 0	
1 X	X X	← o' is unknown, because o is unknown
X 0	X 1	← i can initially be X, and o' is still defined
X 1	X X	← o' is unknown, because i is unknown
X X	X X	

A critical requirement for trajectory evaluation is that the next-state function be monotonic, which is captured in Axiom 1.

Axiom 1. *Monotonicity of* **Y**

For all s, s'. $(s \sqsubseteq s')$ implies $(\mathbf{Y}s \sqsubseteq \mathbf{Y}s')$

Monotonicity can be seen in the NAND-DELAY circuit by comparing a transition in which one of the current state variables (e.g., o) is X with a transition in which o is either 0 or 1. A bit of the algorithmic efficiency of trajectory evaluation is illustrated here. The initial value for some of the circuit nodes can be X and a meaningful result can still be verified. In this way, the lattice often allows trajectory evaluation to prove results with fewer BDD variables than would otherwise be needed.

A sequence σ is said to be *in the language* of a circuit (Definition 2) if the set of behaviors that the sequence encodes is a subset of the behaviors that the circuit can exhibit. This means that the result of applying $\mathbf{Y}$ to any element of the state stream σ^T is no more specified (with respect to the $\sqsubseteq$ ordering) than the succeeding element of σ^T.

Definition 2. *Sequence is in the language of a circuit*

$$\sigma \in \mathcal{L} \, ckt \ \overset{def}{=} \ \text{For all } t \geq 0. \ (\mathbf{Y} \, ckt \, (\sigma^\mathsf{T} \, t)) \ \sqsubseteq \ (\sigma^\mathsf{T} \, (t+1))$$

2.3 Trajectory Evaluation Logic

Trajectory evaluation correctness statements (known as *trajectory assertions*) are written as:

$$\models_{ckt} [ant \Longrightarrow cons]$$

where *ant* and *cons* are *trajectory formulas*. The intuition is that the antecedent *ant* provides stimuli to nodes in the circuit and the consequent *cons* specifies the values expected on nodes in the circuit. Before further describing trajectory assertions, we define trajectory formulas (Definition 3) and what it means for a sequence to satisfy a trajectory formula (Definition 4).

Definition 3. *Trajectory formulas*

$$
\begin{aligned}
f \overset{def}{=} \ & n \text{ is } 0 && /\!/ \ n \text{ has value 0} \\
| \ & n \text{ is } 1 && /\!/ \ n \text{ has value 1} \\
| \ & f_1 \text{ and } f_2 && /\!/ \ \text{conjunction of formulas} \\
| \ & f \text{ when } g && /\!/ \ f \text{ is asserted only when } g \text{ is true} \\
| \ & \mathsf{N} \, f && /\!/ \ f \text{ holds in the next time step}
\end{aligned}
$$

where f, f_1, f_2 *range over formulas;* n *ranges over the node names of the circuit; and* g *is a Boolean expression, commonly called a* guard.

Trajectory formulas are guarded expressions defining values on nodes in the sequence. Guards may contain free variables. In fact, guards are the only place that free variables are allowed in the primitive definition of trajectory formulas. Syntactic sugar for is is commonly defined to allow guards in the value field as well. This is illustrated in Figure 1 and Section 5.2.

Definition 4 describes when a sequence σ satisfies a trajectory formula f. Satisfaction is defined with respect to an assignment ϕ of Boolean values to the variables that appear in the guards of the formula.

Definition 4. *Sequence satisfies a trajectory formula*

$(\phi, \sigma) \models_{STE} (n \text{ is } 0) \overset{def}{=} \sigma\, n\, 0 \sqsupseteq 0$

$(\phi, \sigma) \models_{STE} (n \text{ is } 1) \overset{def}{=} \sigma\, n\, 0 \sqsupseteq 1$

$(\phi, \sigma) \models_{STE} (f_1 \text{ and } f_2) \overset{def}{=} ((\phi, \sigma) \models_{STE} f_1) \text{ and } ((\phi, \sigma) \models_{STE} f_2)$

$(\phi, \sigma) \models_{STE} (f \text{ when } g) \overset{def}{=} (\phi \models g) \text{ implies } ((\phi, \sigma) \models_{STE} f)$

$(\phi, \sigma) \models_{STE} (Nf) \overset{def}{=} (\phi, \sigma_1) \models_{STE} f$

Where $\phi \models g$ means that the assignment that ϕ makes to the free variables in g renders g true.

We now have sufficient notation to define a *trajectory assertion* (Definition 5). In trajectory evaluation, correctness criteria are formulated as trajectory assertions.

Definition 5. *Trajectory assertion*

$\phi \models_{ckt} [ant \Longrightarrow cons] \overset{def}{=}$
For all σ. $(\sigma \in \mathcal{L}\, ckt)$ implies $((\phi, \sigma) \models_{STE} ant)$ implies $((\phi, \sigma) \models_{STE} cons)$

The fundamental theorem of trajectory evaluation [16] says the trajectory evaluation algorithm (STE *ckt ant cons*) computes the Boolean condition e on the free variables in *ant* and *cons* if and only if any assignment ϕ satisfying e also proves the trajectory assertion $\phi \models_{ckt} [ant \Longrightarrow cons]$ (Theorem 1).

Theorem 1. *Correctness of STE algorithm*

For all circuits ckt, antecedents ant, and consequences cons, the implementation of the STE algorithm returns the value e (that is, $e = STE\ ckt\ ant\ cons$) if and only if:

$$\text{For all } \phi.\ \phi \models e \text{ implies } \phi \models_{ckt} [ant \Longrightarrow cons]$$

3 Trajectory Logic over Boolean Streams

In this section we give a definition of trajectory logic over Boolean streams (as opposed to the standard lattice-valued streams in Section 2) and prove that trajectory evaluation results that hold over the four valued lattice also hold over Boolean streams. Boolean identifiers (e.g., next-state relations, sequences, and languages) will be distinguished from their lattice valued counterparts by marking them with a ∘, as in $\overset{\circ}{Y}$, $\overset{\circ}{\sigma}$, and $\overset{\circ}{\mathcal{L}}$.

3.1 Definitions and Axioms

In the Boolean world, circuit behavior is modeled as a *relation* between current and next states. In contrast, circuit behavior in trajectory evaluation is modeled as a next state *function*. The Boolean next state relation, denoted $\overset{\circ}{Y}$, has the type:

$$\overset{\circ}{Y} :: ckt \rightarrow \texttt{bool state} \rightarrow \texttt{bool state} \rightarrow \texttt{bool}$$

We write $\overset{\circ}{Y}$ *ckt* as an infix operator, as in: $s(\overset{\circ}{Y}$ *ckt*$)s'$.

As a concrete example, the next state relation for the NAND-DELAY circuit of Figure 1 is defined by Table 2, where the vectors $\langle i, o \rangle$ and $\langle i', o' \rangle$ denote the current and next states of the input and output. Note that the non-determinism that was represented by Xs in Y (Table 1) appears as multiple next states with the same current state in Table 2.

$\langle\ i\ ,\ o\ \rangle$		$\langle\ i'\ ,\ o'\ \rangle$	
0	0	0	1
0	0	1	1
0	1	0	1
0	1	1	1
1	0	0	1
1	0	1	1
1	1	0	0
1	1	0	1

Table 2. Boolean next-state relation for NAND-DELAY

Given a circuit's next state relation $\overset{\circ}{Y}$, we say that a Boolean sequence $\overset{\circ}{\sigma}$ is *in the language* of the circuit when consecutive states in the state stream $\overset{\circ}{\sigma}{}^{\mathsf{T}}$ are included in the next-state relation (Definition 6).

Definition 6. *Boolean sequence is in the language of a circuit:*

$$\overset{\circ}{\sigma} \in \overset{\circ}{\mathcal{L}}\ ckt \overset{def}{=} \text{For all } t \geq 0.\ (\overset{\circ}{\sigma}{}^{\mathsf{T}}t)\ (\overset{\circ}{Y}\ ckt)\ (\overset{\circ}{\sigma}{}^{\mathsf{T}}(t+1))$$

We now define when a Boolean sequence $\overset{\circ}{\sigma}$ satisfies a trajectory formula f (Definition 7). The only distinction between $\overset{\circ}{\models}$ and satisfaction over lattice sequences ($\models_{\overline{\mathsf{STE}}}$) is that for the formulas $(n \text{ is } 0)$ and $(n \text{ is } 1)$, satisfaction is defined in terms of values in the Boolean domain rather than the lattice domain.

Definition 7. *Boolean sequence satisfies a trajectory formula*

$$(\phi, \overset{\circ}{\sigma}) \overset{\circ}{\models} (n \text{ is } 0) \overset{def}{=} (\overset{\circ}{\sigma}\ n\ 0) = \mathrm{F}$$

$$(\phi, \overset{\circ}{\sigma}) \overset{\circ}{\models} (n \text{ is } 1) \overset{def}{=} (\overset{\circ}{\sigma}\ n\ 0) = \mathrm{T}$$

$$(\phi, \overset{\circ}{\sigma}) \overset{\circ}{\models} (f_1 \text{ and } f_2) \overset{def}{=} ((\phi, \overset{\circ}{\sigma}) \overset{\circ}{\models} f_1)\ \textbf{and}\ ((\phi, \overset{\circ}{\sigma}) \overset{\circ}{\models} f_2)$$

$$(\phi, \overset{\circ}{\sigma}) \overset{\circ}{\models} (f \text{ when } g) \overset{def}{=} (\phi \models g)\ \textbf{implies}\ ((\phi, \overset{\circ}{\sigma}) \overset{\circ}{\models} f)$$

$$(\phi, \overset{\circ}{\sigma}) \overset{\circ}{\models} (\mathsf{N}\ f) \overset{def}{=} (\phi, \overset{\circ}{\sigma}_1) \overset{\circ}{\models} f$$

3.2 Correctness of Boolean Valued Trajectory Evaluation

To link the worlds of lattice and Boolean based trajectory evaluation, we require that the two models of the circuit behavior (Y and $\overset{\circ}{Y}$) describe the same behavior. Axiom 2 says that if two Boolean states s and s' satisfy the next-state relation $\overset{\circ}{Y}$, then the result of applying the next-state function Y to the dropped versions of s results in a state that is no higher in the lattice than $s'\!\downarrow$ (**Y** is a ternary extension of $\overset{\circ}{\mathbf{Y}}$).

Axiom 2. *Relating next-state relation and next-state function*

For all ckt, s, s'. $s\ (\overset{\circ}{\mathbf{Y}}\ ckt)\ s'$
implies
$(\mathbf{Y}\ ckt\ (s\!\downarrow)) \sqsubseteq (s'\!\downarrow)$

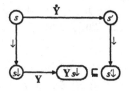

Fig. 4. Illustration of Axiom 2

Axiom 2, which is illustrated in Figure 4, says that any Boolean next state s' possible in the relational model is consistent with the next state in the lattice-valued model. It also constrains $\overset{\circ}{\mathbf{Y}}\ ckt$ to return F whenever it is applied to two states s, s' that are inconsistent with $\mathbf{Y}ckt$ (e.g., unreachable states). Inconsistency is manifested by $\mathbf{Y}ckt\ s$ returning the lattice value T (top).

Theorem 2 makes the connection between trajectory assertions over lattice "values" and Boolean trajectory assertions. If a trajectory assertion holds over lattice-valued streams, then the same antecedent leads to the same consequent over Boolean-valued streams. This is the crux of connection from the lattice world to the Boolean world.

Theorem 2. *Translate trajectory logic from lattice to Boolean sequences.*

For all ckt, ant, $cons$.
   For all ϕ, σ.
      $\sigma \in \mathcal{L}\ ckt$ **implies**
         For all $t \geq 0$. $((\phi, \sigma_t) \models_{\overline{\text{STE}}} ant)$ **implies** $((\phi, \sigma_t) \models_{\overline{\text{STE}}} cons)$
   implies
   For all $\phi, \overset{\circ}{\sigma}$.
      $\overset{\circ}{\sigma} \in \overset{\circ}{\mathcal{L}}\ ckt$ **implies**
         For all $t \geq 0$. $((\phi, \overset{\circ}{\sigma}_t) \models^{\circ} ant)$ **implies** $((\phi, \overset{\circ}{\sigma}_t) \models^{\circ} cons)$

The proof of Theorem 2 relies on Lemmas 1 and 2. Lemma 1 says that if a Boolean sequence $\overset{\circ}{\sigma}$ is in the Boolean language of a circuit then the dropped version $\overset{\circ}{\sigma}\!\downarrow$ is in the lattice-valued language of the circuit. The proof of Lemma 1 is done by unfolding the definition of $\in \overset{\circ}{\mathcal{L}}$ and using Axiom 2.

Lemma 1. *Relationship between Boolean and ternary sequences*

For all $\overset{\circ}{\sigma}$. $\overset{\circ}{\sigma} \in \overset{\circ}{\mathcal{L}}\ ckt$ **implies** $\overset{\circ}{\sigma}\!\downarrow \in \mathcal{L}\ ckt$

Lemma 2 relates satisfaction over Boolean sequences to satisfaction over lattice-valued sequences. Its proof is by induction over the structure of trajectory formulas, unfolding the definitions of $\models^{\circ}$ and $\models_{\overline{\text{STE}}}$, and employing properties of the drop operator (Lemma 3).

Lemma 2. *Satisfaction over Boolean and ternary sequences*

$$\text{For all } \phi, f, \mathring{\sigma}. \ (\phi, \mathring{\sigma}) \models f \ \text{ iff } \ (\phi, \mathring{\sigma}{\downarrow}) \models_{\overline{\text{STE}}} f$$

Lemma 3 says that the value of an element of a Boolean sequence $\mathring{\sigma}$ is F (T) if-and-only-if the value of the dropped sequence at that point is higher in the lattice than 0 (1). The lemma holds because $\text{F}{\downarrow} = 0$, $0 \sqsupseteq 0$ (similarly, $\text{T}{\downarrow} = 1$, and $1 \sqsupseteq 1$).

Lemma 3. *Properties of drop*

$$\text{For all } \mathring{\sigma}. \ (\mathring{\sigma} \ n \ 0 \ = \ \text{F}) \ \text{ iff } \ ((\mathring{\sigma}{\downarrow}) \ n \ 0 \ \sqsupseteq 0)$$
$$\text{For all } \mathring{\sigma}. \ (\mathring{\sigma} \ n \ 0 \ = \ \text{T}) \ \text{ iff } \ ((\mathring{\sigma}{\downarrow}) \ n \ 0 \ \sqsupseteq 1)$$

Theorem 2 is an implication and not an if-and-only-if result. The reason stems from Lemma 3, which causes Lemma 2 to be universally quantified over Boolean sequences, as opposed to lattice-valued sequences. Examining the case in which the trajectory formula f contains both n is 1 and n is 0 illustrates why Lemma 2 does not hold for all lattice-valued sequences. There are no Boolean sequences that satisfy both n is 1 and n is 0, but a lattice valued sequence in which $\sigma \ n \ 0 = \text{T}$ would satisfy the formula.

4 Trajectory Logic as Relations over Streams

This section begins with a description of a shallow embedding of trajectory assertions in a higher-order logic[2] version of Boolean sequences. In the shallow embedding, trajectory formulas are predicates over Boolean sequences. In Section 4.2, we link the shallow embedding with a deep embedding of trajectory formulas. The deep embedding definitions mirror those in the metalogic from Section 3, and so we do not include them. Later, in Section 5 we use the deep embedding as an intermediate representation to connect the shallow embedding to trajectory assertions over Boolean streams from Section 3.

The decision to develop our shallow embedding via an intermediate deep embedding was made consciously. While we much prefer the shallow embedding for reasoning about properties of circuits stated as trajectory assertions, the deep embedding enables reasoning about trajectory logic itself (in particular, it allows quantification over trajectory formulas). We consider both activities important.

4.1 Definitions and Axioms

Circuits are implemented as relations over streams in our shallow embedding. Checking that a sequence is in the language of a circuit is done by simply applying the circuit relation to the sequence. We axiomatize the relationship between the language of circuits in our shallow embedding, the language of circuits in our deep embedding (in_lang), and the language of circuits over Boolean sequences

---

[2] This is not a mechanized implementation, but rather a paper description that could be implemented in a higher-order logic proof system.

($\in \overset{\circ}{\mathcal{L}}$) in Axiom 3. This is an axiom, rather than a lemma, because in this paper we do not give interpretations of circuits. Indeed, for much practical work we only require the ability to distinguish verification results obtained on different circuits, and for this purpose it is sufficient to leave circuits uninterpreted. A complete implementation of the work described here would need to prove that the implementation of circuits satisfies Axiom 3.

Axiom 3. *Relating languages in deep and shallow embeddings.*
A sequence $\overset{\circ}{\sigma}$ is in the language of a deeply-embedded circuit ckt
($\overset{\circ}{\sigma}$ in_lang ckt) if and only if $\overset{\circ}{\sigma} \in \overset{\circ}{\mathcal{L}}$ ckt.
A sequence $\overset{\circ}{\sigma}$ is in the language of a shallowly-embedded circuit
ckt (ckt $\overset{\circ}{\sigma}$) if and only if $\overset{\circ}{\sigma}$ in_lang ckt.

Definition 8 presents the trajectory formula type and the trajectory formula constructs is, and, when, and N for the shallow embedding in a higher-order logic. In the shallow embedding, a sequence satisfies a formula if applying the formula to the sequence yields true.

Definition 8. *Shallow embedding of trajectory formulas in Boolean streams*

$$\text{traj\_form} \overset{type}{=} (\text{bool traj}) \to \text{bool}$$

$$
\begin{aligned}
n \text{ is } 1 &\overset{def}{=} \lambda \overset{\circ}{\sigma}.\ \overset{\circ}{\sigma}\ n\ 0 = 1 \\
n \text{ is } 0 &\overset{def}{=} \lambda \overset{\circ}{\sigma}.\ \overset{\circ}{\sigma}\ n\ 0 = 0 \\
f_1 \text{ and } f_2 &\overset{def}{=} \lambda \overset{\circ}{\sigma}.\ (f_1\ \overset{\circ}{\sigma}) \wedge (f_2\ \overset{\circ}{\sigma}) \\
f \text{ when } g &\overset{def}{=} \lambda \overset{\circ}{\sigma}.\ g \Longrightarrow (f\ \overset{\circ}{\sigma}) \\
\text{N } f &\overset{def}{=} \lambda \overset{\circ}{\sigma}.\ f \overset{\circ}{\sigma}_1
\end{aligned}
$$

4.2 Verification of Shallow Embedding Against Deep Embedding

As mentioned previously, the deep embedding of trajectory formulas and satisfaction (is, and, when, N and sat) is not shown because it is a direct implementation of the metalogical presentation in Section 3. We identify the deeply embedded operators by underlining them.

Definition 9 defines a translation [·] from deeply-embedded to shallowly-embedded trajectory formulas.

Definition 9. *Translation from deep to shallow embedding*

$$
\begin{aligned}
[\![(n \underline{\text{ is }} 1)]\!] &\overset{def}{=} n \text{ is } 1 \\
[\![(n \underline{\text{ is }} 0)]\!] &\overset{def}{=} n \text{ is } 0 \\
[\![(f_1 \underline{\text{ and }} f_2)]\!] &\overset{def}{=} [\![f_1]\!] \text{ and } [\![f_2]\!] \\
[\![(f \underline{\text{ when }} g)]\!] &\overset{def}{=} [\![f]\!] \text{ when } g \\
[\![(\underline{\text{N }} f)]\!] &\overset{def}{=} \text{N } [\![f]\!]
\end{aligned}
$$

The core of the relationship between trajectory formulas in the deep and shallow embeddings is captured in Theorem 3. The theorem says that translating a deeply embedded formula f' to a shallow embedding via $[\![\cdot]\!]$ (Definition 9) results in a trajectory formula that is satisfied by exactly the same set of sequences as f'. The proof is done by induction over the structure of trajectory formulas in the deep embedding.

Theorem 3. *Translate trajectory logic over Booleans in logic from deep to shallow embedding*

$$\forall f',\ \mathring{\sigma}.\ ([\![f']\!]\ \mathring{\sigma}) \Longleftrightarrow (\mathring{\sigma}\ \underline{\text{sat}}\ f')$$

5 Wrapping It All Up

In this section we gather together the various theorems proved in Sections 3 and 4 to produce our final result. We then demonstrate the use of this result on the simple NAND-DELAY circuit first introduced in Section 1.

5.1 Gluing the Pieces Together

The focus of this paper is on the connection from trajectory evaluation over lattice values to relations over Boolean streams. Formalizing this connection forces us to reason about three different worlds: trajectory assertions with both lattice and Boolean values, and higher-order logic. A completely formal representation of these worlds (in particular, the semantics of higher-order logic [7]) and mappings between them would obfuscate the focus of our work. To maintain focus, we gloss over some of the semantic mappings between these different worlds. In particular, we use the same representation of trajectory formulas for both the metalogical results and a deep embedding of trajectory formulas in logic.

We link our shallow embedding of trajectory formulas in Section 4 with the results from Section 3 via an intermediate representation of trajectory logic that is deeply embedded in a higher-order logic. Theorem 4 says that a trajectory evaluation result over Boolean streams holds if and only if it holds in logic using a deep embedding of trajectory formulas.

Theorem 4. *Translate trajectory logic over Booleans to deep embedding in higher-order logic*
    For all circuits ckt, antecedents ant, and consequences cons, if

> For all ϕ.
> $\phi \models e$ implies
>   For all $\mathring{\sigma}$.
> $\phi \models \left(\mathring{\sigma} \in \mathring{\mathcal{L}}\ ckt \right)$ implies
>   For all $t \geq 0$. $((\phi, \mathring{\sigma}_t) \models ant)$ implies $((\phi, \mathring{\sigma}_t) \models cons)$
> *then the following is a true formula in HOL:*

$$\models_{\text{HOL}} \left(e \Longrightarrow \forall \mathring{\sigma}.\ \mathring{\sigma}\ \underline{\text{in\_lang}}\ ckt \Longrightarrow \forall t \geq 0.\ (\mathring{\sigma}_t\ \underline{\text{sat}}\ ant) \Longrightarrow (\mathring{\sigma}_t\ \underline{\text{sat}}\ cons) \right)$$

We now have the pieces in place to prove a relationship between the standard trajectory logic and our shallow embedding of trajectory logic as predicates over Boolean streams in HOL (Theorem 5). This is proved by connecting Theorems 1, 2, 3, and 4.

Theorem 5. *Translate STE result to shallow embedding of Boolean streams. For all circuits ckt, antecedents ant, and consequences cons, if an implementation of the STE algorithm returns e:*

$$e = \text{STE } ckt \; ant \; cons$$

then we can introduce the following axiom in HOL:

$$\vdash \left(e \implies \forall \mathring{\sigma} . \; ckt \; \mathring{\sigma} \implies \forall t \geq 0. \; [\![ant]\!] \; (\mathring{\sigma}_t) \implies [\![cons]\!] \; (\mathring{\sigma}_t) \right)$$

The proof of Theorem 5 requires going through some simple transitivity reasoning to work from the STE algorithm to trajectory assertions (Theorem 1), unfolding $\models_{ckt} [ant \implies cons]$ (Definition 5) and using Theorem 2 to arrive at trajectory assertions over Boolean streams, and then to a deep embedding of trajectory logic in HOL (Theorem 4). We then use Theorem 3 to unify the right-hand-sides of Theorem 4 and Theorem 5. The unification is done by instantiating Theorem 3 first with "f'" as ant and then with "f'" as $cons$. Finally, we use the axioms relating the various representations of the circuit to prove that the sequence $\mathring{\sigma}$ is in the language of the circuit.

5.2 Simple Example Revisited

We now revisit the simple NAND-DELAY circuit first introduced in Section 1. When we introduced the circuit, we showed an intuitive, but not quite technically correct, translation from a standard trajectory assertion to a relations over Boolean streams result. We now demonstrate how the semantic link provided by our principal theorem (Theorem 5) induces a formally correct and intuitively satisfying connection between user-level verifications in trajectory evaluation and higher-order logic.

We begin with the trajectory formula that first appeared in Figure 2.

$$i \overset{def}{=} "i"$$
$$o \overset{def}{=} "o"$$
$$\text{NAND-DELAY} \models \left[\begin{matrix} (i \text{ is } b_1) \text{ and} \\ (o \text{ is } b_2) \end{matrix} \implies (\text{N} \; (o \text{ is } \neg(b_1 \land b_2))) \right]$$

Running the STE algorithm establishes that the NAND-DELAY circuit does satisfy the specification, and indeed, satisfies it for all valuations of the Boolean variables b_1 and b_2. We instantiate e with T in Theorem 5, and, after some unfolding and beta-reduction, we are left with the following.

$$\vdash \forall \mathring{\sigma}.$$
$$\text{NAND-DELAY } \mathring{\sigma} \implies$$
$$\forall t \geq 0.$$
$$\left(\begin{array}{l} (\mathring{\sigma} \text{ "}i\text{" } t) = b_1 \wedge \\ (\mathring{\sigma} \text{ "}o\text{" } t) = b_2 \end{array} \right) \implies$$
$$((\mathring{\sigma} \text{ "}o\text{" } (t+1)) = \neg(b_1 \wedge b_2))$$

Because b_1 and b_2 are free variables, we can universally quantify over them, and then use the equalities in the antecedent to substitute for b_1 and b_2 through the consequent. This finally leaves us with the result we had hoped for, namely the following intuitive user-level HOL theorem.

$$\vdash \forall \mathring{\sigma}.$$
$$\text{NAND-DELAY } \mathring{\sigma} \implies$$
$$\forall t \geq 0.$$
$$((\mathring{\sigma} \text{ "}o\text{" } (t+1)) = \neg((\mathring{\sigma} \text{ "}i\text{" } t) \wedge (\mathring{\sigma} \text{ "}o\text{" } t)))$$

6 Voss Implementation of Trajectory Evaluation

In this section we briefly touch on the Voss implementation of trajectory logic. Information on Voss has been published previously [8], we include this section to make the connection to an implementation of trajectory evaluation more concrete. In Voss, the four-valued lattice is represented by a pair of Booleans, called a dual-rail value (Definition 10). In Voss, Booleans are implemented as BDDs and may therefore be symbolic. A Boolean value v is translated to a dual-rail value by putting v on the high rail and its negation on the low rail $(v\!\downarrow \stackrel{def}{=} (v, \neg v))$.

Definition 10. *Implementation of dual-rail lattice in Voss*

$$\top \stackrel{def}{=} (F, F)$$
$$1 \stackrel{def}{=} (T, F)$$
$$0 \stackrel{def}{=} (F, T)$$
$$X \stackrel{def}{=} (T, T)$$

Trajectory formulas are represented in Voss by a very simple, but slightly indirect, deep embedding. Rather than representing trajectory formulas by a data type that mirrors the abstract syntax tree of the formulas, a trajectory formula is a list of "5-tuples" (Figure 5). The five elements of each tuple are the guard, a node name, a Boolean value (which may be a variable), a natural-number start-time, and a natural-number end-time. The meaning is "if the guard is true, then the node takes on the given value from the start-time to the end-time."

As the definitions in Figure 5 show, every formula in the trajectory logic can be represented quite simply by a 5-tuple list. Moreover, it is clear that the list representation does not add anything new to the expressive power of our formulas: any 5-tuple whose duration spans more than one-time unit can be expressed by an appropriate conjunction of applications of the next time operator.

$$
\begin{array}{lllll}
// & guard & node & value & start & end \\
\texttt{traj\_form} \overset{type}{=} & (\texttt{bool} & \times \texttt{string} & \times \texttt{bool} & \times \texttt{nat} & \times \texttt{nat}) \texttt{list}
\end{array}
$$

$$
\begin{aligned}
n \text{ is } v \quad &\overset{def}{=} [(\text{T}, n, v, 0, 1)] \\
f_1 \text{ and } f_2 \quad &\overset{def}{=} f_1 \text{ append } f_2 \\
f \text{ when } g \quad &\overset{def}{=} \text{map } (\lambda(g', n, v, t_0, t_1). \ (g' \wedge g, n, v, t_0, t_1)) \ f \\
\text{N } f \quad &\overset{def}{=} \text{map } (\lambda(g, n, v, t_0, t_1). \ (g, n, v, t_0 + 1, t_1 + 1)) \ f \\
\\
f \text{ from } t_0 \quad &\overset{def}{=} \text{map } (\lambda(g, n, v, z, t_1). \ (g, n, v, t_0, t_1)) \ f \\
f \text{ to } t_1 \quad &\overset{def}{=} \text{map } (\lambda(g, n, v, t_0, z). \ (g, n, v, t_0, t_1)) \ f
\end{aligned}
$$

Fig. 5. Implementation of trajectory formulas in Voss

7 Conclusion

The motivation of our work is to combine the automatic deductive power of finite-state model checking with the expressive power and flexibility of general-purpose theorem proving. Key elements of such an architecture are semantic links between the general logic of the theorem prover and the more specialised model-checker logic. To this end, we have established a semantic link between correctness results in the temporal logic of symbolic trajectory evaluation and relational hardware specifications in higher order logic.

Ultimately, our aim is to combine results from several different model checkers in a single system. The expressive power of typed higher-order logic, which is a foundational formalism, and the generality of the relational approach are an excellent "glue logic" for this. For many other linear-time temporal logics, at least, it seems quite reasonable to expect smooth embeddings into the relational world. Indeed, there have been several such embeddings in the literature [18, 4]. Branching time logics may be more challenging, but the PVS results cited earlier are encouraging. More general temporal logics, such as Extended Temporal Logic [17] will provide a further challenge.

Acknowledgments

This paper benefited from the helpful comments of Carl Seger, Ching-Tsun Chou, and the anonymous referees.

References

1. M. D. Aagaard, R. B. Jones, and C.-J. H. Seger. Formal verification using parametric representations of Boolean constraints. In *ACM/IEEE Design Automation Conference*, July 1999.
2. S. Agerholm and H. Skjødt. Automating a model checker for recursive modal assertions in HOL. Technical Report DAIMI IR-92, Computer Science Department, Aarhus University, 1990.
3. A. Cheng and K. Larsen, editors. *Program and Abstracts of the BRICS Autumn School on the Verification*, Aug. 1996. BRICS Notes Series NS-96-2.
4. C.-T. Chou. Predicates, temporal logic, and simulations. In C.-J. H. Seger and J. J. Joyce, editors, *HOL User's Group Workshop*, pages 310–323. Springer Verlag; New York, Aug. 1994.
5. C.-T. Chou. The mathematical foundation of symbolic trajectory evaluation. In *Workshop on Computer-Aided Verification*. Springer Verlag; New York, 1999. *To appear.*
6. G. C. Gopalakrishnan and P. J. Windley, editors. *Formal Methods in Computer-Aided Design*. Springer Verlag; New York, Nov. 1998.
7. M. J. C. Gordon and T. F. Melham, editors. *Introduction to HOL: a theorem proving environment for higher order logic*. Cambridge University Press, New York, 1993.
8. S. Hazelhurst and C.-J. H. Seger. Symbolic trajectory evaluation. In T. Kropf, editor, *Formal Hardware Verification*, chapter 1, pages 3–78. Springer Verlag; New York, 1997.
9. A. J. Hu and M. Y. Vardi, editors. *Computer Aided Verification*. Springer Verlag; New York, July 1998.
10. J. Joyce and C.-J. Seger. Linking BDD based symbolic evaluation to interactive theorem proving. In *ACM/IEEE Design Automation Conference*, June 1993.
11. M. Newey and J. Grundy, editors. *Theorem Proving in Higher Order Logics*. Springer Verlag; New York, Sept. 1998.
12. J. O'Leary, X. Zhao, R. Gerth, and C.-J. H. Seger. Formally verifying IEEE compliance of floating-point hardware. *Intel Technical Journal*, First Quarter 1999. Online at http://developer.intel.com/technology/itj/.
13. The omega project, 1999. http://www.ags.uni-sb.de/projects/deduktion/.
14. Proof and specification assisted design environments, ESPRIT LTR project 26241, 1999. *http://www.dcs.gla.ac.uk/prosper/*.
15. S. Rajan, N. Shankar, and M. Srivas. An integration of model checking automated proof checking. In *Workshop on Computer-Aided Verification*. Springer Verlag; New York, 1996.
16. C.-J. H. Seger and R. E. Bryant. Formal verification by symbolic evaluation of partially-ordered trajectories. *Formal Methods in System Design*, 6(2):147–189, Mar. 1995.
17. M. Y. Vardi and P. Wolper. Reasoning about infinite computations. *Information and Computation*, 115:1–37, 1994.
18. J. von Wright. Mechanising the temporal logic of actions in HOL. In M. Archer, J. J. Joyce, K. N. Levit, and P. J. Windley, editors, *International Workshop on the HOL Theorem Proving System and its Applications*, pages 155–159. IEEE Computer Society Press, Washington D.C., Aug. 1991.

Verification of Infinite State Systems by Compositional Model Checking

K. L. McMillan

Cadence Berkeley Labs

Abstract. A method of compositional verification is presented that uses the combination of *temporal case splitting* and *data type reductions* to reduce types of infinite or unbounded range to small finite types, and arrays of infinite or unbounded size to small fixed-size arrays. This supports the verification by model checking of systems with unbounded resources and *uninterpreted functions*. The method is illustrated by application to an implementation of Tomasulo's algorithm, for arbitrary or infinite word size, register file size, number of reservation stations and number of execution units.

1 Introduction

Compositional model checking reduces the verification of a large system to a number of smaller verification problems that can be handled by model checking. This is necessary because model checkers are limited with respect to state space size. Compositional methods are implemented in, for example, the SMV system [13] and the Mocha system [1]. The typical proof strategy using these systems is to specify *refinement relations* between an abstract model and certain variables or signals in an implementation. This allows components of the implementation to be verified in context of the abstract model. This basic approach is limited in two respects. First, it does not reduce data types with large ranges, such as addresses or data words. For example, it is ineffective for systems with 32-bit or 64-bit memory address spaces. Second, the approach can verify only a fixed configuration of a design, with fixed resources. It cannot, for example, verify a parameterized design for all values of the parameter (such as the number of elements in an array).

Here, we present a method based on *temporal case splitting* and a form of *data type reduction*, that makes it possible to handle types of arbitrary or infinite range, and arrays of arbitrary or infinite size. Temporal case splitting breaks the correctness specification for a given data item into cases, based on the path the data item has taken through the system. For each case, we need only consider a small, fixed subset of the elements of the large data structures. The number of cases, while potentially very large, can be reduced to a small number by existing techniques based on symmetry [13]. Finally, for any given case, a data type reduction can reduce the large or infinite types to small finite types. The reduced types contain only a few values relevant to the given case, and an abstract value representing the remaining values in the original type. Thus, we reduce the large or infinite types and structures to small finite types and structures for the purpose of model checking.

Together, these methods also allow specification and verification using uninterpreted functions. For example, we can use model checking to verify the correctness of an instruction set processor independent of its arithmetic functions. This separates the verification of data and control flow from the verification the the arithmetic units.

The techniques described in this paper have been implemented in a proof assistant which generates model checking subgoals to be discharged by the SMV model checker.

Related work Data type reduction, as applied here, can be viewed as a special case of *abstract interpretation* [5]. It is also related to reductions based on *data independence* [18] in that a large data type is reduced to a small finite one, using a few representative values and an extra value to represent everything else. The technique used here is more general, however, in that it does not require control to be independent of data. For example, it allows control to depend on comparisons of data values. The technique may therefore be applied to reduce addresses, tags and pointers, which are commonly compared to determine control behavior (an example of this appears later in the paper). Also, the technique reduces not only the data types in question, but also any arrays indexed by these types. This makes it possible to handle systems with unbounded memories, FIFO buffers, *etc.*.

Lazic and Roscoe [11] also describe a technique for reducing unbounded arrays to finite ones for verification, under certain restrictions. Their technique is a complete procedure for verifying a particular property (determinism). It works by identifying a finite configuration of a system, whose determinism implies determinism of of any larger configurations. The technique presented here, on the other hand, is not restricted to a particular property. More importantly, the method of [11] does not allow equality comparison of values stored in arrays, nor the storage of array indices in arrays. Thus, for example, it cannot handle unbounded cache memories, content-addressable memories, or the example presented in this paper, an out-of-order processor that stores tags (*i.e.*, array indices) in arrays, and compares them for equality. Note that comparing values stored in an unbounded array, or even including one bit of status information in the elements of an unbounded array, is sufficient to make reachability analysis undecidable. Unfortunately, these conditions are ubiquitous in hardware design. Thus, while the technique presented here is incomplete, being based on a conservative abstraction, this incompleteness should be viewed as inevitable if we wish to verify hardware designs for unbounded resources.

Data type reduction has also been used by Long [12] in his work on generating abstractions using BDD's. However, that work applied only to concrete finite types. Here, types of arbitrary or infinite size are reduced to finite types. Also, Long's work did not treat the reduction of arrays. What makes it possible to do this here is the combination of data type reductions with temporal case splitting and symmetry reductions, a combination which appears to be novel.

The use of uninterpreted functions here is also substantially different from previous applications, both algorithmically and methodologically. The reason for using uninterpreted functions is the same – to abstract away from the actual functions computed on data. However, existing techniques using uninterpreted functions, such as [4, 10, 7, 14, 17, 2] are based essentially on symbolic simulation. The present method allows the combination of uninterpreted functions with model checking. This distinction has significant practical consequences for the user. That is, the existing methods are all based on proving commutative diagrams. In the simplest case, one shows that, from any state, applying an abstraction function and then a step of the specification model is equivalent to applying a step of the implementation model and then the abstraction function. However, since not all states are reachable, the user must in general provide an *inductive invariant*. The commutative diagram is proved only for those states satisfying the invariant. By contrast, in the present technique, there is no need to provide an inductive invariant, since the model checker determines the strongest invariant

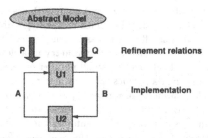

Fig. 1. Compositional refinement verification

by reachability analysis. This not only saves the user a considerable effort, but also improves the re-usability of proofs, as we will observe later.

In general, when uninterpreted functions with equality are added to temporal logic, the resulting logic is undecidable. The present method is not a decision procedure for such a logic, but rather a user-guided reduction to the propositional case that is necessarily incomplete. Note, an earlier semi-decision procedure for such a logic [8], is sound only in a very restricted case; for most problems of practical interest, the procedure is not sound, and can only be used to find counterexamples. Of the various non-temporal techniques using uninterpreted functions, the present method is most similar to [17], since it is also based on *finite instantiation*. However, the methods are not similar algorithmically.

Thus, the methods presented here are novel in three aspects: first the particular techniques of data type reduction and of handling uninterpreted functions are novel. Second, the combination of these techniques with existing compositional methods and techniques of exploiting symmetry is novel. Finally, the implementation of all these techniques into a mechanical proof assistant based on symbolic model checking is novel.

Outline of the article Section 2 is a brief overview of earlier work on compositional methods and symmetry, on which the current work is based. Section 3 then describes temporal case splitting and its implementation in the SMV system. Section 4 covers data type reduction and its implementation in SMV. Finally, in section 5, these techniques are illustrated by applying them to an implementation of Tomasulo's algorithm. This is the same example used in [13], however in this case the new techniques substantially simply the proof, decrease the run-time of the prover, and allow verification for unbounded or infinite resources.

2 Compositional verification and symmetry

The SMV system uses compositional model checking to support refinement verification – proving that an abstract model, acting as the system specification, is implemented by some more detailed system model. Correctness is usually defined by *refinement relations* that specify signaling behavior at suitable points in the implementation in terms of events occurring in the abstract model (see fig. 1). Typically, the abstract model, the implementation and the refinement relations are all expressed in the same HDL-like language, as sets of equations that may involve time delay. Formally, however, we can view them as simply linear temporal logic properties.

The refinement relations decompose the system structurally into smaller parts for separate verification. This relies on a method of *circular compositional proof* whereby we may assume that one temporal property P holds true while verifying property Q, and *vice versa*. In the figure, for example, we can assume that signal A is correct w.r.t.

the abstract model when verifying signal B, and assume that signal B is correct w.r.t. the abstract model when verifying signal A. This makes it possible to compositionally very systems that have cyclic data flow, such as instruction set processors.

In addition, the SMV system can exploit symmetry in a design to reduce a large number of symmetric proof obligations to a small number of representative cases. This is based on the use of symmetric data types called *scalarsets*, borrowed from the Murphi language [9]. To exploit the symmetry of a given type, we must guarantee that values of that type are only used in certain symmetric ways. For example, they may be compared for equality, or used as indices of arrays. SMV enforces these conditions by static type checking. For further details of these methods, the reader is referred to [13].

3 Temporal case splitting

Hardware designs typically contain large arrays, such as memories, FIFO buffers, content-addressable memories (CAM's) and so forth. Their state space is often intractably large, and thus we usually cannot apply model checking to them directly. However, we can often confine the verification problem to one or two elements of the array by means of *temporal case splitting*. Using this approach, we verify the correctness of only those data items that have passed through a given fixed element of an array. Thus, we consider individually each path that a data item might take through a given system, and reduce the number of state variables accordingly.

Temporal case splitting breaks the proof of temporal property $G\phi$ (ϕ at all times) into cases based on the value of a given variable v. For each possible value i of v, we show that ϕ is true at just those times when $v = i$. Then, since at all times v must have some value, we can infer that ϕ must be true at all times. This inference is based on the following fairly trivial fact about temporal logic:

Theorem 1. *If, for all i in the range of variable v, $\models G(v = i \Rightarrow \phi)$, then $\models G\phi$.*

Typically, v is an auxiliary variable recording the location in some array that was used to store the data item currently appearing at a unit output (see [13] for the use of auxiliary variables in compositional proofs). To prove a give case $v = i$, it is commonly only necessary to refer to element i of the array. The other elements of the array can be abstracted from the model by replacing them with an "unknown" value $\bot$, much as in ternary symbolic simulation [3]. Thus, in effect, we decompose a large array into its elements for the purposes of model checking. Several examples of this can be found in section 5.

In the SMV system, a specification s of the form Gp is split into cases using a declaration of the following form:

```
forall (i in TYPE) subcase c[i] of s for v = i;
```

Here, s is the name of the original specification, and TYPE is the type of variable v. This generates an array of specifications c where each specification c[i] is the formula $G((v = i) \Rightarrow p)$. If every element of c can be separately proved, then SMV infers the original specification s.

4 Data type reductions

Although temporal case splitting may reduce a large array to a small number of elements, the model will still have types with large ranges, such as addresses or data words. In this case, *data type reduction* can reduce a large (perhaps unbounded or infinite) type to a small finite one, containing only one or two values relevant to the case being verified. The remaining values are represented by a single abstract value. Corresponding to this reduction, we have an abstract interpretation of constructs in the logic. This abstraction is conservative, in that any property that is true in the reduced model is also true in the original.

For example, let t be an arbitrary type. Regardless of the actual range of t, we can reduce the range to a set containing a distinguished value i and an abstract value (which we will denote $t \setminus i$) representing the remaining values. Thus, in the reduced model, all variables of type t range over the set $\{i, t \setminus i\}$. Now, consider, for example, the equality operator. In order to obtain a conservative abstraction, we use the following truth table for equality:

$=$	i	$t \setminus i$
i	1	0
$t \setminus i$	0	$\perp$

That is, the specific value i is equal to itself, and not equal to $t \setminus i$. However, two values not equal to i may themselves be equal or unequal. Thus, the result of comparing $t \setminus i$ and $t \setminus i$ for equality is an "unknown" value $\perp$.

Similarly, an array reference $a[x]$, where a has index type t, yields the value of signal $a[i]$ if $x = i$ and otherwise $\perp$. As a result, in the reduced verification problem, only one element, $a[i]$, of array a is referenced.

Using such an abstract interpretation, any formula that is true in the reduced model will be true in the original. On the other hand, in some cases the truth value of a formula in the abstraction will be $\perp$. In this case we cannot infer anything about the truth of the formula in the original model. In practice, an appropriate data type reduction for a given type can often be inferred automatically, given the particular case being verified. For example, if the case has two parameters, i and j, both of type t, then by default SMV would reduce the type t to the two values i and j, and the abstract value $t \setminus \{i, j\}$. Some examples of this will appear in section 5.

Formalizing data type reductions Data type reductions, as used here, are a particular instance of abstract interpretation [5]. This, of course, is an old subject, however the particular abstract interpretation used here is believed by the author to be novel. Formalizing the notion of data type reduction for the complete logic used by SMV would require at the very least introducing the entire logic, which is well beyond the scope of this paper. However, we can easily formalize data type reductions in a general framework, for an arbitrary logic, and show in some special cases how this relates to SMV's logic.

To begin with, suppose that we are given a set U of *values*, a set V of *variables*, a set T of *types* and a function $\mathcal{T} : V \to T$, assigning types to variables. Suppose also that we are given a language $\mathcal{L}$ of *formulas*. Formulas are built from a set of *constructors* C, where each constructor $c \in C$ has a *arity* $n_c \geq 0$. Examples of constructors in SMV's logic are function symbols, constants, variables and quantifiers. An *atomic formula* is $c()$, where $c \in C$ is of arity $n_c = 0$. A *formula* is defined to be either an atomic formula, or $c(\psi_1, \ldots, \psi_n)$, where $c \in C$ is a constructor of arity $n_c = n \geq 1$, and $\psi_1, \ldots, \psi_n$ are formulas. Now, let a *structure* be a triple $M = (\mathcal{R}, \mathcal{N}, \mathcal{F})$, where $\mathcal{R}$ is a function $T \to \mathcal{P}(U)$, assigning a range of values to every type, $\mathcal{N}$ is a set of *denotations*, and $\mathcal{F}$ is an *interpretation*, assigning to each constructor $c \in C$, a function $\mathcal{F}(c) : \mathcal{N}^n \to \mathcal{N}$. The *denotation* of a formula f in structure M will be written f^M. This is defined inductively. That is, for formula $\phi = c(\psi_1, \ldots, \psi_n)$, where $c \in C$ is a constructor, $\phi^M = (\mathcal{F}(c))(\psi_1^M, \ldots, \psi_n^M)$.

We will assume that the set of denotations $\mathcal{N}$ admits a pre-order, $\leq$, and that every function $\mathcal{F}(c)$ where $c \in C$, is monotonic with respect to this pre-order. That is, for all $c \in C$, if $x_1 \leq y_1, \ldots, x_n \leq y_n$, where $x_i, y_i \in \mathcal{N}$, then $(\mathcal{F}(c))(x_1, \ldots, x_n) \leq$

$(\mathcal{F}(c))(y_1, \ldots, y_n)$. Given two structures $M = (\mathcal{R}, \mathcal{N}, \mathcal{F})$ and $M' = (\mathcal{R}', \mathcal{N}', \mathcal{F}')$, we will say that a function $h : \mathcal{N} \to \mathcal{N}'$ is a *homomorphism* from M to M' when, for every $c \in C$,

$$h(\mathcal{F}(c)(x_1, \ldots, x_n)) \geq (\mathcal{F}'(c)(h(x_1), \ldots, h(x_n))$$

If h is a homomorphism from M to M', we will write $M \xrightarrow{h} M'$.

Theorem 2. *If $M \xrightarrow{h} M'$ then for all formulas $\phi \in \mathcal{L}$, $h(\phi^M) \geq \phi^{M'}$*

Proof. By induction over the structure of ϕ, using monotonicity.

Now, let us fix a structure $M = (\mathcal{R}, \mathcal{N}, \mathcal{F})$. Let a *data type reduction* be any function $r : T \to \mathcal{P}(U)$, assigning to each type $t \in T$ a set of values $r(t) \subseteq \mathcal{R}(t)$. That is, a data type reduction maps every type to a subset of its range in M. We wish to define, for every data type reduction r, a reduced structure M_r, and a map h_r, such that $M \xrightarrow{h_r} M_r$.

As an example, let us say that $\mathcal{N}$ is the set of functions $\mathcal{M} \to \mathcal{Q}$, where $\mathcal{M}$ is a set of *models* and $\mathcal{Q}$ is a set of *base denotations*. For now, let $\mathcal{Q}$ be simply the set U of values. A model is a function $\sigma : V \to \mathcal{Q}$, assigning to every variable $v \in V$ a value $\sigma(v)$ in $\mathcal{R}(\mathcal{T}(v))$, the range of its type in M. For any data type reduction $r : T \to \mathcal{P}(U)$, let M_r be a structure $(\mathcal{R}_r, \mathcal{N}_r, \mathcal{F}_r)$. For every type $t \in T$, let $\mathcal{R}_r(t)$ be the set $\{\{x\} \mid x \in r(t)\} \cup \{U \setminus r(t)\}$. That is, the range of type t in the reduced model is the set consisting of the singletons $\{x\}$, for x in $r(t)$, and the set containing all the remaining values. Let $\mathcal{N}_r$, the set of denotations of the reduced structure, be as above, except that $\mathcal{Q}_r$, the set of base denotations in the reduced model, is $\mathcal{P}(U)$. For any $x, y \in \mathcal{Q}_r$, we will say that $x \leq y$ when $x \supseteq y$, and we will equate $\perp_r$ with the set U of all values. The pre-order $\leq$ on $\mathcal{N}_r$ is the point-wise extension of this order to denotations. That is, for all $x, y \in \mathcal{N}$, $x \leq y$ iff, for all models $\sigma \in \mathcal{M}_r$, $x(\sigma) \leq y(\sigma)$. Finally, the map h_r is defined as follows:

$$h_r(x)(\sigma') = \{x(\sigma) \mid \sigma \in \sigma'\}$$

where for any $\sigma \in \mathcal{M}$ and $\sigma' \in \mathcal{M}_r$, we say $\sigma \in \sigma'$ iff for all variables $v \in V$, $\sigma(v) \in \sigma'(v)$. Now, for every constructor c, we must define an abstract interpretation $\mathcal{F}_r(c)$, corresponding to the original interpretation $\mathcal{F}(c)$, such that $M \xrightarrow{h_r} M_r$.

Definition 1. *A constructor $c \in C$ is safe for r iff, for all $\sigma \in \mathcal{M}$ and $\sigma' \in \mathcal{M}_r$ for all $x \in \mathcal{N}^n$, $y \in \mathcal{N}_r^n$ if $\sigma \in \sigma'$ and $x_i \sigma \in y_i \sigma'$ for $i = 1 \ldots n$, then $(\mathcal{F}(c))(x_1, \ldots, x_n)\sigma \in (\mathcal{F}_r(c))(y_1, \ldots, y_n)\sigma'$.*

Theorem 3. *If every $c \in C$ is safe for r, then $M \xrightarrow{h_r} M_r$.*

Proof. Suppose that for all $\sigma \in \sigma'$, $(\mathcal{F}(c))(x_1, \ldots, x_n)\sigma \in (\mathcal{F}_r(c))(y_1, \ldots, y_n)\sigma'$. Then,

$$h_r(\mathcal{F}(c)(x_1, \ldots, x_n))(\sigma') = \{\mathcal{F}(c)(x_1, \ldots, x_n)(\sigma) \mid \sigma \in \sigma'\}$$
$$\subseteq \mathcal{F}_r(c)(y_1, \ldots, y_n)\sigma'$$

Hence by definition $h_r(\mathcal{F}(c)(x_1, \ldots, x_n)) \geq (\mathcal{F}_r(c))(y_1, \ldots, y_n)$.

We will consider here a few atomic formulas and constructors in the SMV logic of particular interest. For example, every variable $v \in V$ is a atomic formula in the logic. In the original interpretation we have, somewhat tautologically,

$$\mathcal{F}(v)()(\sigma) = \sigma(v)$$

In the abstract interpretation, we have the same definition:

$$\mathcal{F}_r(v)()(\sigma') = \sigma'(v)$$

We can immediately see that v is safe for any data type reduction r. That is, if $\sigma \in \sigma'$, then by definition, for every variable v, $\sigma(v) \in \sigma'(v)$. Thus, $\mathcal{F}(v)()(\sigma) \in \mathcal{F}_r(v)()(\sigma')$.

Now let us consider the equality operator. The concrete interpretation of this operator is:

$$\mathcal{F}(=)(x_1, x_2)(\sigma) = \begin{cases} 1 & ; \; x_1\sigma = x_2\sigma \\ 0 & ; \; \text{else} \end{cases}$$

The abstract interpretation is:

$$\mathcal{F}_r(=)(y_1, y_2)(\sigma') = \begin{cases} \{1\} & ; \; y_1\sigma' = y_2\sigma', \; |y_1\sigma'| = 1 \\ \{0\} & ; \; y_1\sigma' \cap y_2\sigma' = \emptyset \\ \bot_r & ; \; \text{else} \end{cases}$$

That is, if the arguments are equal singletons, they are considered to be equal, if they are disjoint sets, they are considered to be unequal, and otherwise the result is $\bot_r$. This constructor is clearly safe for any r. That is, a simple case analysis will show that if $x_1\sigma \in y_1\sigma'$ and $x_2\sigma \in y_2\sigma'$, then $\mathcal{F}(=)(x_1, x_2)(\sigma) \in \mathcal{F}_r(=)(y_1, y_2)(\sigma')$.

Finally, we consider array references. An *array* a is an n-ary constructor. To each argument position $0 \leq i < n$, we association an *index type* $\mathcal{T}_i(a) \in T$. We also associate with a a collection of variables $\{a[x_1] \cdots [x_n] \mid x_i \in \mathcal{R}(\mathcal{T}_i(a))\} \subseteq V$. The interpretation of the array constructor a is:

$$\mathcal{F}(a)(x_1, \ldots, x_n)(\sigma) = \begin{cases} \sigma(a[x_1\sigma] \cdots [x_n\sigma]) & ; \; x_i\sigma \in \mathcal{R}(\mathcal{T}_i(a)) \\ \bot & ; \; \text{else} \end{cases}$$

That is, if all of the indices are in the range of their respective index types, then the result is the value of the indexed array element in σ, else it is $\bot$ (the "unknown" value in the concrete model). The abstract interpretation is

$$\mathcal{F}_r(a)(y_1, \ldots, y_n)(\sigma') = \begin{cases} \sigma'(a[m_1] \cdots [m_n]) & ; \; y_i\sigma' = \{m_i\}, \; m_i \in r(\mathcal{T}_i(a)) \\ \bot_r & ; \; \text{else} \end{cases}$$

That is, if all of the indices are *singletons* in the *reduced* range of their respective index types, then the result is the value of the indexed array element in σ', else it is $\bot_r$. Thus, if a given array index has index type t, then the abstract interpretation depends only on array elements whose indices are in $r(t)$. This is what allows us to reduce unbounded arrays to a finite number of elements for model checking purposes. It is a simple exercise in case analysis to show that array references are safe for r.

Most of the remaining constructors in the SMV logic are trivially safe in that they return $\bot_r$ when any argument is not a singleton. Since all the constructors are safe, we have $M \overset{h}{\rightsquigarrow} M_r$. Now, suppose that we associate with each structure $M = (\mathcal{R}, \mathcal{N}, \mathcal{F})$ a distinguished denotation $\textsc{True} \in \mathcal{N}$, the denotation of valid formulas. We will say that a homomorphism h from M to M' is *truth preserving* when $h(x) \geq \textsc{True}'$ implies $x = \textsc{True}$. This gives us the following trivial corollary of theorem 2:

Corollary 1. *If $M \overset{h}{\rightsquigarrow} M'$ and h is truth preserving, then $\phi^{M'} = \textsc{True}'$ implies $\phi^M = \textsc{True}$.*

In the case of a data type reduction r, we will say that $\textsc{True}$ in structure M is the denotation that maps every model to 1, while $\textsc{True}_r$ in the reduced model M_r maps every model to the singleton $\{1\}$. In this case, h_r is easily shown to be truth preserving, hence every formula that is valid in M_r is valid in M. Note that, if r maps every type to a finite range, then the range of all variables is finite in M_r. Further, because of the abstract interpretation of array references, the number of variables that ϕ^{M_r} depends on is finite for any formula ϕ. Thus, we can apply model checking techniques to evaluate ϕ^{M_r} even if the ranges of types in M are unknown, which also allows the possibility that they are infinite.

The above treatment is simplified somewhat relative the the actual SMV logic. For example, the SMV logic is a linear temporal logic. To handle temporal operators, we must extend the type of base denotations Q from values in U to infinite sequences in U^ω. Further extensions of Q are required to handle nondeterministic choice and quantifiers. However, the basic theory presented above is not substantially changed by these extensions.

5 Verifying a version of Tomasulo's algorithm

Combining the above methods – compositional reasoning, symmetry reduction, temporal case splitting and data type reduction – we can reduce the verification of a complex hardware system with unbounded resources to a collection of finite state verification problems, each with a small number of state bits. When the number of state bits in each subproblem is sufficiently small, verification can proceed automatically by model checking. As an example, we apply the above techniques to the verification of an implementation of Tomasulo's algorithm for out-of-order instruction execution.

Tomasulo's algorithm Tomasulo's algorithm [15] allows an instruction set processor to execute instructions in data-flow order, rather than sequential order. This can increase the throughput of the unit, by allowing instructions to be processed in parallel. Each pending instruction is held in a "reservation station" until the values of its operands become available. It is then issued to an execution unit. The flow of instructions in our implementation is pictured in figure 2. Each instruction, as it arrives, fetches its two operands from a special register file. Each register in this file holds either

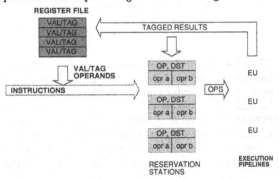

Fig. 2. Flow of instructions in Tomasulo's algorithm

an actual value, or a "tag" indicating the reservation station that *will* produce the register value when it completes. The instruction and its operands (either values or tags) are stored in a reservation station (RS). The RS watches the tagged results returning from the execution unit. When a tag matches one of its operands, it records the value in place of the tag. When the RS has the values of both of its operands, it may issue its instruction to an execution unit. When the result returns from the execution unit, the RS is cleared, and the result value, if needed, is stored in the destination register.

In addition to ALU instructions, our implementation includes instructions that read register values to an external output and write values from an external input. There is also a "stall" output, indicating that an instruction cannot currently be received. A stall can happen either because there is no available RS to store the instruction, or because the value of the register to be read to an output is not yet available.

Structural decomposition The implementation is modeled in the SMV language at the RTL level. This machine is in turn specified with respect to an abstract model. This is a simple implementation of the instruction set that executes instructions sequentially. The input and output signals of the abstract model and the implementation are the same, so there is no need to write refinement relations between them.[1]

We begin the proof by using refinement relations to break the verification problem into tractable parts. In [13], the circular compositional rule was used decompose the arrays (e.g., the register file, and reservation station array). Here, a substantially simpler proof is obtained using temporal case splitting and data type reductions. Essentially, we break the verification problem into two lemmas. The first lemma specifies the operand values stored in the reservation stations, while the second specifies the values returning on the result bus from the execution units, both in terms of the abstract model. We apply circular compositional reasoning, using operand correctness to prove result correctness, and result correctness to prove operand correctness.

To specify the operand and result values, we need to know what the correct values for these data items actually are. We obtain this information by adding auxiliary state to the model (exactly as in [13]). In this case, our auxiliary state variables record the correct values of the operands and the result of each instruction, as computed by the abstract model. These values are recorded at the time an instruction enters the machine to be stored in a reservation station. The SMV code for this is the following:

```
if(~stallout & opin = ALU){
  next(aux[st_choice].opra) := opra;
  next(aux[st_choice].oprb) := oprb;
  next(aux[st_choice].res)  := res;
}
```

That is, if the machine does not stall, and we have an ALU operation, then we store in array aux the correct values of the two operands (opra and oprb) and the result res from the abstract model. The variable st_choice indicates the reservation station to be used. Storing these values will allow us to verify that the actual operands and results we eventually obtain are correct.

The refinement relations themselves are also written as assignments, though they are treated as temporal properties to be proved. For example, here is the operand correctness specification (for operand opra):

```
layer lemma1 :
  forall(k in TAG)
    if(st[k].valid & st[k].opra.valid)
      st[k].opra.val := aux[k].opra;
```

For present purposes, the declaration "layer lemma1:" simply attaches the name lemma1 to the property. TAG is the type of RS indices. Thus, the property must hold for all reservation stations k. If station st[k] is is valid (contains an instruction) and its opra operand is a value (not a tag), then the value must be equal to the correct

---

[1] Here, we prove only safety. For the liveness proof, see "Circular compositional reasoning about liveness", in this volume.

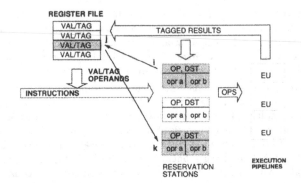

Fig. 3. Path splitting in Tomasulo's algorithm.

operand value stored in the auxiliary array aux. Note that semantically, the assignment operator here simply stands for equality. Pragmatically, however, it also tells the system that this is a specification of signal st[k].opra.val, and that this specification depends on signal aux[k].opra. The SMV proof system uses this information when constructing a circular compositional proof.

The result correctness lemma is just as simply stated:

```
forall (i in TAG)
  layer lemma2[i] :
    if(pout.tag = i & pout.valid)
      pout.val := aux[i].res;
```

That is, for all reservation stations i, if the tag of the returning result on the bus pout is i, and if the result is valid, then its value must be the correct result value for reservation station i.

Using temporal case splitting The refinement relations divide the implementation into two parts for the purpose of verification (operand forwarding logic and instruction execution logic). However, there remain large arrays in the model that prevent us from applying model checking at this point. These are the register file, the reservation station array and the execution unit array. Therefore, we break the verification problem into cases, as a function of the path a data item takes when moving from one refinement relation to another.

Consider, for example, a value returning on the result bus. It is the result produced by a reservation station i (the producer). It then (possibly) gets stored in a register j. Finally it is fetched as an operand for reservation station k (the consumer). This suggests a case split which reduces the verification problem to just two reservation stations and one register. For each operand arriving at consumer RS k, we split the specification into cases based on the producer i (this is indicated by the "tag" of the operand) and the register j (this is the source operand index of the instruction). To prove just one case, we need to use only reservation stations i and k, and register j. The other elements of these arrays are automatically abstracted away, replacing them with the "unknown" value $\perp$. The effect of this reduction is depicted in figure 3.

To apply temporal case splitting in SMV, we use the following declaration (for operand opra):

```
forall (i,k in TAG; j in REG)
  subcase lemma1[i][j]
```

```
of st[k].opra.val//lemma1
for st[k].opra.tag = i & aux[k].srca = j;
```

That is, for all consumer reservation stations k, we break lemma1 into an array of cases (i, j), where i is the producer reservation station and j is the source register. Note, we added an auxiliary variable auk[k].srca to record the source operand register srca, since the implementation does not store this information. Verifying each case requires only one register and two reservation stations in the model. Thus, we have effectively broken the large arrays down into their elements for verification purposes. For the result lemma a similar case splitting declaration can be specified; we split cases on the producing reservation station of the result on the bus, and the execution unit that computed it.

To verify operand correctness, we now have one case to prove for each triple (i, j, k) where i, k are reservation stations and j is an element of the register file. However, if all the registers are symmetric to one another, and all the reservation stations are similarly symmetric, then two representative cases will suffice: one where $i = k$ and one where $i \neq k$. To exploit the symmetry of the design in this way in SMV, we simply declare the types of register indices and reservation station indices to be scalarsets. SMV verifies the symmetry and automatically chooses a set of representative cases. In fact, it chooses the cases $(i = 0, j = 0, k = 0)$ and $(i = 0, j = 0, k = 1)$. All other cases reduce to one of these by permuting the scalarset types. Thus, we have reduced $O(n^3)$ cases to just two.

Infinite state verification Up to this point we have defined refinement relations, used path splitting to decompose the large structures, and applied symmetry to reduce the number of cases to a tractable level. There remain, however, the large types, *i.e.* the data values and possibly the index types. To handle these, we use data type reduction to reduce these types to small sets consisting of a few relevant values and an abstract value representing the rest. In fact, using data type reduction, we can verify our implementation for an arbitrary (or infinite!) number of registers, reservation stations, and execution units. To do this, we simply declare the index types to be scalarsets with undefined range, as follows:

```
scalarset REG undefined;
scalarset TAG undefined;
```

This declares both REG (the type of register indices) and TAG (the type of reservation station indices) to be symmetric, but does not declare ranges for these types. This is possible because the verification process, using symmetry and data type reductions, is independent of the range of these types. For example, when verifying operand correctness, for a given case (i, j, k), SMV by default reduces the type TAG to just three values: i, k and an abstract value. Similarly, the type REG is reduced to just two values: j and an abstract value. This has the side effect of eliminating all the reservation stations other than i and k, and all the registers other that j, by substituting the value $\perp$.

Further, due to symmetry, we only need to verify a fixed set of cases for i, j and k, regardless of the actual range of the types. Thus, we can verify the system generically, for any range, finite or infinite, of these types.

Uninterpreted functions Finally, we come to the question of data values. Suppose, for example, that the data path is 64 bits wide. Although model checkers can handle some arithmetic operations (such as addition and subtraction) for binary values of this width, they cannot handle some other operations, such as multiplication. Moreover, it would be better to verify our implementation generically, regardless of the arithmetic operations in the instruction set. This way, we can isolate the problem of binary arithmetic verification. This is done by introducing an *uninterpreted function symbol* f for the ALU. Assuming only that the abstract model and the implementation execution units compute the same function f, we can prove that our implementation is correct for all ALU functions. The uninterpreted function appraoch also has the advantage that the symmetry of data values is not broken. Thus, we can apply symmetry reductions to data values. As a result, we use only a few representative data values rather than all 2^{64} possible values.

Interestingly, the techniques described above are sufficient to handle uninterpreted functions, without introducing any new logical constructs or decision procedures. To introduce an uninterpreted function in SMV, we simply observe that an *array* in SMV is precisely an uninterpreted function (or, if you prefer, the "lookup table" for an arbitrary function). Thus, to introduce an uninterpreted function symbol in SMV, we simply declare an array of the appropriate type. For example:

```
forall (a,b in WORD) f[a][b] : WORD;
```

This declares a binary function f that takes two words a and b and returns a word f[a][b]. Since we want our arithmetic function to be invariant over time, we declare:

```
next(f) := f;
```

We replace ALU operations in both the abstract model and implementation with lookups in the the array f. We can exploit the symmetry of data words by declaring the type of data words to be a scalarset. That is:

```
scalarset WORD undefined;
```

In fact, since the actual range of the type is undeclared, in principle we are verifying the implementation for any size data word. We then use case splitting on data values to make the problem finite state. That is, we verify result correctness for the case when the operands have some particular values a and b, and where the result $f[a][b]$ is some particular value c. Since we have three parameters a, b and c of the same type, the number of cases needed to have a representative set is just $3! = 6$. Here is SMV declaration used to split the problem into cases:

```
forall(i in TAG; a,b,c in WORD)
    subcase lemma2[i][a][b][c]
    of pout.val//lemma2[i]
    for aux[i].opra = a & aux[i].oprb = b & f[a][b] = c;
```

SMV automatically applies symmetry reduction to reduce an infinite number of cases to 6 representatives. By default, it uses data type reduction to reduce the (possibly infinite) type of data words to the specific values a, b and c, and an abstract value. Thus, in the worst case, when a, b and c are all different, only two bits are needed to encode data words. We have thus reduced an infinite state verification problem to a finite number of finite state problems.

Applying model checking Applying the above proof decomposition, the SMV system produces a set of finite state proof subgoals for the model checker. When all are checked, we have verified our implementation of Tomasulo's algorithm for an arbitrary (finite or infinite) number of registers and reservation stations, for an arbitrary (finite or infinite) size data word, and for an arbitrary ALU function. Note that we have not yet applied data type reduction to execution unit indices. Thus, our proof still applies only to a fixed number of execution units. For one execution unit, there are 11 model checking subgoals, with a maximum of 25 state variables. The overall processing time (including generation of proof goals and model checking) is just under 4 CPU seconds on a SPARC Ultra II server. Increasing the number of execution units to 8, the processing time increases to roughly one minute. We will discuss shortly how to generalize the proof to an arbitrary number of execution units.

Perhaps a more important metric for a technique such as this is the user effort required. The time required for an experienced user of SMV (its author!) to write, debug and verify the proof was approximately an hour and ten minutes. Note that the design itself was already debugged and was previously formally verified using an earlier methodology [13]. The time required to write and debug the design was far greater than that required to effect the proof.[2]

Proving noninterference We can also verify the design for an arbitrary number of execution units. To do this, as one might expect, we split the result lemma into cases based on the execution unit used to produce the result, eliminating all the other units. This, however, requires introducing a "noninterference lemma". This states that no other execution unit spuriously produces the result in question. Such interference would confuse the control logic in the RS and lead to incorrect behavior. The noninterference lemma is stated in temporal logic as follows:

```
lemma3 : assert G (pout.valid -> (complete_eu = aux[pout.tag].eu));
```

In effect, if the result bus pout is returning a value from an execution unit complete_eu, and if that value is tagged for reservation station i, then complete_eu must be the execution unit from which reservation station i is expecting a result. Note, an auxiliary variable aux[i].eu is used to store the index of this execution unit. We can prove this lemma by splitting cases on pout.tag and complete_eu:

```
forall(i in TAG) forall(j in EU)
  subcase lemma3[i][j] of lemma3 for pout.tag = i & complete_eu = j;
```

In this case, the data types TAG and EU (the type of execution unit indices) are reduced to just the values i and j respectively, plus an abstract value. Thus, we can prove the lemma for an arbitrary number of execution units. However, here an interesting phenomenon occurs: the lemma for a given execution unit j holds at time t only if it is true for all the other units up to time $t - 1$. In effect, we must prove that unit j is not the *first* unit to violate the lemma. This is done using the circular compositional method, with the following declaration:

```
forall(i in TAG) forall(j in EU)
  using (lemma3) prove lemma3[i][j];
```

The parentheses around lemma3 tell SMV to assume the general lemma up to time $t - 1$ when proving case i, j at time t. This is typical of noninterference lemmas, where the first unit to violate the lemma may cause others to fail in the future. We can now use lemma3 to prove result correctness for any reservation station i and execution unit j, for any number of execution units. The resulting model checking subgoals require only a few seconds to discharge.

Adding a reorder buffer Now, suppose that we modify the design to use a "reorder buffer". That is, instead of writing results to the register file when they are produced, we store them in a buffer and write them to the register file in program order. This might be done so that the processor can be returned to a consistent state after an "exceptional" condition occurs, such as an arithmetic overflow. The simplest way to

---

[2] Details of this example can be found in a tutorial on SMV, included with the SMV software. At the time of this writing, the software and tutorial can be downloaded from the following URL:

    http://www-cad.eecs.berkeley.edu/~kenmcmil/smv.

do this in the present implementation is to store the result in an extra field res of the reservation station, and then modify the allocation algorithm so that reservation stations are allocated and freed in round-robin order. The result of an instruction is written to the register file when its reservation station is freed.

Interestingly, after this change, the processor can be verified without modifying one line of the proof! This is because our three lemmas (for operands, results and noninterference) are not affected by the design change. This highlights an important difference between the present methodology and techniques such as [4, 10, 7, 14, 17, 2], which are based on symbolic simulation. Because we are using model checking, it is not necessary to write inductive invariants. Instead, we rely on model checking to compute the strongest invariant of an abstracted model. Thus, our proof only specifies the values of three key signals: the source operands in the reservation stations, the value on the result bus and the tag on the result bus. Since the function of these signals was not changed in adding the reorder buffer, our proof is still valid. On the other hand, if we had to an inductive invariant, this would involve in some way all of the state holding variables. Thus, after changing the control logic and adding data fields, we would have to modify the invariants. Of course, in some cases, such as very simple pipelines, almost all states will be reachable, so the required invariant will be quite simple. However in the case of a system with more complex control (such as an out-of-order processor), the invariants are nontrivial, and must be modified to reflect design changes. While this is not an obstacle in theory, in practice a methodology that requires less proof maintenance is a significant advantage.

6 Conclusions and future work

Within a compositional framework, a combination of case splitting, symmetry, and data type reductions can reduce verification problems involving arrays of unbounded or infinite size to a tractable number of finite state subgoals, with few enough state variables to be verified by model checking. This is enabled by a new method of data type reduction and a method of treating uninterpreted functions in model checking. These techniques are part of an overall strategy for hardware verification, that can be applied to such diverse hardware applications as out-of-order instruction set processors, cache coherence systems [6] and packet buffers for communication systems [16]. Note that the model checking, symmetry reduction, temporal case splitting, and data type reduction are tightly interwoven in this methodology. All are used, for example to support uninterpreted functions. Their integration into a mechanical proof assistant means that the proof does not rely in any way on reasoning "on paper".

One possible form of data type reduction is described here. There are, however, many other possibilities. For example, an inductive data type has been added, which allows incrementation (i.e., a successor function). This can be used, for example, to show by induction that a FIFO buffer delivers an infinite sequence of packets in the correct order.

The methodology used here is an attempt to combine in a practical way the strengths of model checking and theorem proving. The refinement relation approach, combined with various reductions and model checking, makes it possible to avoid writing assertions about all state holding component of the design, and also to avoid interactively generated proof scripts. In this way, the manual effort of proofs is reduced. Such proofs, since they specify fewer signals than do proofs involving inductive invariants, can be less sensitive to design changes, as we saw in the case of adding a reorder buffer. On the other hand, the basic ability of theorem proving to break large proofs down into smaller ones is exploited to avoid model checking's strict limits on model size. Thus, by combining the strengths of these two methods, we may arrive at a scalable methodology for formal hardware verification.

References

1. R. Alur, T. A. Henzinger, F. Mang, S. Qadeer, S. K. Rajamani, and S. Tasiran. Mocha: Modularity in model checking. In A. J. Hu and M. Y. Vardi, editors, *CAV '98*, number 1427 in LNCS, pages 521–25. Springer-Verlag.

2. S. Berezin, A. Biere, E. Clarke, and Y. Zhu. Combining symbolic model checking with uninterpreted functions for out-of-order processor verification. In *FMCAD '98*, number 1522 in LNCS, pages 351–68. Springer, 1998.

3. R. E. Bryant and C.-J. Seger. Formal verification of digital circuits using symbolic ternary system models. In R. Kurshan and E. M. Clarke, editors, *Workshop on Computer-Aided Verification*, New Brunswick, New Jersey, June 1990.

4. J. R. Burch and D. L. Dill. Automatic verification of pipelined microprocessor control. In *Computer-Aided Verification (CAV '94)*. Springer-Verlag, 1994.

5. P. Cousot and R. Cousot. Abstract interpretation: a unified lattice model for static analysis of programs by construction or approximation of fixpoints. In *4th POPL*, pages 238–252. ACM Press, 1977.

6. A. Eiriksson. Formal design of 1M-gate ASICs. In *FMCAD '98*, number 1522 in LNCS, pages 49–63. Springer, 1998.

7. R. Hojati and R. K. Brayton. Automatic datapath abstraction of hardware systems. In *CAV '95*, number 939 in LNCS, pages 98–113. Springer-Verlag, 1995.

8. R. Hojati, A. Isles, D. Kirkpatrick, and R. K. Brayton. Verification using uninterpreted functions and finite instantiations. In *FMCAD '96*, volume 1166 of *LNCS*, pages 218–32. Springer, 1996.

9. C. Ip and D. Dill. Better verification through symmetry. *Formal Methods in System Design*, 9(1-2):41–75, Aug. 1996.

10. R. B. Jones, D. L. Dill, and J. R. Burch. Efficient validity checking for processor verification. In *ICCAD '95*, 1995.

11. R. S. Lazić and A. W. Roscoe. Verifying determinism of concurrent systems which use unbounded arrays. Technical Report PRG-TR-2-98, Oxford Univ. Computing Lab., 1998.

12. D. E. Long. Model checking, abstraction, and compositional verification. Tecnical report CMU-CS-93-178, CMU School of Comp. Sci., July 1993. Ph.D. Thesis.

13. K. L. McMillan. Verification of an implementation of tomasulo's algorithm by compositional model checking. In *CAV '98*, number 1427 in LNCS, pages 100–21. Springer-Verlag, 1998.

14. J. U. Skakkabaek, R. B. Jones, and D. L. Dill. Formal verification of out-of-order execution using incremental flushing. In *CAV '98*, number 1427 in LNCS, pages 98–109. Springer-Verlag, 1998.

15. R. M. Tomasulo. An efficient algorithm for exploiting multiple arithmetic units. *IBM J. of Research and Development*, 11(1):25–33, Jan. 1967.

16. T. E. Truman. *A Methodology for the Design and Implementation of Communication Protocols for Embedded Wireless Systems*. PhD thesis, Dept. of EECS, University of CA, Berkeley, May 1998.

17. M. Velev and R. E. Bryant. Bit-level abstraction in the verification of pipelined microprocessors by correspondence checking. In *FMCAD '98*, number 1522 in LNCS, pages 18–35. Springer, 1998.

18. P. Wolper. Epressing interesting properties of programs in propositional temporal logic. In *13th ACM POPL*, pages 184–193, 1986.

Formal Verification of Designs with Complex Control by Symbolic Simulation

Gerd Ritter, Hans Eveking, and Holger Hinrichsen

Dept. of Electrical and Computer Engineering
Darmstadt University of Technology, D-64283 Darmstadt, Germany
ritter/eveking/hinrichsen@rs.tu-darmstadt.de

Abstract. A new approach for the automatic equivalence checking of behavioral or structural descriptions of designs with complex control is presented. The verification tool combines symbolic simulation with a hierarchy of equivalence checking methods, including decision-diagram based techniques, with increasing accuracy in order to optimize overall verification time without giving false negatives. The equivalence checker is able to cope with different numbers of control steps and different implementational details in the two descriptions to be compared.

1 Introduction

Verifying the correctness of designs with complex control is crucial in most areas of hardware design in order to avoid substantial financial losses. Detecting a bug late in the design cycle can block important design resources and deteriorate the time-to-market. Validating a design with high-confidence and finding bugs as early as possible is therefore important for chip design.

"Classical" simulation with test-vectors is incomplete since only a non-exhaustive set of cases can be tested and costly as well in the simulation itself as in generating and checking the tests. Formal hardware verification covers all cases completely, and gives therefore a reliable positive confirmation if the design is correct.

The formal verification technique presented in this paper uses *symbolic* simulation. Employing symbolic values makes the complete verification of *all* cases possible. One symbolically simulated path corresponds in general to a large number of "classical" simulation runs. During symbolic simulation, relationships between symbolic terms, e.g., the equivalence of two terms are detected and recorded. A given verification goal like the equivalence of the contents of relevant registers, is checked at the end of every symbolic path. If it is not demonstrated, more time-consuming but also more accurate procedures including decision diagram based techniques are used to derive undetected relationships.

Currently, the approach is used to check the *computational equivalence* of two descriptions but it is also applicable to the verification of properties which is planned for future research. Two descriptions are computationally equivalent if both produce the same final values on the same initial values relative to a set of relevant variables. For instance, the two descriptions in Fig. 1 are computationally equivalent with respect to the final value of the relevant variable r. The equivalence checker simulates symbolically all possible paths. False paths

Specification	Implementation
`x←a;`	`(x←a, y←b);`
`if opcode(m)=101;`	`z←opcode(m);`
`    then r← b ⊕ x;`	`if z=101`
`    else r← ¬b ∨ ¬x;`	`    then r← x ⊕ y;`
	`    else r← ¬(x ∧ y);`

Fig. 1. Example of two computationally equivalent descriptions

are avoided by making only consistent decisions at branches in the description. A case split is performed if a condition is reached which cannot be decided but depends on the initial register and memory values, e.g., `opcode(m)=101` in Fig. 1. The symbolic simulation of the specification and of the implementation is executed in parallel. The example in Fig. 1 requires, therefore, the symbolic simulation of two paths. Note that both symbolic paths represent an important number of "classical" symbolic runs.

Each symbolically executed assignment establishes an equivalence between the destination variable on the left and the term on the right side of an assignment. Additional equivalences between terms are detected during simulation. Equivalent terms are collected in equivalence classes. During the path search, only relationships between terms that are fast to detect or are often crucial for checking the verification goal are considered on the fly. Some functions remain uninterpreted while others are more or less interpreted to detect equivalences of terms, which is considered by unifying the corresponding equivalence classes.

Having reached the end of both descriptions with consistent decisions, a complete path is found and the verification goal is checked for this path, e.g., if both produce the same final values of r. This check is trivial for the *then*-branches in Fig. 1 since the equivalence of b⊕x and x⊕y is detected on the fly.

Using only a selection of function properties for equivalence detection which are fast to compute during the path search, we may fail to prove the equivalence of two terms at the end of a path, e.g., the equivalence of ¬b ∨ ¬x and ¬(x ∧ y) in the *else*-branches of Fig. 1 (application of De Morgan's Law on *bit-vectors*). In these cases the equivalence of the final values of r is checked using *decision diagrams* as described in section 5. If this fails, it is verified whether a false path is reached, since conditions may be decided inconsistently during the path search due to the limited equivalence detection. If the decisions are sound, the counterexample for debugging is reported. Relevant details about the symbolic simulation run can be provided since all information is available on every path in contrast to formula based verification. Our automatic verification process does not require insight of the designer into the verification process.

Some related work is reviewed in section 2. Section 3 shows the preparation of the initial data structure. Section 4 describes the path search itself, the symbolic simulation in the proper sense, and gives an overview of the algorithm. Section 5 presents the more powerful, but less time-efficient algorithms that are used if the verification goal can not be demonstrated at the end of a path. The use of *vectors* of OBDD's as canonizer is discussed and compared to other approaches. Experimental results are presented in section 6. Finally, section 7 gives a conclusion and directions for future work.

2 Related Work

Several approaches have been proposed for formal verification of designs with complex control. Theorem provers were used to verify the control logic of processors requiring extensive user guidance from experts which distinguishes the approach from our automated technique. Prominent examples are the verification of the FM9001 microprocessor [4] using Nqthm, of the Motorola CAP processor [5] using ACL2 and the verification of the AAMP5 processor using PVS [23]. [18] proposed an interesting approach to decompose the verification of pipelined processors in sub-proofs. However, the need for user guidance especially for less regular designs remains.

The idea of symbolic state-space representation has already been applied by [3, 12] for equivalence checking or by [10] for traversing automata for model checking. Their methods use decision diagrams for state-space representation, and are therefore sensitive to graph explosion. [24] developed an encoding technique for uninterpreted symbols, i.e., the logic of uninterpreted functions with equality is supported, and they abstracted functional units by memory models. The complexity of their simulation increases exponentially if the memory models are addressed by data of other memory models, therefore the processor they verified contains no data-memory or branching.

[11] proposed an approach to generate a logic formula that is sufficient to verify a pipelined system against its sequential specification. This approach has also been extended to dual-issue, super-scalar architectures [19, 9, 25] and with some limitations to out-of-order execution by using incremental flushing [22, 20]. SVC (the Stanford Validity Checker) [1, 2, 19] was used to automatically verify the formulas. SVC is a proof tool requiring for each theory to add that functions are canonizable and algebraically solvable, because every expression must have a unique representation. If a design is transformed by using theories, that are not fast to canonize/solve or that are not supported, SVC can fail to prove equivalence. Verifying bit-vector arithmetic [2], which is often required to prove equivalence in control logic design, is fast in SVC if the expressions can be canonized without slicing them into single bits, otherwise computation time can increase exponentially. Our approach does not canonize expressions in general. Only if the verification goal can not be demonstrated at the end of a path, formulas are constructed *using previously collected information* and are checked for equivalence using *vectors* of OBDD's. The efficiency of vectors of OBDD's in our application area is compared with SVC and *BMD's in section 5. Another problem of building formulas first and verifying them afterwards is the possible

term-size explosion which may occur if the implementation is given at the structural rt-level or even gate-level (see section 4 and 6). In addition, the debugging information given by a counter-example is restricted to an expression in the initial register values.

Symbolic simulation of executable formal specifications as described in [21] uses ACL2 without requiring expert interaction. Related is the work in [13], where pre-specified microcode sequences of the JEM1 microprocessor are simulated symbolically using PVS. Expressions generated during simulation are simplified on the fly. Multiple "classical" simulation runs are also collapsed but the intention of [21] is completely different since concrete instruction sequences at the machine instruction level are simulated symbolically. Therefore a *fast* simulation on *some* indeterminate data is possible for debugging a specification. Our approach checks equivalence for *every* possible program, e.g., not only some data is indeterminate but also the control flow. Indeterminate branches would lead in [21] to an exponentially grow of the output to the user. Furthermore, insufficient simplifications on the fly can result in unnecessary case splits or/and term-size explosion.

3 The Internal Data Structure

Our equivalence checker compares two *acyclic* descriptions at the rt-level. For many cyclic designs, e.g., pipelined machines the verification problem can also be reduced to the equivalence check of acyclic sequences, which is shown for some examples in section 6.

The inherent timing structure of the initial descriptions is expressed explicitly by indexing the register names. An indexed register name is called a *RegVal*. A new *RegVal* with an incremented index is introduced after each assignment to a register. An additional upper index s or i distinguishes the *RegVals* of specification and implementation. Only the initial *RegVals* as anchors are identical in specification and implementation, since the equivalence of the two descriptions is tested with regard to arbitrary but identical initial register values. Fig. 2 gives a simple example written in our experimental rt-level language LLS [15]. Parenthesis enclose synchronous parallel transfers. The sequential composition operator "." separates consecutive transfers. "Fictive" assignments (italic in Fig. 2) have to be generated, if a register is assigned in only one branch of an *if-then-else* clause in order to guarantee that on each possible path the sequence of indexing is complete. Checking computational equivalence consists of verifying that the final *RegVals*, e.g., adr_2 or pc_1 are equivalent to the according final *RegVals* in the other description. The introduction of *RegVals* makes all information about the sequential or parallel execution of assignments redundant which is, therefore, removed afterwards. Finally every distinct term and subterm is replaced for technical reasons by an arbitrary chosen distinct variable. A new variable is introduced for each term where the function type or at least one argument is distinct, e.g., pc_1+2 and pc_2+2 are distinguished.

Formula based techniques like SVC do not use distinct *RegVals*, because they

```
adr←pc;                                adr₁ ←pc;
ir←mem(adr);                           ir₁ ←mem(adr₁);
if ir[0:5]=000111                      if ir₁[0:5]=000111
then    (pc←pc+1, adr←ir[6:15]);       then    (pc₁ ←pc+1, adr₂ ←ir₁[6:15]);
        mi←mem(adr);                           mi₁ ←mem(adr₂);
        ac←ac+mi;                              ac₁ ←ac+mi₁;
else    pc←pc+2;                       else    pc₁ ←pc+2;
                                               adr₂ ←adr₁;
                                               mi₁ ←mi;
                                               ac₁ ←ac;
```

Fig. 2. Numbering registers after each new assignment

represent the changes of registers in the term-hierarchy implicitly. Expressing the timing structure explicitly has several advantages. Term size explosion is avoided, because terms can be expressed by intermediate *RegVals*, see also section 4. We do not loose information about intermediate relationships by rewriting or canonizing so that arbitrary additional techniques can be used at the end of a path to establish the verification goal. In addition, support of debugging is improved.

4 Symbolic Simulation

4.1 Identifying Valid Paths

One subgoal of our symbolic simulation method is the detection of equivalent terms.

Definition 1 (Equivalence of terms). *Two terms or RegVals are equivalent* $\equiv_{term}$, *if under the decisions* $C_0, ..., C_n$ *taken preliminary on the path, their values are identical for all initial RegVals. The operator* $\downarrow$ *denotes that each case-split, leading to one of the decisions* $C_0, ..., C_n$, *constrains the set of possible initial RegVals.*

$$term_1 \equiv_{term} term_2 \Leftrightarrow \forall RegVal_{initial} : (term_1 \equiv term_2) \downarrow (C_0 \wedge C_1 ... \wedge C_n)$$

Equivalent terms are detected along valid paths, and collected in *equivalence classes*. We write $term_1 \equiv_{sim} term_2$ if two terms are in the same equivalence class established during simulation. If $term_1 \equiv_{sim} term_2$ then $term_1 \equiv_{term} term_2$. Initially, each *RegVal* and each term gets its own equivalence class. Equivalence classes are unified in the following cases:

- two terms are identified to be equivalent;
- a condition is decided; if this condition is
  - a test for equality $a = b$, the equivalence class of both sides are unified *only if the condition is asserted*,
  - otherwise (e.g., $a < b$ or a status-flag) the equivalence class of the condition is unified with the equivalence class of the constant 1 or 0 if the condition is asserted or denied;

- after every assignment. Practically, this union-operation is significantly simpler because the equivalence class of the *RegVal* on the left-hand side of the assignment was not modified previously.

Equivalence classes permit to keep also track about unequivalences of terms:

Definition 2 (Unequivalence of terms). *Two terms or RegVals are unequivalent $\not\equiv_{term}$, if under the decisions $C_0, ..., C_n$ taken preliminary on the path their values are never identical for arbitrary initial RegVals:*

$$term_1 \not\equiv_{term} term_2 \Leftrightarrow \neg \exists RegVal_{initial} : (term_1 \equiv term_2) \downarrow (C_0 \wedge C_1 ... \wedge C_n)$$

We write $term_1 \not\equiv_{sim} term_2$ if two terms are identified to be $\not\equiv_{term}$ during simulation. Equivalence classes containing $\not\equiv_{sim}$ terms are unequivalent. This is the case

- if they contain different constants;
- if a condition with a test for equality (e.g., $a = b$) is decided to be false.

Implementing a path enumeration requires a decision algorithm each time an *if-then-else* is reached. Identifying *CondBits* in the conditions accelerates this decision procedure. *CondBits* replace

(a) tests for equality of bit-vectors, i.e., terms or *RegVals* (e.g., $r_3^s = x_2^s + y_1^s$);
(b) all terms with Boolean result (e.g., $r_3^s < x_2^s$) except the connectives below;
(c) single-bit registers (e.g., status-flags).

After the replacement, the conditions of the *if-then-else*-structures contain only propositional connectives (NOT, AND, IOR, XOR) and *CondBits*. Because identical comparisons might be done multiple times in one path, this approach avoids multiple evaluation of a condition by assigning one of three values (UNDEFINED, TRUE, FALSE) to the *CondBits*. If a *CondBit* appears the first time in a path, its value is UNDEFINED. Therefore, its condition is checked by comparing the equivalence classes of two terms or *RegVals*: In case (a), we have to check the terms on the left-hand and right-hand side, whereas in cases (b) and (c) the equivalence class of the term is compared to the equivalence class of the constant 1. There are three possible results:

i. The two terms to be compared are in the same equivalence class. Then the *CondBit* is asserted or TRUE *in this path* for arbitrary initial register values;
ii. The equivalence classes of the terms have been decided preliminary to be unequivalent or contain different constants. The *CondBit* is always denied or FALSE;
iii. Otherwise the *CondBit* may be true or false, depending on the initial register and memory values. Both cases have to be examined in a case split. Denying/Asserting a *CondBit* leads to a decided unequivalence/union-operation.

Fig. 3 (a) gives an example of the symbolic simulation of one path during the equivalence check of the example in Fig. 1. The members of the equivalence classes after every simulation step are given in Fig. 3 (b). Initially all terms and

(a)   **Specification**   **Implementation**

Specification	Implementation
S1 $x_1^s \leftarrow a$;	I1 $(x_1^i \leftarrow a,\ y_1^i \leftarrow b)$;
S2 if opcode(m)=101;	I2 $z_1^i \leftarrow$ opcode(m);
S3 then $r_1^s \leftarrow b \oplus x_1^s$	I3 if $z_1^i = 101$
else ...	I4 then $r_1^i \leftarrow x_1^i \oplus y_1^i$
	else ...

(b)

	x_1^s	a	x_1^i	y_1^i	b	z_1^i	opcode(m)	101	r_1^s	$b \oplus x_1^s$	$x_1^i \oplus y_1^i$	r_1^i
S1	x_1^s	a	x_1^i	y_1^i	b	z_1^i	opcode(m)	101	r_1^s	$b \oplus x_1^s$	$x_1^i \oplus y_1^i$	r_1^i
I1	x_1^s	a	x_1^i	y_1^i	b	z_1^i	opcode(m)	101	r_1^s	$b \oplus x_1^s$	$x_1^i \oplus y_1^i$	r_1^i
I2	x_1^s	a	x_1^i	y_1^i	b	z_1^i	opcode(m)	101	r_1^s	$b \oplus x_1^s$	$x_1^i \oplus y_1^i$	r_1^i
I3	x_1^s	a	x_1^i	y_1^i	b	z_1^i	opcode(m)	101	r_1^s	$b \oplus x_1^s$	$x_1^i \oplus y_1^i$	r_1^i
I4	x_1^s	a	x_1^i	y_1^i	b	z_1^i	opcode(m)	101	r_1^s	$b \oplus x_1^s$	$x_1^i \oplus y_1^i$	r_1^i
S3a	x_1^s	a	x_1^i	y_1^i	b	z_1^i	opcode(m)	101	r_1^s	$b \oplus x_1^s$	$x_1^i \oplus y_1^i$	r_1^i
S3b	x_1^s	a	x_1^i	y_1^i	b	z_1^i	opcode(m)	101	r_1^s	$b \oplus x_1^s$	$x_1^i \oplus y_1^i$	r_1^i

Fig. 3. Example of a simulation run

Reg Vals are in distinct equivalence classes. S1 is simulated first. When symbolic simulation reaches S2, the condition of S2 depends on the initial *Reg Vals* (case iii) and the simulation is blocked. Paths are searched simultaneously in specification and implementation. After the simulation of I1 and I2, I3 requires also a case split. Decisions in the normally more complex implementation have priority in order to facilitate a parallel progress. Therefore, a case split on the condition in I3 is performed. Only the case with the condition asserted is sketched in Fig. 3, where the equivalence classes of z_1^i and the constant 101 are then unified and I4 is simulated. The condition of S2 is now decidable in the given context since both sides of the condition are in the same *EqvClass* (case i), i.e., no additional case split is required. First the equivalence of $b \oplus x_1^s$ and $x_1^i \oplus y_1^i$ is detected (S3a) and then the assignment to r_1^s is considered (S3b). Finally r_1^s and r_1^i are in the same equivalence class, therefore, computational equivalence is satisfied at the end of this path. If they were in different equivalence classes, equivalence would be denied. Note that simultaneous progress in implementation and specification avoids simulating S1 again for the denied case.

4.2 Identifying Equivalent Terms

Ideally, all $\equiv_{term}$ equivalent terms and *Reg Vals* are in the same equivalence class, but it is too time consuming to search for all possible equivalences on the fly. Therefore, no congruence closure is computed *during the path search*, i.e., building eventually incomplete equivalence classes is accepted in favor of a fast path search. If congruence closure or undetected equivalences are required to check the verification goal, the algorithms described in section 5 are used.

In order to speed up the path search, the following simplifications are made with respect to completeness of equivalence detection:

- Some functions, e.g., user-defined functions are always treated as uninterpreted.
- Only fast to check or "crucial" properties of interpreted functions are considered. Some examples are:
  - If a bit or a bit-vector of a term or a *RegVal* is selected which is in an equivalence class with a constant, the (constant) result is computed (e.g., from $IR\equiv_{sim} 011$ follows $IR[1]\equiv_{sim} 1$). If at least one argument of a Boolean function is $\equiv_{sim}$ to 1 or 0 then it is checked whether the function is also $\equiv_{sim}$ to one of these constants.
  - Functions representing multiplexers, i.e., structures where N control signals select one of $M = 2^N$ data-words, have to be interpreted. A transformation into an adequate *if-then-else*-structure is feasible, but blows up the descriptions. Note that, therefore, multiplexers can lead to term-size explosion, if the overall formula is build in advance and verified afterwards (e.g., if a big ROM is used). This can be avoided in symbolic simulation by using intermediate carriers and evaluating expressions on the fly.
  - Symmetric functions are equivalent, if every argument has an equivalent counter-part (e.g., $(a \equiv d) \wedge (b \equiv c) \Rightarrow (a + b) \equiv (c + d)$). Note that preliminary sorting of the arguments can not always tackle this problem because different terms can be assigned to *RegVals*.
- The transformation steps done during preprocessing preserve the timing structure. In general, equivalence of the arguments of two terms is already obvious, when the second term is found on the path. Therefore, it is sufficient to check only at the first occurrence of a term whether it is equivalent to terms previously found.
- In most cases the equivalence of terms can be decided by simply testing if the arguments are $\equiv_{sim}$ or $\not\equiv_{sim}$ which avoids the expansion of the arguments.
- Equivalence checking for a term is stopped after the first union operation, since all equivalent terms are (ideally) already in the same equivalence class.

This procedure fails in two cases:

- Equivalence cannot be detected by the incomplete function interpretation.
- A decision about the relationship of the initial *RegVals* is done *after* two terms are found on the path and equivalence of the terms is given only considering this decision.

The last situation occurs especially in the case of operations to memories. Similar to [11, 1], two array operations are used to model memory access: `read(mem,adr)` reads a value at the address `adr` of memory `mem` while `store(mem,adr,val)` stores `val` corresponding without changing the rest of the memory. In the example of Fig. 4, the order of the `read` and the `store` operations is reversed in the implementation. Thus, `val` is forwarded if the addresses are identical. The problem is to detect, that in the opposite case the final values of `x` are identical, which is only obvious *after* the case split (setting $adr1 \not\equiv_{sim} adr2$) and not already after the assignments to `x`. The example indicates, that it is important to check `read`

Specification	Implementation

```mem_1^s[adr1]←val;```
```x_1^s ←mem_1^s[adr2];```
```z_1^s ← ⎡x_1^s⎤+y;```
	```x_1^i ←mem[adr2];```
	```mem_1^i[adr1]←val;```
	```if adr1=adr2```
	``` then z_1^i ←val+y;```
	``` else z_1^i ← ⎡x_1^i⎤+y;```

Fig. 4. Forwarding example

and **store**-terms whenever the equivalence classes of the related addresses are modified. Note that our address comparison uses only the information of the *EqvClasses* and does not evaluate boolean expressions as in [24].

4.3 Overview of the Algorithm

Lines 3 to 10 in Fig. 5 summarize the path search. For every case split due to a condtition **to_decide**, first the denied case is examined (line 9) while the asserted case is stored in **rem_cases** (line 8). Initially **rem_cases** contains the whole specification and implementation with a dummy-condition (line 1). Note that only those parts of the descriptions, that are not simulated yet in this path, are examined after case splits, i.e., **remain(act_case**$_{spec/impl}$**)** (line 8). Lines 12 to 22 describe the case where computational equivalence is not reported at the end of a path (line 11), and are explained in the next section in full detail.

5 Examining Differences of the Descriptions

5.1 Overview

In the first hierarchy-level of the checking algorithm, arbitrary function properties can be considered in order to detect term-equivalence. Adding the check of function properties during the path search is a trade-off: the accuracy increases, therefore less false negatives to be checked afterwards occur. But the additional checks can be time-consuming since every time a term is found a check is done which may actually be necessary only in few cases or not at all.

According to *Algorithm Equivalence Check*, if the verification goal is not given in a path (line 11), then the first step is to consider additional function properties which are less often necessary or more time consuming to check.

If then the verification goal is not yet reported for all pairs of final *RegVals* an attempt is made to decide the equivalence. Formulas are built considering knowledge about path-dependent equivalence/unequivalence of intervenient terms which are sufficient for the equivalence of the final *RegVals* (line 14). A pre-check follows, which applies some logic minimization techniques and which checks whether a formula was built previously and stored in a hash-table. Before hashing, all *RegVals* and cut-points (see below) in the formulas are replaced in order of their appearance in the formula by auxiliary variables T1, T2,...,Tn,

```
INPUT spec, impl;
 1. rem_cases := {(dummy_cond,spec,impl)};
 2. WHILE rem_cases ≠ ∅ DO
 3.    act_case := pop(rem_cases);
 4.    assert(act_case_{to_decide});
 5.    REPEAT
 6.        to_decide := simulate_parallel(act_case);
 7.        IF to_decide THEN
 8.            push(to_decide,remain(act_case_{spec}),remain(act_case_{impl}))
                  rem_cases;
 9.            deny(to_decide);
10.    UNTIL to_decide not found;
11.    IF ∃i : R^{spec}_{i_final} ≢_{sim} R^{impl}_{i_final} THEN
12.        check_additional_properties;
13.        IF ∃i : R^{spec}_{i_final} ≢_{sim} R^{impl}_{i_final} THEN
14.            ∀i : (R^{spec}_{i_final} ≢_{sim} R^{impl}_{i_final}) : LET F_j ⇒ R^{spec}_{i_final} ≡_{sim} R^{impl}_{i_final} ;
15.            IF ∃j : F_j ≡ TRUE THEN
16.                mark_new_relations;
17.                return_in_path;
18.            ELSIF ∃k : inconsistent(decision_k)
19.                mark_new_relations;
20.                return_to_wrong_decision;
21.            ELSE report_debug_information;
22.                RETURN(FALSE);
23. ENDWHILE;
24. RETURN(TRUE);
```

Fig. 5. Algorithm Equivalence Checking

because the same formula may appear with only different *RegVals* or cut-points with regard to a previously computed formula. New formulas are checked using binary decision diagrams. This is the first time a canonical form is built.

If none of the formulas is satisfiable, all decided *CondBits*, i.e., conditions for which a case-split was done, are checked in order of their appearance to search for a contradictory decision due to the incomplete equivalence detection on the fly. Using the information about the equivalence classes again facilitates considerably building the required formulas.

If at least one formula is valid (line 15) or if a contradictory decision has been detected (line 18), the path is backtracked and the relationship is marked so that it is checked during further path search on the fly. This is done since the probability is high, that also in other paths the more time consuming algorithms are invoked unnecessarily again due to this relationship.

Otherwise the descriptions are not equivalent and the counterexample is reported for debugging (line 21). A complete error trace for debugging can be generated since all information about the symbolic simulation run on this path is available. For example, it turned out that a report is helpful which summarizes different microprogram-steps or the sequence of instructions carried through the pipeline registers. Note that if formulas were canonized only a counterexample in the initial *RegVals* would be available. Simulation-information can also be useful if the descriptions are equivalent. For instance, a report of never taken branches *in one if-clause* indicates redundancy which may not be detected by logic minimizers.

5.2 Building Logic Formulas without Sequential Content

For each unsatisfied goal (equivalence of two *RegVals*), a formula is built. The knowledge about equivalence/unequivalence of terms, which is stored in the equivalence classes, is used in order to obtain formulas which are easy to check. It is possible to obtain formulas in terms of the initial *RegVals* without term-size explosion by backward-substitution because a specific path is chosen. In many cases, however, less complex formulas can be derived by using intermediate cut points already identified to be equivalent in specification and implementation during symbolic simulation. A greedy algorithm guides the insertion of cut-points in our prototype version. Validating the formulas may be infeasible if these cut-points are misplaced or hide necessary function properties. Therefore, a failed check is repeated without cut-points.

5.3 Checking Formulas by means of Decision Diagrams

A *Multiple-Domain Decision Diagram Package* (TUDD-package) [16, 17] developed at TU Darmstadt with an extension for vectors of OBDD's [6] is used to prove the formulas. Another possibility is to use word-level decision diagrams like *BMD's [7, 8]. In practical examples of control logic, bit-selection functions are used frequently, either explicitly, e.g., R[13:16], or implicitly, e.g., storing the result of an addition in a register without carry. Using *BMD's, terms are represented by one single *BMD. Bit-selection, therefore, requires one or two *modulo*-operations which are worst-case exponentially with *BMD.

Bit-selection is quasi for free, if terms are expressed as *vectors* of OBDD's, where each graph represents one bit. Bit-selection can then be done by simply skipping the irrelevant bits, i.e., the corresponding OBDD's and by continuing the computation with the remaining OBDD's. Checking equivalence just consists of comparing each bit-pair of the vectors.

The initial *RegVals* and the cut-points are represented by a vector of decision diagrams, where each of the diagrams represents exactly one bit of the *RegVal* or cut-point. There is no fixed assignment of vectors of decision diagrams to initial *RegVals*/cut-points, but association is done dynamically after a formula is built. Decision diagrams with a fixed variable ordering (interleaved variables) are built during pre-processing since reordering would be too time consuming.

All formerly applied algorithms are (fairly) independent of the bit-vector length. Results obtained during symbolic simulation are used to simplify formulas before OBDD-vector construction. But even without simplification even large bit-vectors can be handled by OBDD-vectors in acceptable computation time. In [2] the results of SVC on five bit-vector arithmetic verification examples are compared to the results of the *BMD package from Bryant and Chen and also to

Laurent Arditi's *BMD [2] implementation which has special support for bit-vector and Boolean expressions. We verified these examples also with OBDD-vectors. Tab. 1 summarizes the results. All our measurements are on a Sun Ultra II with 300 MHz. Various orderings of the variables for our *BMD-measurements are used. The line DM contains the verification results for a bit-wise application of De Morgan's law to two bit-vectors a and b, i.e., $\overline{a_0 \wedge b_0}\#...\#\overline{a_n \wedge b_n} \equiv (\overline{a_0} \vee \overline{b_0})\#...\#(\overline{a_n} \vee \overline{b_n})$, and the ADD-example is the verification of a ripple-carry-adder. Note that the input is also one *word* for the two last examples and not a vector of inputs (otherwise *BMD-verification is of course fast since no slicing or modulo operation is required). The inputs may represent some intermediate cut-points for which, e.g., the *BMD is already computed.

Obviously, *BMD-verification suffers from the modulo-operations in the examples. According to [2], the results of example 1 to 4 are independent of the bit-vector length for SVC, but the verification times with OBDD-vectors are also acceptable even for large bit-vectors. These times can be reduced especially for small bit-vectors by optimizing our formula parsing. In example 5, SVC ends

	SVC[1]		*BMD Bryant/Chen[1]		*BMD Arditi[1]		OBDD-vector TUDD				
	200MHz Pentium		200MHz Pentium		300MHz UltraSparc 30		300MHz Sun Ultra II				
Bits	16	32	16	32	16	32	16	32	64	128	256
1	N/A	0.002	N/A	N/A	N/A	0.04	0.14	0.27	0.38	0.68	1.38
2	N/A	0.002	N/A	N/A	N/A	1.10	0.13	0.20	0.25	0.44	0.93
3	0.002	0.002	265.0	>500	0.07	0.18	0.21	0.32	0.51	0.95	1.95
4	0.002	0.002	26.4	>500	0.72	8.79	0.24	0.40	0.71	1.53	4.38
5	0.111	0.520	22.7	>500	0.39	3.78	0.14	0.21	0.31	0.57	1.15
	Measured at TUDD		Measured with TUDD *BMD-package								
Bits	16	32	16	32	64		16	32	64	128	256
DM	>5min		>5min				0.12	0.22	0.28	0.48	1.03
ADD	-[2]		5.19	37.2	282.7		0.21	0.31	0.48	0.98	1.90

[1] Measurements reported in [2].
[2] 2 Bit: 1.01s; 4 Bit: 9.47s; 5 Bit 44.69s; Verification with more than 5 Bit was not feasible with the current version of SVC.

Table 1. Comparison of SVC, *BMD and OBDD-Vectors. Times are in seconds.

up slicing the vector and thus the execution time depends on the number of bits and shows, therefore, a significant increase, whereas the computation time for OBDD-vectors increases only slightly. The increase in *this* example may be eliminated in a future version of SVC [2], but the general problem is that slicing a vector has to be avoided in SVC. This can be seen for the examples DM and ADD, where verification is only practical with OBDD-vectors.

Note that functions that are worst-case exponentially with OBDD's or have no representation, e.g., multiplication, are only problematic in the rare cases where special properties of the functions are necessary to show equivalence. Normally, the terms are replaced by cut-points during the formula-construction since we use information from the simulation-run.

6 Experimental Results

DLX-Processor Descriptions

Two implementations of a subset of the DLX processor [14] with 5-pipeline-stages have been verified, the first from [18], initially verified in [11], and a second one designed at TU Darmstadt. The latter contains more structural elements, e.g., the multiplexers and corresponding control lines required for forwarding are given.

For both descriptions, acyclic sequences are generated by using the flushing approach of [11]; i.e., execution of the inner body of the pipeline loop followed by the flushing of the pipeline is compared to the flushing of the pipeline followed by one serial execution. Different to [11] (see also [9]), our flushing schema guarantees, that one instruction is fetched *and* executed in the first sequence, because otherwise it has to be communicated between specification and implementation if a instruction has to be executed in the sequential processor or not (e.g., due to a load interlock in the implementation). [9] describes this as keeping implementation and specification in sync; using their flushing approach with communication reduces our number of paths to check and verification time, too. Verification is done automatically, only the (simple) correct flushing schema, guaranteeing that one instruction is fetched and executed, has to be provided by the user. In addition, some paths are collapsed by a simple annotation that can be used also for other examples. Forwarding the arguments to the ALU is obviously redundant, if the EX-stage contains a NO_OP or a branch. The annotation expresses, that in these cases the next value of these arguments can be set to a distinct unknown value. The verification remains *complete*, because the equivalence classes of the *RegVals* to check would always be different, if one of these final *RegVals* depends on such a *distinct* unknown value. Note that verification has been done for both cases also without this annotation, but with $\approx$ 90% more paths to check. Two errors introduced by the conversion of the data-format used

Version	paths	aver. time per path	total time
DLX from [18]	310,312	12.6 ms	1h 5min 13s
DLX with multiplexers	259,221	19.5 ms	1h 24min 14s

Table 2. Verification results for DLX-implementations

by [18] and several bugs in our hand crafted design have been detected automatically by the equivalence checker. Verification results of the correct designs are given in Tab. 2. Measurements are on a Sun Ultra II with 300 MHz. Note that the more detailed and structural description of the second design does not blow up verification time: the average time per path increases acceptable, but the number of paths remains nearly the same (even decreases slightly due to a minor different realization of the WB-stage).

Verifying the DLX-examples does not require the more complex algorithms, especially the decision diagrams, because with exception of the multiplexers in the second design, the pipelined implementation can be derived from a sequential specification using a small set of simple transformations (in [15] a formally correct automatic synthesis approach for pipelined architectures using such a set is presented). Verifying examples like the DLX is not the main intention of our approach since the capabilities of the equivalence checker are only partly used, but it demonstrates that also control logic with a complex branching can be verified by symbolic simulation.

Microprogram-Control with and without Cycle Equivalence

In this example, two behavioral descriptions of a simple architecture with microprogram control are compared to a structural implementation. In both behavioral descriptions, the microprogram control is performed by simple assignments and no information about the control of the datapath-operations, e.g., multiplexer-control, is given. The structural description of the machine compromises an ALU, 7 registers, a RAM, and a microprogram-ROM. All multiplexers and control lines required are included. The two behavioral descriptions differ in the number of cycles for execution of one instruction:

- The first is cycle-equivalent to the structural description; i.e., all register-values are equivalent in every step. Generating the finite sequences consists of simply comparing the loop-bodies describing one micro-program step.
- The second is less complex than the first and more intuitive for the designer. It contains an instruction fork in the decode phase. No cycle equivalence is given, therefore, the sequences to be compared are the complete executions of one instruction. The only annotation of the user is the constant value of the microprogram counter, that indicates the completion of one instruction.

The ROM is expressed as one multiplexer with constant inputs. In this example, the read/write-schema used also in SVC would not work, since the ROM has constant values on all memory-places. The ROM-accesses and the other multiplexers would lead to term-size explosion if they are interpreted as functions (canonizing!) as well as if they are considered as *if-then-else*-structures, since symbolic simulation goes over several cycles in this example. Results are given in Tab. 3.

Example	paths*	ext. checks	false paths	time
with cycle equivalence	291	56	39	24.53s
different number of cycles	123	41	16	19.58s

* including false paths

Table 3. Verification of microprogram-controller

Measurements are on a Sun Ultra II with 300 MHz, verification times include the construction of decision diagrams. The third column indicates how often the extended checks of section 5 are used either to show equivalence or to detect an inconsistent decision, i.e., one of the false path reported in the fourth column is reached. The in principle more difficult verification without cycle equivalence requires less paths since the decisions in the behavioral description determines the path in the structural description. Note that again no insight into the automatic verification process is required.

7 Conclusion and Future Work

A new approach for equivalence checking of designs with complex control using symbolic simulation is presented. The number of possible paths to simulate can be handled even for complex examples since symbolic values are used. All indeterminate branches, that depend on initial register values, are considered by case splits to permit a complete verification for an arbitrary control flow.

Our equivalence detection on the fly is not complete to permit a fast simulation. If the verification goal is not given at the end of a path, additional and more powerful algorithms including decision-diagram based techniques are used to review the results of the simulation run.

The approach is flexible to integrate various equivalence detection algorithms which are applied either finally or during simulation on the fly. There are no special requirements like canonizability to integrate new theories since we keep track about equivalences of terms by assembling them in equivalence classes. Therefore, all information about the simulation run is available at the end of a path. This is also useful for debugging: simulation *"is a natural way engineers think"*.

First experimental results demonstrate the applicability to complex control logic verification problems. The equivalence checker supports different number of control steps in specification and implementation. Structural descriptions with implementational details can be compared with their behavioral specification. By using intermediate carriers, term-size explosion is avoided which can occur in formula-based techniques when implementational details are added.

The approach has so far only been used to check the computational equivalence of two descriptions. An application to designs, where relationships between intermediate values or temporal properties have to be verified, is planned in future work. Another topic is to parallelize the verification on multiple workstations in order to reduce the overall computational time.

Acknowledgement

The authors would like to thank the anonymous reviewers for helpful comments.

References

[1] C. W. Barrett, D. L. Dill, and J. R. Levitt: Validity checking for combinations of theories with equality. In Proc. FMCAD'96, Springer LNCS 1166, 1996.

[2] C. W. Barrett, D. L. Dill, and J. R. Levitt: A decision procedure for bit-vector arithmetic. In Proc. DAC'98, 1998.

[3] D. L. Beatty and R. E. Bryant: Formally verifying a microprocessor using a simulation methodology. In Proc. DAC'94, 1994.

[4] B. Brock, W. A. Hunt, and M. Kaufmann: The FM9001 microprocessor proof. Technical Report 86, Computational Logic Inc., 1994.

[5] B. Brock, M. Kaufmann, and J. S. Moore: ACL2 theorems about commercial microprocessors. In Proc. FMCAD'96, Springer LNCS 1166, 1996.

[6] R. E. Bryant: Graph-based algorithms for Boolean function manipulation. In IEEE Trans. on Computers, Vol. C-35, No. 8, pages 677-691, 1986.

[7] R. E. Bryant and Y.-A. Chen: Verification of arithmetic functions with binary moment diagrams. Technical Report CMU-CS-94-160, Carnegie Mellon University, 1994.

[8] R. E. Bryant and Y.-A. Chen: Verification of arithmetic circuits with binary moment diagrams. In Proc. DAC'95, 1995.

[9] J. R. Burch: Techniques for verifying superscalar microprocessors. In Proc. DAC'96, 1996.

[10] J. R. Burch, E. Clarke, K. McMillan, and D. Dill: Sequential circuit verification using symbolic model checking. In Proc. DAC'90, 1990.

[11] J. R. Burch and D. L. Dill: Automatic verification of pipelined microprocessor control. In Proc. CAV'94. Springer LNCS 818, 1994.

[12] O. Coudert, C. Berthet, and J.-C. Madre: Verification of synchronous sequential machines based on symbolic execution. In Proc. Automatic Verification Methods for Finite State Systems, Springer LNCS 407, 1989.

[13] D. A. Greve: Symbolic simulation of the JEM1 microprocessor. In Proc. FM-CAD'98, Springer LNCS 1522, 1998.

[14] J. L. Hennessy, D. A. Patterson: Computer architecture: a quantitative approach. Morgan Kaufman, CA, second edition, 1996.

[15] H. Hinrichsen, H. Eveking, and G. Ritter: Formal synthesis for pipeline design. In Proc. DMTCS+CATS'99, Auckland, 1999.

[16] S. Höreth: Implementation of a multiple-domain decision diagram package. In Proc. CHARME'97, pp. 185-202, 1997.

[17] S. Höreth: Hybrid Graph Manipulation Package Demo. URL : http://www.rs.e-technik.tu-darmstadt.de/~sth/demo.html, Darmstadt 1998.

[18] R. Hosabettu, M. Srivas, and G. Gopalakrishnan: Decomposing the proof of correctness of pipelined microprocessors. In Proc. CAV'98, Springer LNCS 1427, 1998.

[19] R. B. Jones, D. L. Dill, and J. R. Burch: Efficient validity checking for processor verification. In Proc. ICCAD'95, November 1995.

[20] R. B. Jones, J. U. Skakkebæk, and D. L. Dill: Reducing manual abstraction in formal verification of out-of-order execution. In Proc. FMCAD'98, Springer LNCS 1522, 1998.

[21] J. S. Moore: Symbolic simulation: an ACL2 approach. In Proc. FMCAD'98, Springer LNCS 1522, 1998.

[22] J. U. Skakkebæk, R. B. Jones, and D. L. Dill: Formal verification of out-of-order execution using incremental flushing. In Proc. CAV'98, Springer LNCS 1427, 1998.

[23] M. Srivas and S. P. Miller: Applying formal verification to a commercial microprocessor. In Computer Hardware Description Language, August 1995.

[24] M. N. Velev and R. E. Bryant: Bit-level abstraction in the verification of pipelined microprocessors by correspondence checking. In FMCAD'98, Springer LNCS 1522, 1998.

[25] P. J. Windley and J. R. Burch: Mechanically checking a lemma used in an automatic verification tool. In Proc. FMCAD'96, November 1996.

Hints to Accelerate Symbolic Traversal*

Kavita Ravi[1] and Fabio Somenzi[2]

[1] Cadence Design Systems
kravi@cadence.com
[2] Department of Electrical and Computer Engineering,
University of Colorado at Boulder
Fabio@Colorado.EDU

Abstract. Symbolic model checking is an increasingly popular debugging tool based on Binary Decision Diagrams (BDDs). The size of the diagrams, however, often prevents its application to large designs. The lack of flexibility of the conventional breadth-first approach to state search is often responsible for the excessive growth of the BDDs. In this paper we show that the use of *hints* to guide the exploration of the state space may result in orders-of-magnitude reductions in time and space requirements. We apply hints to *invariant checking*. The hints address the problems posed by difficult image computations, and are effective in both proving and refuting invariants. We show that good hints can often be found with the help of simple heuristics by someone who understands the circuit well enough to devise simulation stimuli or verification properties for it. We present an algorithm for guided traversal and discuss its efficient implementation.

1 Introduction

Great strides have been made in the application of formal methods to the verification of hardware The most successful technique so far has been model checking [19]. Model checking exhaustively explores the state space of a system to ascertain whether it satisfies a property expressed in some temporal logic.

Given the exponential growth of the number of states with the number of state variables, several techniques have been devised to turn model checking into a practical approach to verification. The most fundamental technique is abstraction: Verification is attempted on a simplified model of the system, which is meant to preserve the features related to the property of interest [11, 16]. Compositional verification [1, 21, 15] in particular applies abstraction to hierarchically defined systems so that the environment of each subsystem is summarized by the properties that it guarantees.

Abstraction and compositional verification can be used to reduce the cost of model checking experiments. The modeling effort required of the user, however, grows with the degree of abstraction. Other techniques are therefore needed to increase the intrinsic efficiency of the state exploration. Explicit model checkers use clever hashing schemes and external storage. Implicit model checkers, on the other hand, rely on Binary Decision Diagrams (BDDs) [5] to represent very large sets of states and transitions. BDD-based model checkers can sometimes analyze models with over 5000 state variables

* This work was supported in part by SRC contract 98-DJ-620.

without resorting to abstraction. In other cases, however, even models with 40 state variables prove intractable. This large variability is ultimately due to the fact that only functions with high degrees of regularity possess compact BDDs. In many instances, however, the occurrence of large BDDs in model checking experiments is an artifact of the specific state search strategy and can be avoided. The analysis of the conditions that cause the occurrence of large BDDs during the exploration of the state space, and the description of a strategy to prevent those conditions are the topics of this paper.

Our discussion will focus on the form of model checking known as *invariant checking*, which consists of proving that a given predicate holds in all reachable states of a system. This form of verification is the most commonly applied in practice. We aid invariant checking by guiding reachability analysis with *hints*. Hints specifically address the computational bottlenecks in reachability analysis, attempting to avoid the memory explosion problem (due to large BDDs) and accelerate reachability analysis. Hints may depend on the property to be verified, but they are successful at speeding up the proof as well as the refutation of invariants.

Hints are applied by constraining the transition relation of the system to be verified. They specify possible values for (subsets of) the primary inputs and state variables. The constrained traversal of the state space proceeds much faster than the standard breadth-first search (BFS), because the traversal with hints is designed to produce smaller BDDs. Once all states reachable following the hints have been visited, the system is unconstrained and reachability analysis proceeds on the original system. Given enough resources, the algorithm will therefore search all reachable states, unless a state that violates the invariant is found. In model checking, constraints sometimes arise from assumptions on the environment. Consequently, these constraints need to be validated on the environment. In our algorithm, since hints are eventually lifted, they leave no proof obligations.

Hints are reminiscent of simulation stimuli, but the distinguishing feature of our application is that the state space is exhaustively explored. Simulation is a partial exploration of the state space and can only disprove invariants—both concrete and symbolic simulation suffer this limitation. Using hints is similar to applying stimuli to the system, but temporarily. Moreover, hints are designed to make symbolic computations with BDDs easier, whereas simulation stimuli are only chosen to exercise the circuit with respect to a property.

It has been observed in the reachability analysis of many systems that the BDDs at completion are smaller than the intermediate ones. The BFS curve of Fig. 2 illustrates this phenomenon for the traversal of the circuit Vsa (described in Section 6). The constrained traversal sidesteps the intermediate size explosion. When the hints are removed, the unconstrained traversal is expected to be past the intermediate size explosion. Our algorithm also takes advantage of the information gathered in the constrained traversal.

Some invariants can be checked directly by induction (if the predicate holds in all initial states and in all successors of the states where it holds). In general, however, proving invariants entails performing reachability analysis. Our algorithm is designed to benefit the most general case. It is compatible with abstraction techniques like *localization reduction* [16], which is particularly useful when the invariants describe local properties of the system.

The algorithmic core of BDD-based invariant checking is the computation of least fixpoints. Our guided search approach is applicable in general to least fixpoint computations and hence to a wider class of verification procedures, like those of [14, 3]. The rest of this paper is organized as follows. In Section 2 we discuss background material and define notation. In Section 3 we discuss the main computational problems in image computation—the step responsible for most resource consumption in invariant checking. In Section 4 we introduce guided symbolic traversal by discussing case studies and presenting our algorithm. Section 5 is devoted to a review of the relevant prior art in relation with our new approach. Experimental results are presented in Section 6 and conclusions are drawn in Section 7.

2 Preliminaries

Binary Decision Diagrams (BDDs): (BDDs) represent boolean functions. A BDD is obtained from a binary decision tree by merging isomorphic subgraphs and eliminating redundant nodes. For a given variable order, this reduction process leads to a canonical representation. Therefore, equivalence tests are efficient, and, thanks to the extensive use of memoization, the algorithms that operate on BDDs are fast. Large sets can be manipulated via their characteristic functions, which in turn can be represented by BDDs. BDDs are used to represent sets of states and transitions in symbolic verification.

Though almost all functions have optimal BDDs of size exponential in the number of variables, the functions encountered in several applications tend to have well-behaved BDDs. This is not necessarily the case in model checking: Sometimes the sets of states or transitions that a model checker manipulates are irregular and have large BDDs.

The sizes of the BDDs for many functions depend critically on the variable orders. Good variable orders are hard to predict *a priori*. Heuristic algorithms like sifting [27] have been devised to dynamically change the order and reduce BDD sizes during the computation. However, the high cost of the reordering and short-sighted optimization (which may be bad for a later stage) impedes its effectiveness.

Finite State Machine: A sequential circuit is modeled as a finite state machine $(S, \Sigma, O, T, \lambda, I)$, where S is the set of states, Σ is the input alphabet, O is the output alphabet, $T = S \times \Sigma \times S$ is the state transition relation, $Z = S \times \Sigma \times O$ is the output relation, $I \subseteq S$ is the set of initial states.

S is encoded with a set of variables x. T is encoded with three sets of variable—the set of present state variables x (same as S), a set of next state variables y, and a set of primary input variables w. $T(x, w, y) = 1$ if and only if a state encoded by y is reached in one step from a state encoded by x under input w.

For deterministic circuits, the transition relation is customarily constructed as a product of the *bit relations* of each state variable, $\prod_{i=1}^{n}(y_i \equiv \delta_i(x, w))$, where y_i is the next state variable corresponding to the state variable x_i, δ_i is the next state function of the i-th state variable. When T is represented by a single BDD, it is said to be *monolithic*. The monolithic transition relation may have a large BDD, even for mid-size circuits. In practice, a more efficient *partitioned* representation [7, 28] is used as an implicit conjunction of blocks of bit relations.

$$T(w, x, y) = \prod_{i=0}^{m} T_i(w, x, y),$$

Image Computation: With BDDs it is possible to compute the successors (predecessors) of a set of states symbolically (without enumerating them). This computation is referred to as an *image (preimage)* computation, and is defined as follows:

$$\text{Image}(T, C) = [\exists_{x,w} T(x, w, y) \wedge C(x)]_{y=x},$$

where $\text{Image}(T, C)$ is the image of a set of states $C(x)$ (expressed in terms of the x variables) under the one-step transition relation $T(x, w, y)$. Further details are discussed in Section 3.

BFS: The traversal of the state space of a circuit is accomplished by a series of image computations, starting from the initial states and continuing until no new states are acquired. This is a BFS of the state graph: All the states at a given minimum distance from the initial states are reached during the same image computation. Some states are considered multiple times: A commonly applied optimization is to compute only the image of the states that were first reached during the previous iteration. This optimization still guarantees a BFS of the graph. Such a set of states is called the *frontier*.

3 Image Computation

And-Exists: For a monolithic relation, image computation is carried out with a single BDD operation called *And-Exists*. *And-Exists*(f, g, v) produces $\exists_v(f \wedge g)$. The complexity of this operation has no known polynomial bound [19]. In the worst case, the known algorithms are exponential in the number of variables in the first two operands.

We adopt the approach of Ranjan *et al.*, [23] for image computation. A linear order is determined for the blocks of the transition relation, $T_i(x, y, w)$. A series of *And-Exists* operations are performed. At each step, the operands are the current partial product, next block of the transition relation $T_j(w, x, y)$ and the quantifiable variables. The initial partial product is set to $C(x)$.

Quantification Schedule: *And-Exists* has the property that $\exists_v(f \wedge g) = g \wedge \exists_v f$, if g is independent of v. This property can be used to quantify x and w variables earlier than the last *And-Exists* in image computation to reduce the worst-case sizes by reducing the peak number of variables. This has been observed to produce significant practical gains. The application of this property is called *early quantification*. Several researchers have proposed heuristics to optimally order the operands of the *And-Exists*, $T_i(w, x, y)$, $C(x)$ and scheduling the quantification of variables [13, 23]. In Ranjan's approach, the lifespan of variables during image computation is minimized—variables are quantified out as early as possible and introduced into the partial product as late as possible. However, the proposed techniques are far from being optimal.

Issues in Image Computation: Image computation comprises the largest fraction of traversal time, typically 80-90%. Most of the BDD size explosion problems in symbolic traversal are observed during image computation. The main causes for size explosion are complex functions for the next state relations, insufficient early quantification, and conflicting variable ordering requirements.

Three sets of variables—x, y, and w are involved in image computation. Several ordering heuristics have been experimented with—the most common one being the interleaving of x and y variables (as done in VIS [4]). This controls the intermediate

BDD sizes in the substitution of the y variables with the x variables at the end of image computation but may be sub-optimal for the partial products.

In the monolithic transition relation, the interleaving order may cause a size blowup since all the x and y variables interact in the same BDD. In a partitioned transition relation, normally the size stays under control since every block $T_i(x, w, y)$ has few variables and the blocks are kept separate. However, during image computation, depending on the set whose image is computed and the quantification schedule, all the x and y variables may interact. The situation may arise due to two reasons—the heuristics used may not find the optimal quantification schedule or, a good quantification schedule may not exist when many partitions depend on most of the variables.

Example: An ALU is an example of function that causes large intermediate sizes. ALUs usually have operations such as ADD, SHIFT that make the output bits depend on many of the input bits in complex functions, resulting in a bad quantification schedule and large BDDs. In Section 4 we show how the use of hints can address this problem.

4 Guided Symbolic Traversal Using Hints

We address the main problem of traversal—large intermediate BDD sizes during image computation. We propose to simplify the transition relation in order to reduce these sizes. The simplification addresses the issues discussed in Section 3—reducing the peak number of variables involved in image computation and (or) reducing the complexity of the transition relation by removing some transitions.

The simplification is achieved by the use of *hints*. As explained in Section 1, applying hints to traversal is equivalent to constraining the environment or the operating mode of the FSM. Breath-first search (BFS) explores all possible states at each iteration of traversal i.e., all possible transitions from the current set of states are considered. With *hints*, only a subset of these transitions considered.

Hints are expressed as constraints on the primary inputs and the states of the FSM. The transition relation simplified with hints (conjoined with the constraints) may have fewer variables in its support or fewer nodes or both. This leaves room for a better clustering of the bit relations based on support and size considerations. A improved quantification schedule may result from this modified clustering. Every iteration of traversal performed with the simplified transition relation benefits in terms of smaller intermediate BDD sizes and faster image computations.

In invariant checking, all reachable states are checked for violation of the property. The order in which the states are checked for the violation is irrelevant for the correctness of invariant checking. Traditional methods employ BFS, starting at the initial states, as their search strategy. Applying hints results in a different search order. Removing the hints at the end ensures that the invariant is checked on all states. Counterexamples can be produced for invariant checking with hints in the same manner as invariant checking without hints. An error trace, though correct, may not be the shortest possible error trace since hints do not perform a BFS.

We use the following grammar to express hints.

$$hint ::= atomic\_hint \mid hint : hint \mid repeat(hint, n)$$

where an *atomic_hint* is a predicate over the primary inputs and states of the FSM. The ":" operator stands for concatenation. The *repeat* operator allows the application of hints for n (finite) iterations or infinitely many times ($n = \infty$). repeat(*hint*, ∞) results in exploration of all states reachable from the given set in the hint-constrained FSM.

We illustrate the effectiveness of hints by studying their application to some circuits.

Am2901: This model is a bit-sliced ALU and contains sixteen 4-bit registers organized into a register file, along with a 4-bit shift register. The ALU operations include LOAD, AND, OR, XOR, and ADD. The registers are fed by the output of the ALU. The operands of the ALU are either primary inputs or registers.

The hint given to this circuit restricts the instruction set to LOAD. Traversal without hints runs out of memory whereas with hints completes in 3.93 seconds. The LOAD instruction is sufficient to generate the entire reachable set. The hint results in every register being fed by primary inputs and removes the dependence on other bit registers. The size of each bit relation decreases, the number of blocks in the transition relation reduces due to better clustering and the quantification schedule improves, allowing more variables to be quantified earlier.

Another hint that enhances traversal is the sequencing of destination registers. Since there is only one write port, only one register gets written into in each cycle. In BFS, at the k-th iteration, sets of k registers are filled with data values of the ALU output and $17 - k$ registers retain their initial value. The circuit takes 17 iterations to complete reachability analysis. In sequencing the destination address, a specific register is chosen as the destination for the k-th iteration. At the end of 17 iterations, the effect is the same as that of not sequencing addresses. The time for this traversal is 53 seconds.

The two hints mentioned above indicate that the transition relation can be simplified in different ways. In reducing the instruction set to LOADs only, the peak number of variables in the computation decreases. Sequencing the addresses resulted in only one set of registers being fed by the ALU output while the rest retain their original values.

Pipelined ALU (PALU): This model is a simple ALU with a three-stage pipeline. The ALU allows boolean, arithmetic and shift operations. It also reads from and writes to a register bank. The register bank is 4-bits wide and has 4 registers. There are three pipeline stages—fetch, execute and write-back stage. A register bypass is allowed if the destination address in the write-back stage matches the source address in the fetch stage. The pipeline is stalled if one of the inputs is asserted. On stalling, the data and addresses are not allowed to move through the pipeline. This creates dependencies between the latches in different stages of the pipeline.

The traversal of this circuit takes 1560 seconds. With the hint that disallows pipeline stalls, traversal completes in 796 seconds. In this case, the hint chooses a common mode of operation of the circuit. Disabling stalls changes the quantification schedule since many registers in the original FSM depend on the stall input. Additionally, the constrained FSM explores a denser [25] set of states due to the absence of simultaneous dependencies introduced by stall and addresses. There is no instruction such as LOAD to generate enough data values to explore a large portion of the reachable states.

Summary: The above examples demonstrate the usefulness of hints in traversal. Hints may make traversal possible where it was not or may speed up traversal considerably. In

```
Guided_Traversal (hints,T, P)
        switch type (hints)
        case "atomic_hint"
            T_H(x, w, y) = T(x, w, y) ∧ H(w, x)
            P = P ∨ Image(T_H, P)
        case ":"
            P = Guided_Traversal (left_child(hints),T, P)
            P = Guided_Traversal (right_child(hints),T, P)
        case "repeat"
            for (counter =0; counter < read_n(hints); counter++)
                P_prev = P
                P = Guided_Traversal (left_child(hints),T, P)
                if (P_prev == P) break
        return P
```

Fig. 1. Guided traversal algorithm.

the following section, we describe how these hints are applied and a modified algorithm for hint-enhanced traversal.

4.1 Algorithm

Our algorithm for traversal is illustrated in Fig. 1. *hints* is a parse tree of hints expressed in the grammar described at the beginning of Section 4. Every node of this parse tree has a type, that may be *"atomic_hint"*, ":" or *"repeat"*. An *"atomic_hint"* node has no children, a ":" node has two children and a *"repeat"* node has one child and an associated field, n, for the number of repetitions.

The algorithm is called with P set to I and T set to the transition relation of the FSM. The hints are automatically suffixed with *:repeat* $(1, \infty)$. This serves to restore the original transition after applying all the hints. (The first argument here implies no constrain on the transition relation.) The algorithm in Fig. 1 recurs on the parse tree of *hints*. The leaf of the recursion is the case of the atomic hint $(H(w, x))$. In this case, the algorithm computes a constrained transition relation $T_H(x, w, y)$ with respect to the hint and computes the image of the current set P. In the case of a node of type ":" (concatenation), the procedure recurs on each subgraph (concatenated hint) and updates P. With a *"repeat"* node, the procedure recurs as many times as required by n or until P converges, whichever comes first. The *atomic_hint* produces new successors while ":" and *repeat* order the application of hints.

The case of *repeat* with $n > 1$ involves repeatedly computing the image of P using the same transition relation $T_H(x, w, y)$ (generated using this *atomic_hint*). The implementation of *repeat(atomic_hint, n)* can be made more efficient by using the frontier states (as mentioned in Section 2). Our implementation makes use of this optimization. The correctness of the algorithm in Fig. 1 is established by the following theorem.

Theorem 1. *1. Algorithm* Guided_Traversal *terminates.*
2. The set of states computed by Guided_Traversal *is contained in the reachable states and no less than the reachable states are computed.*

Proof Sketch: (1) is proved by induction on the depth of *hints*. The proof of (2) relies on the fact that application of hints to the transition is monotonic and the operation $P \vee$

$Image(T_H, P)$ is monotonic. Finally, since the original transition relation is restored, all reachable states are computed.

Optimization. Touati [28] proved that the BDD *Constrain* operator has the property

$$\exists_{x,w}T(x,y,w) \wedge C(x) = \exists_{x,w}T(x,y,w) \downarrow C(x).$$

In applying the hint, the same result can be used as

$$P(x) \vee (\exists_{x,w}(T(x,w,y) \wedge P(x)) \downarrow H(w,x))|_{y=x}.$$

If the hint depends on primary inputs only, the computation can be further simplified to

$$P(x) \vee (\exists_{x,w}(T(x,w,y) \downarrow H(w)) \wedge P(x))|_{y=x}.$$

Images of Large Sets. Every iteration in the algorithm computes the image of P. If the size of the BDD of P increases, computing the image of P is likely to cause a memory blowup. To address this problem, we decompose P disjunctively and compute the image of each individual part [9, 22, 26]. Specifically, we adopt the *Dead-End* computation approach of [26]—a two-way recursive disjunctive decomposition of P is done and an image is computed only when the size of a disjunct are below a certain threshold. A tradeoff may occur in terms of CPU time when the decomposition does not produce enough reduction in size and many disjuncts are generated before the allowed size for image computation is attained. However, the image computation of a large BDD is avoided and in many cases this allows traversal to complete.

4.2 Identification of Good Hints

Hints to aid traversal fall in two main categories: Those that depend on the invariants being checked and those that capture knowledge of the design at hand, independent of the property. Invariant checking may profit from both property-dependent and machine-dependent hints. False invariants may be negatively impacted by hints that further the distance of the violating states from the initial states. Property-dependent hints will tend to avoid this situation. Counterexamples, using property-dependent hints, will tend to be shorter than those with general purpose hints.

In Section 4, we presented two circuits and appropriate hints that accelerated the traversal of these circuits. The hints fall in the category of machine-dependent hints. The acceleration was achieved due to reduced BDD sizes in image computation and exploration of a dense set of states for a majority of iterations. Figure 2 shows the comparison of BDD sizes for the reached set between BFS and traversal using hints of the circuit Vsa (described in Section 6). In this section, we try to summarize the kinds of hints that are easy to identify *and* achieve the desired effect. We believe that these hints are easily extracted from a reasonable knowledge of the circuit. (Sometimes it may even be possible to guess them.) We expect that traversal of circuits containing features (ALUs, register files) similar to the ones described in this paper will profit from the same kinds of hints.

The hints we have utilized can be classified into four categories.

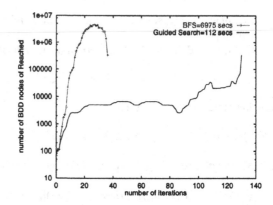

Fig. 2. Comparison of BDD sizes on a log scale of the reached states in the BFS traversal and hint-enhanced traversal of Vsa.

1. Pick the most common operation mode of the circuit. One example is disabling stalls. If there is a counter in the circuit, the counter should be enabled first so that counting iterations have small BDDs.
2. Simplify complex functions such as ALUs: Pick an opcode that produces small intermediate sizes and many states. Loads, in this case, tend to be ideal as they reduce variables in the support of the ALU outputs and produce many next states.
3. Disabling latches: Disable output latches or latches with a small transitive fanin.
4. In the presence of addresses to access banks of registers, pick a sequence of addresses. This prevents the simultaneous interaction of variables belonging to independent registers. These registers tend to be quasi-symmetric in the circuit. Symmetry has traditionally been viewed problematic in model-checking because of its effects on variable order. Several researchers have studied the effects of symmetry [10, 18]. The proposed hint breaks the symmetry in the circuit.

A combination of the above hints can also enhance traversal time. In the Vsa example (described in Section 6), a combination of opcode hints and sequencing of addresses reduced the BDD sizes and traversal time dramatically. For property dependent hints, it is useful to pick an input that is likely to cause the violation. While simulation requires the specification of all input stimulus, the property dependent hints require only a partial specification.

So far, we have discussed the extraction and application of hints by the user. It may also be possible to extract these hints automatically and apply them to traversal. The above mentioned guidelines may prove useful in the automatic detection of hints. In particular, the last two categories may be easy to extract by static analysis of the circuit. A property dependent hint for the same circuit decreased the

Effect on dead-end computations. Dead-end computations are expensive in comparison with image computations of frontier sets with small BDDs. Since repeated dead-end computations may slow down the overall traversal, a good choice of hints should try to minimize the number of dead-end computations. In this regard, retaining a frontier is desirable. The hint *repeat(atomic_hint, n)* is implemented to use frontier sets. Hints should also be chosen to produce dense sets of states, to ease dead-end computations. A particular hint may also be abandoned when the density of the reached state set begins to deteriorate.

5 Comparison to Prior Work

In this section we review the techniques that share significant algorithmic features with our approach.

High-density reachability analysis [25] is based on the observation that BDDs are most effective at representing sets of states when their *density* is high. The density of a BDD is defined as the ratio of the number of minterms of the function to the number of nodes of the BDD. By departing from pure BFS, one can try to direct the exploration of the state space so that the state sets have dense BDDs. Dense sets are obtained by applying subsetting algorithms [24] to the frontier states when the BDDs grow large. The increased density of the frontier sets often improves the density of the other sets as well. When this is not the case, corrective measures can be attempted [26].

Both high density reachability and guided traversal analysis depart from BFS to avoid its inefficiencies. The latter adopts the dead-end computation algorithm of the former. However, the search strategies differ in several respects. High density reachability is a fully automated approach, which can be applied regardless of the knowledge of the system, but occasionally exhibits erratic behavior. Also, it addresses directly the density of sets of *states*, but only indirectly the difficulties of image computation that may be connected to the representation of *transitions*. (When the intermediate products of image computation grow too large, they are subsetted. This leads to the computation of partial images [26].)

Disjunctive partitioning of the transition relation has been proposed in [9, 8, 22, 20]. In the original formulation, partitioning is applied to image computation only. This keeps the test for convergence to the fixpoint simple (no dead-end computation is required,), is effective in reducing the image computation cost, and is fully automated. However, it provides for no departure from BFS, and is therefore ineffective at controlling the sizes of the state sets. When the sub-images computed for the partition blocks have substantial overlap, the disjunctive approach may do unnecessary work.

[9] also introduces the notions of λ-latches and partial iterative squaring. These are latches on which no other state variables depend; they are constrained at their initial value until no new states are achieved. Guided traversal subsumes this technique. Freezing λ-latches is indeed a good hint. Partial iterative squaring may be effective when the sequential depth of the circuit is high.

Biasing the search towards reaching particular states is the subject of [30, 2, 29]. A technique called *saturated simulation* is proposed in [30, 2]. It consists of dividing the state variables into control and data variables, and selectively discarding reached states to allow the exploration of a large number of values for the control variables. Since this technique concentrates on "control" states, its usefulness is mostly in disproving invariants in circuits with a clear separation between control and data.

In [29] the idea of *guidepost* is presented in the context of explicit state search. Given a set of target states, a guidepost is a set of states that the search should try to reach as an intermediate step towards reaching the targets. There is a clear relation between guideposts and the hints used in guided symbolic traversal formulated in terms of state variables. However, there are important differences. In explicit state enumeration, guideposts only help if the target states are indeed reachable. By contrast, the hints used in the symbolic algorithm do not have this limitation. Apart from pointing the way towards the target states, the hints used in guided traversal try to address the difficulties of image computation—not addressed by explicit search.

Constraints on the inputs of the system being verified are also used in symbolic trajectory evaluation [6]. However, full-fledged reachability analysis is not the objective of that technique. Constraints are also used in [15] to implement an assume/guarantee approach to verification. The constraining of the transition relation proceeds in the same way as in guided traversal, but the constraints are not chosen to speed up verification; they are imposed by the environment. One consequence is that there is no need to lift the constraints, and therefore no dead-end computation.

A similar situation occurs in *forward* model checking [14, 3]. The traversal of the state space is constrained by the formula being verified. This has been reported as one of the reasons why forward model checking often outperforms the classical algorithm based on backward analysis. However, as in the previous cases, the constraints are not deliberately chosen to improve performance of the algorithm.

6 Experimental Results

We implemented the algorithm described in Fig. 1 in VIS [4]. Experiments were conducted on a 400MHz Pentium II machine with 1GB of RAM running Linux. We conducted two sets of experiments: One for reachability analysis; the other for invariant checking. We report the results of these experiments in Tables 1 and 2. We used thirteen circuits in our experiments. Of these, Am2901, and PALU are described in Section 4. The Am2910 is a microprogram sequencer. It has a 5-deep stack to store 12-bit addresses, a stack counter and a micro-program counter. CRC computes a 32-bit cyclic redundancy code of a stream of bytes. Fabric is an abstracted version of an ATM network switching fabric [17]. BPB is a branch prediction buffer that predicts whether a branch should be taken depending on the correctness of the previous predictions. Rotator is a barrel shifter sandwiched between registers. Vsa is a very simple non-pipelined microprocessor that executes 12-bit instructions—ALU operations, loads, stores, conditional branch—in five stages: fetch, decode, execute, memory access, and write-back. It has four registers, with one always set to zero. DAIO is a digital audio input output receiver. Soap is a model of a distributed mutual exclusion protocol [12]. CPS is a flat description of a landing gear controller. s1269 and s1512 belong to the ISCAS89 Addendum benchmarks.

The set of atomic hints belong to one of the four categories described in Section 4.2. We guessed some hints for CPS, s1269, and s1512 as we had no information regarding their behavior. In all these circuits, constraints on the primary inputs alone were effective. The hints were expressed as a concatenation of *repeat (atomic_hint,* ∞) and computed with frontier sets. We conducted our experiments without dynamic reordering in order to compare without variability introduced by it.

Table 1 compares reachability analysis with hints against BFS runs. Columns 1, 2, and 3 give the name of the circuit, number of flip-flops (state variables) and number of reachable states of the circuit. A memory limit of 1GB was set on each traversal run.

Table 1. Experimental results for reachability analysis.

Circuits	FFs	Reachable States	Peak Live Nodes		Times in seconds		
			without hints	with hints	without hints	with hints	constr. traversal
CRC	32	4.295e+09	>42,384,405	16,312	Mem. out	0.11	0.11
BPB	36	6.872e+10	1,884,853	46,904	236.9	1.04	1.04
PALU	37	2.206e+09	29,134,676	25,753,334	1560.78	796.48	40.9
s1269	37	1.131e+09	31,225,168	3,287,777	2686.08	47.07	25.31
s1512	57	1.657e+12	23,527,792	26,210,543	5036.9	2372.7	2372.05
Rotator	64	1.845e+19	>12,891,752	16,071	Mem. out	0.18	0.18
Vsa	66	1.625e+14	25,061,852	6,858,369	6974.7	111.8	23.2
Am2901	68	2.951e+20	>38,934,128	349,781	Mem. out	3.65	3.65
DAIO	83	3.451e+14	6,390,705	3,746,631	24584.3	1752.12	1591.36
Fabric	87	1.121e+12	14,220,404	16,197,453	340.12	178.65	117.15
Am2910	99	1.161e+26	>36,696,241	26,238,783	Mem. out	1674.39	14.97
Soap	140	4.676e+08	1,011,972	628,959	36.93	13.87	9.67
CPS	231	1.108e+10	4,648,226	4,032,671	108.86	105.9	68.9

Columns 4 and 5 show the peak number of live nodes during traversal. These numbers are reflective of the peak intermediate sizes during image computation. Columns 6 and 7 compare run times for reachability analysis without and with hints. Column 8 shows the portions of the times in Column 7 spent in traversing the constrained machines.

The circuits in this table are mid-sized, but four of these circuits—CRC, Rotator, Am2901, and Am2910 run out of memory for BFS. The hints described in Section 4 provide dramatic improvements to the traversal of these circuits, enabling completion times of a few seconds. With the remaining circuits in Table 1, BPB, s1269, DAIO, and Vsa demonstrate 1-2 orders of magnitude improvement. For Am2901 and s1269, different hints give comparable reduction in traversal times. The best times are reported in the table. With PALU, Fabric and s1512, the improvement in traversal time is roughly a factor of 2. In cases of dramatic improvement in traversal time, the difference peak number of live nodes support the notion that small computation times result from manipulation of small BDDs (see Fig. 2). CRC, BPB, s1269, Am2901, and Rotator also display this behavior.

The difference in traversal times for CPS is small. The traversal of the constrained machine takes only 69 seconds. Most of the remaining traversal time is taken up in the dead-end computation (only one additional image computation after the dead-end computation is required to prove convergence). This is an example of a case where hints speed up traversal but the dead-end computation offsets these gains. It illustrates the importance of a hint producing a dense set of reachable states in addition to simplifying each image computation.

Table 2 shows comparisons for invariant checking runs without and with hints. Both passing and failing invariants were tried on the various circuits. Entries in Column 1 indicate the circuit name and the number of distinct invariants tried on them. Column 2 indicates whether an invariant failed or passed. Some invariants led to a reduced FSM based on the transitive fanin of the variables involved in the invariant property. Such invariants required exploration of fewer states to pass or fail. For a fair comparison the

Table 2. Experimental results for invariant checking.

Circuits	Invariant TRUE/ FALSE	Reached States		Times in seconds	
		without hints	with hints	without hints	with hints
CRC-1	FALSE	Mem. out	4.295e+09	Mem. out	0.2
BPB-1	FALSE	2.76347e+08	2.41592e+09	20.0	0.9
PALU-1	FALSE	6.05235e+06	5.27424e+06	2.3	1.3
PALU-2	FALSE	5.50225e+08	5.05979e+08	71.0	8.6
PALU-3	FALSE	1.50252e+09	1.33688e+09	164.8	19.2
PALU-4	TRUE	2.20562e+09	2.20562e+09	1529.9	824.5
Rotator-1	FALSE	Mem. out	1.84467e+19	Mem. out	0.3
Vsa-1	FALSE	5.81677e+12	1.23589e+11	806.8	5.6
Vsa-2	FALSE	2.92565e+11	9.47068e+13	173.2	40.5
Vsa-3	TRUE	1.62485e+14	1.62485e+14	6813.5	111.1
Am2901-1	FALSE	Mem. out	2.95148e+20	Mem. out	19.6
Fabric-1	FALSE	4.37925e+09	2.73703e+08	10.4	8.0
Fabric-2	TRUE	6.88065e+10	6.88065e+10	7.0	14.2
Fabric-3	TRUE	1.11004e+12	1.11004e+12	77.9	65.6
Am2910-1	FALSE	Mem. out	3.45961e+18	Mem. out	872.6
Am2910-2	TRUE	24576	24576	3.5	2.8
Am2910-3	TRUE	Mem. out	1.161e+26	Mem. out	1734.8
Soap-1	TRUE	4.676e+08	4.676e+08	31.5	16.1

same reduced FSM was provided to the algorithm without and with hints. Columns 3 and 4 of the table indicate the number of reachable states that needed to be explored to prove these invariants. The last two columns give time comparisons. Failing invariants are reported as soon as they are detected.

The results show that hints may help both passing and failing invariants. For failing invariants where using hints was faster, hints steered the search toward the violating states. For example, in PALU, invariants involving the register values are checked faster when stalls are disabled. Invariants PALU-2 and PALU-3 benefit from this. On the other hand, a user should be careful that the hints provided do not conflict with the invariants being checked, thereby resulting is slower computation times. Invariant checking with hints on CRC, Rotator, Am2901, Am2910, and Soap enjoys the same benefits as reachability analysis. Sometimes hints have less of an impact since the violating states are close to the initial states, or the reduced FSM is small, like in the case of Fabric.

In the case of the false invariant for Vsa-2, the 40.5 second run-time was generated with hint based on the knowledge of the circuit. A property-dependent hint is more effective, completes traversal in 6.5 seconds and generates the shortest error trace. This hint also works for the other invariants, but is about three times slower than the general-purpose one. So general purpose and property-dependent hints may be also useful in the reverse situations.

7 Conclusions

In this paper we have shown that state traversal guided by hints can substantially speed up invariant checking. Orders-of-magnitude improvement have been obtained is several cases, and visible gains have been achieved in most experiments we have run. The hints prescribe constraints for the input and state variables of the system. Deriving the hints requires some knowledge of the circuit organization and behavior, at the level a verification engineer needs to devise simulation stimuli or properties to be checked. Simple heuristics often allow one to find hints that are effective for both properties that fail and properties that pass. There is no guarantee, however, that a hint will be beneficial, as we have discussed in Section 6. We are therefore investigating methods that would allow guided traversal to recover from an ineffective hint with minimum overhead.

The success obtained in deriving hints from a rather small catalog of hint types seems to suggest the automatic derivation of hints as a fruitful area of investigation. We are interested in both methods that work on the structural description of the system (e.g., the RTL code), and methods that analyze the transition relation for information concerning function support and symmetries. We are also investigating the application of hints to model checking more general properties than invariants.

References

[1] M. Abadi and L. Lamport. The existence of refinement mappings. *Theoretical Computer Science*, 82(2):253–284, 1991.

[2] A. Aziz, J. Kukula, and T. Shiple. Hybrid verification using saturated simulation. In *Proc. of the Design Automation Conference*, pages 615–618, San Francisco, CA, June 1998.

[3] I. Beer, S. Ben-David, and A. Landver. On-the-fly model checking of RCTL formulas. In A. J. Hu and M. Y. Vardi, editors, *Tenth Conference on Computer Aided Verification (CAV'98)*, pages 184–194. Springer-Verlag, Berlin, 1998. LNCS 1427.

[4] R. K. Brayton et al. VIS. In *Formal Methods in Computer Aided Design*, pages 248–256. Springer-Verlag, Berlin, November 1996. LNCS 1166.

[5] R. E. Bryant. Graph-based algorithms for boolean function manipulation. *IEEE Transactions on Computers*, C-35(8):677–691, August 1986.

[6] R. E. Bryant, D. L. Beatty, and C. H. Seger. Formal hardware verification by symbolic ternary trajectory evaluation. In *Proc. of the Design Automation Conference*, pages 397–402, 1991.

[7] J. R. Burch, E. M. Clarke, K. L. McMillan, and D. L. Dill. Sequential circuit verification using symbolic model checking. In *Proc. of the Design Automation Conference*, pages 46–51, June 1990.

[8] G. Cabodi, P. Camurati, L. Lavagno, and S. Quer. Disjunctive partitionining and partial iterative squaring: An effective approach for symbolic traversal of large circuits. In *Proc. of the Design Automation Conference*, pages 728–733, Anaheim, CA, June 1997.

[9] G. Cabodi, P. Camurati, and S. Quer. Improved reachability analysis of large finite state machines. In *Proc. of the International Conference on Computer-Aided Design*, pages 354–360, Santa Clara, CA, November 1996.

[10] E. M. Clarke, E. A. Emerson, S. Jha, and A. P. Sistla. Symmetry reductions in model checking. In A. J. Hu and M. Y. Vardi, editors, *Tenth Conference on Computer Aided Verification (CAV'98)*, pages 147–158. Springer-Verlag, Berlin, 1998. LNCS 1427.

[11] P. Cousot and R. Cousot. Abstract interpretation: A unified lattice model for static analysis of programs by constructions or approximation of fixpoints. In *Proc. of the ACM Symposium on the Principles of Programming Languages*, pages 238–250, 1977.

[12] J. Desel and E. Kindler. Proving correctness of distributed algorithms using high-level Petri nets: A case study. In *International Conference on Application of Concurrency to System Design*, Aizu, Japan, March 1998.

[13] D. Geist and I. Beer. Efficient model checking by automated ordering of transition relation partitions. In D. L. Dill, editor, *Sixth Conference on Computer Aided Verification (CAV'94)*, pages 299–310, Berlin, 1994. Springer-Verlag. LNCS 818.

[14] H. Iwashita, T. Nakata, and F. Hirose. CTL model checking based on forward state traversal. In *Proc. of the International Conference on Computer-Aided Design*, pages 82–87, San Jose, CA, November 1996.

[15] M. Kaufmann, A. Martin, and C. Pixley. Design constraints in symbolic model checking. In A. J. Hu and M. Y. Vardi, editors, *Tenth Conference on Computer Aided Verification (CAV'98)*, pages 477–487. Springer-Verlag, Berlin, 1998. LNCS 1427.

[16] R. P. Kurshan. *Computer-Aided Verification of Coordinating Processes*. Princeton University Press, Princeton, NJ, 1994.

[17] J. Lu and S. Tahar. On the verification and reimplementation of an ATM switch fabric using VIS. Technical Report 401, Concordia University, Department of Electrical and Computer Engineering, September 1997.

[18] G. S. Manku, R. Hojati, and R. K. Brayton. Structural symmetry and model checking. In A. J. Hu and M. Y. Vardi, editors, *Tenth Conference on Computer Aided Verification (CAV'98)*, pages 159–171. Springer-Verlag, Berlin, 1998. LNCS 1427.

[19] K. L. McMillan. *Symbolic Model Checking*. Kluwer, Boston, MA, 1994.

[20] K. L. McMillan. Personal Communication, February 1998.

[21] K. L. McMillan. Verification of an implementation of Tomasulo's algorithm by compositional model checking. In A. J. Hu and M. Y. Vardi, editors, *Tenth Conference on Computer Aided Verification (CAV'98)*, pages 110–121. Springer-Verlag, Berlin, 1998. LNCS 1427.

[22] A. Narayan, A. J. Isles, J. Jain, R. K. Brayton, and A. L. Sangiovanni-Vincentelli. Reachability analysis using partitioned ROBDDs. In *Proc. of the International Conference on Computer-Aided Design*, pages 388–393, November 1997.

[23] R. K. Ranjan, A. Aziz, R. K. Brayton, B. F. Plessier, and C. Pixley. Efficient BDD algorithms for FSM synthesis and verification. IWLS95, Lake Tahoe, CA., May 1995.

[24] K. Ravi, K. L. McMillan, T. R. Shiple, and F. Somenzi. Approximation and decomposition of decision diagrams. In *Proc. of the Design Automation Conference*, pages 445–450, San Francisco, CA, June 1998.

[25] K. Ravi and F. Somenzi. High-density reachability analysis. In *Proc. of the International Conference on Computer-Aided Design*, pages 154–158, San Jose, CA, November 1995.

[26] K. Ravi and F. Somenzi. Efficient fixpoint computation for invariant checking. In *Proc. of the International Conference on Computer Design*, Austin, TX, October 1999. To appear.

[27] R. Rudell. Dynamic variable ordering for ordered binary decision diagrams. In *Proc. of the International Conference on Computer-Aided Design*, pages 42–47, Santa Clara, CA, November 1993.

[28] H. Touati, H. Savoj, B. Lin, R. K. Brayton, and A. Sangiovanni-Vincentelli. Implicit enumeration of finite state machines using BDD's. In *Proc. of the IEEE International Conference on Computer Aided Design*, pages 130–133, November 1990.

[29] C. H. Yang and D. L. Dill. Validation with guided search of the state space. In *Proc. of the Design Automation Conference*, pages 599–604, San Francisco, CA, June 1998.

[30] J. Yuan, J. Shen, J. Abraham, and A. Aziz. On combining formal and informal verification. In O. Grumberg, editor, *Ninth Conference on Computer Aided Verification (CAV'97)*, pages 376–387. Springer-Verlag, Berlin, 1997. LNCS 1254.

Modeling and Checking Networks of Communicating Real-Time Processes[1]

Jürgen Ruf and Thomas Kropf

Institute of Computerdesign and Fault Tolerance (Prof. D. Schmid)
University of Karlsruhe, Kaiserstr. 12, Geb. 20.20, 76128 Karlsruhe, Germany
{ruf,kropf}@ira.uka.de
http://goethe.ira.uka.de/hvg/cats/raven

Abstract. In this paper we present a new modeling formalism that is well suited for modeling real-time systems in different application areas and on various levels of abstraction. These *I/O-interval structures* extend interval structures by a new communication method, where input sensitive transitions are introduced. The transitions can be labeled time intervals as well as with communication variables. For interval structures, efficient model checking techniques based on MTBDDs exist. Thus, after composing networks of I/O-interval structures, efficient model checking of interval structures is applicable. The usefulness of the new approach is demonstrated by various real-world case studies, including experimental results.

1 Introduction

For modeling real-time systems it is necessary to have a formalism which allows the explicit declaration of timing information, e.g., a valve needs 5 seconds to open. Moreover, for real-time systems, typical delay times may vary, e.g., due to fabrication tolerances and thus have to be represented by time intervals indicating minimal and maximal time bounds. Since systems are often described in a modular manner, these modules have to communicate with each other. This means, the modeling formalism must be strong enough for expressing communication.

Many approaches for modeling real-time systems exist. Two main approaches have to be distinguished: those based on timed automata [1] and those extending finite state machines (FSM). Both classes of formalisms have disadvantages.

Timed automata have complex model checking algorithms [2,3,4]. Moreover, timed automata use an event-based communication, i.e., transitions in the automata are labeled with events, and transitions with same labels in different automata are synchronized during composition. This event-based communication is not the usual form of hardware communication. In hardware there exist signals which have a defined value for all time instances and modules use the output signals of other modules as inputs.

FSMs have no possibility to model quantitative timing effects. The expansions of FSMs for modeling real-time systems [5,6] have no intuitive semantics according to the specification logic [7] and there exists no composition strategy which is able to combine many communicating modules.

In [8] we have presented a new formalism called interval structure. This formalism has a proper semantics with regard to the specification logic (CCTL) and it allows the

---

1.This work is sponsored by the German Research Grant (DFG)

declaration of time-intervals. We developed an efficient model checking algorithm based on an MTBDD representation of interval structures [9,10]. In [11] we have presented a method for the composition of interval structures completely working on the MTBDD representation, including two minimization heuristics.

A natural model of a hardware system uses a description of several submodules connected by wires. The description of submodules may contain free input variables. If we now want to use interval structures for the modeling of timed sub-systems, the advantages of a compact time representation gets lost, because at any time instance the structure has to react to the free inputs. This effect destroys all timed edges and splits them into unit delay edges. The resulting interval structure is in principal an FSM with many new intermediate states.

Therefore, in this paper we extend interval structures by a signal-based communication mechanism. In order to model this communication properly, the I/O-interval structures contain additional input signals and support input restrictions on timed transitions. With this concept there exist unconditional transitions only consuming time (input-insensitive) as well as conditional transitions which consume time if they are taken because inputs fulfil a input restriction (input-sensitive). With this formalism, timed transitions are even possible if free input variables are introduced.

Furthermore, this communication allows the integration of untimed FSMs (e.g. a controller) and I/O-interval structures (e.g. the timed environment). With some few changes, the efficient composition algorithms of interval structures [11] are applicable to I/O-interval structures. After composing networks of I/O-interval structures, the efficient model checking algorithms of interval structures are applicable [8].

The following section introduces interval structures and their expansion, called I/O-interval structures. In Section 3 the specification logic CCTL is presented. Section 4 gives a short overview about composition and model checking. Some case studies, which show the flexibility of our new approach will be presented in Section 5. We show that for many real-world examples, one clock per module is sufficient and realistic. In Section 6, we show some experimental results. The last section concludes this paper.

2 Using Structures for Modeling System Behavior

In the following sections we introduce Kripke structures and interval structures [8]. In order to overcome some problems of interval structures that result whenever free inputs are introduced, we extend these structures to I/O-interval structures.

2.1 Interval Structures

Structures are state-transition systems modeling HW- or SW-systems. The fundamental structure is the Kripke structure (unit-delay structure, temporal structure) which may be derived from FSMs.

Definition 2.1. *A Kripke structure (KS) is a tuple* $U = (P, S, T, L)$ *with a finite set* P *of atomic propositions.* S *is a finite set of states,* $T \subseteq S \times S$ *is the transition relation connecting states. We assume that for every state* $s \in S$ *there exists a state* $s' \in S$ *such that* $(s, s') \in T$. *The labeling function* $L: S \to \wp(P)$ *assigns a set of atomic propositions to every state.*

The semantics of a KS is defined by paths representing computation sequences.

Definition 2.2. *Given a KS $U = (P, S, T, L)$ and a starting state $s_0 \in S$. A path is an infinite sequence of states $p = (s_0, s_1, \ldots)$ with $(s_i, s_{i+1}) \in T$.*

The basic models for real-time systems are interval structures, i.e., state transition systems with additional labelled transitions. We assume that each interval structure has exactly one clock for measuring time. The clock is reset to zero if a state is entered. A state may be left if the actual clock value corresponds to a delay time labelled at an outgoing transition. The state must be left if the maximal delay time of all outgoing transitions is reached (Fig. 2.1).

Definition 2.3. *An interval structure (IS) $\mathfrak{S}$ is a tuple $\mathfrak{S} = (P, S, T, L, I)$ with a set of atomic propositions P, a set of states S (i-states), a transition relation between the states $T \subseteq S \times S$ such that every state in S has a successor state, a state labeling function $L: S \to \wp(P)$ and a transition labeling function $I: T \to \wp(I\!N)$ with $I\!N = \{1, \ldots\}$.*

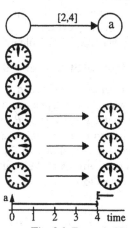

The only difference to KSs are the transitions which are labeled with delay times. Every state of the IS must be left after the *maximal state time*.

Definition 2.4. *The maximal state time of a state s MaxTime:$S \to I\!N_0$ ($I\!N_0 = \{0, \ldots\}$) is the maximal delay time of all outgoing transitions of s, i.e.*

Fig. 2.1. Example IS

$$\text{MaxTime}(s) = max\{t \mid \exists s'.(s, s') \in T \wedge t = max(I(s, s'))\}$$

$$(1)$$

Besides the states, we now also have to consider the currently elapsed time to determine the transition behavior of the system. Hence, the actual state of a system, called the *generalized state*, is given by an i-state s and the actual clock value v.

Definition 2.5. *A generalized state (g-state) $g = (s, v)$ is an i-state s associated with a clock value v. The set of all g-states in an IS $\mathfrak{S} = (P, S, T, L, I)$ is given by:*

$$G = \{(s, v) \mid s \in S \wedge 0 \leq v < \text{MaxTime}(s)\}$$

$$(2)$$

The semantics of ISs is given by runs which are the counterparts of paths in KSs.

Definition 2.6. *Given the IS $\mathfrak{S} = (P, S, T, L, I)$ and a starting state g_0. A run is a sequence of g-states $r = (g_0, g_1, \ldots)$. For the sequence holds $g_j = (s_j, v_j) \in G$ and for all j it holds either*

- $g_{j+1} = (s_j, v_j + 1)$ *with $v_j + 1 < \text{MaxTime}(s_j)$ or*
- $g_{j+1} = (s_{j+1}, 0)$ *with $(s_j, s_{j+1}) \in T$ and $v_j + 1 \in I(s_j, s_{j+1})$.*

The semantics of an IS may also be given in terms of KSs. Therefore in [11] we defined a stutter state expansion operation on ISs, which transforms an IS into a KS. Fig. 2.2 shows an example of expanding timed transitions with stutter states.

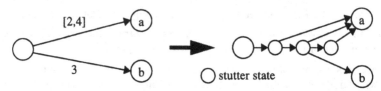

Fig. 2.2. Expansion Semantics of a Timed Structure using Stutter States

IS have no explicit input variables, i.e., if different ISs communicate with each other, they have to share state variables. This communication scheme leads to the following problems:

- one transition may depend on many different valuations of the input variables
- there may exist timed transitions which do not depend on (free) inputs but on the other hand, a variable in an IS may not be declared as don't care.

2.2 Interval Structures with Communication

To understand the second problem, we consider a transition with delay time four in a IS, which is not dependent on an input variable (i). It is however a member of the set of atomic propositions P of the IS. As defined in Def. 2.3, IS are not expressive enough to represent this transition, except by splitting this transition into many unit delay edges as shown in Fig. 2.3. After this splitting operation, the transition may be taken, regardless of the input i, but the advantage of the compact time representation of IS gets lost.

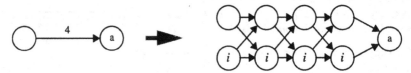

Fig. 2.3. Modeling a free input i with an IS

To overcome these problems which become worse during composition, we introduce a new modeling formalism: I/O-interval structures. These structures carry additional input labels on each transition. Such an input label is a Boolean formula over the inputs. We interpret this formulas as input conditions which have to hold during the corresponding transition times. Input-insensitive edges carry the formula *true*. The formula *false* should not exist, since this transition can never be selected.

In the following definition we formalize Boolean formulas with sets of valuations over the input variables. An element of the set $Inp := \wp(P_I)$ defines exactly one valuation of the input variables: the propositions contained in the set are true, all others are false. An element of the set $\wp(Inp)$ then defines all possible input valuations for one edge. For example, given the inputs a and b, the set $\{\{a\}, \{a, b\}\}$ is represented by the Boolean function $(a \wedge \neg b) \vee (a \wedge b) = a$. This example shows that the variable b does not affect the formula, i.e., the transition labeled with the formula a may be taken independent of the input b.

Definition 2.7. *An I/O-IS is a tuple* $\mathfrak{I}_{I/O} = (P, P_I, S, T, L, I, I_I)$. *For accessing the first (second) component of an element* $x \in S \times Inp$ *we write:* $x[1]$ *(*$x[2]$*). (This access operator is defined to all elements consisting of multiple components)*

- *The components* P, S, L *and* I *are defined analogously to IS*
- P_I *is a finite set of atomic input propositions*
- *The transition relation connects pairs of states and inputs:* $T \subseteq S \times S \times Inp$
- $I_I : T \rightarrow \wp(Inp)$ *is a transition input labeling function*

We assume the following restriction on the input labelling:

$$\forall t_1 \in T. \forall t_2 \in T.\left(\begin{array}{c} (t_1[1] = t_2[1]) \wedge (t_1 \neq t_2) \rightarrow \\ ((I_I(t_1) = I_I(t_2)) \vee (I_I(t_1) \cap I_I(t_2) = \emptyset)) \end{array} \right) \quad (3)$$

This formula ensures that if there exist multiple edges starting in the same, then their input restrictions are either equal or disjoint. This means, that the input valuations on timed edges are clustered.

Definition 2.8. *The cluster function* $C : S \times Inp \rightarrow \wp(Inp)$ *computes all input valuation of a cluster represented by an arbitrary member*

$$C(s, i) := \left\{ \begin{array}{ll} I_I(t) & \text{if } \exists s' \in S. \exists i' \in Inp.t = (s, s', i') \in T \wedge i \in I_I(t) \\ \emptyset & \text{otherwise} \end{array} \right. \quad (4)$$

All input evaluations belonging to the state are clustered. Because of equation (3) all clusters of one state are disjoint, i.e. every evaluation of the input variables represents the cluster it lies in.

Now we describe the semantics of the I/O-ISs by defining runs. Therefore we first need the maximal state time.

Definition 2.9. *The maximal state time* MaxTime:$S \times Inp \rightarrow I\!N$ *is the maximal delay time of all outgoing transitions, i.e.*

MaxTime$(s, i) :=$

$$max\{v|\exists s' \in S. \exists i' \in Inp.t = (s, s', i') \in T \wedge i \in I_I(t) \wedge v = max(I(t))\} \quad (5)$$

G-states in I/O-IS also have to consider the actual inputs besides the i-state and the elapsed time.

Definition 2.10. *An extended generalized state (xg-state)* $g = (s, i, v)$ *is an i-state* s *associated with an input evaluation* $i \in Inp$ *and a clock value* v. *The set of all xg-states in an I/O-IS is given by:*

$$G_I = \left\{ (s, i, v)|s \in S \wedge i \in \bigcup_{i' \in Inp} C(s, i') \wedge 0 \leq v < \text{MaxTime}(s, i) \right\} \quad (6)$$

Definition 2.11. *Given the I/O-IS* $\mathfrak{I} = (P, P_I, S, T, L, I, I_I)$, *a run is a sequence of xg-states* $r = (g_0, g_1, \ldots)$. *For the sequence holds* $g_j = (s_j, i_j, v_j) \in G_I$ *and for all* j *it holds either*

- $g_{j+1} = (s_j, i_{j+1}, v_j + 1)$ *with* $v_j + 1 < \text{MaxTime}(s_j, i_j)$ *and* $i_{j+1} \in C(s_j, i_j)$ *or*
- $g_{j+1} = (s_{j+1}, i_{j+1}, 0)$ *with* $t = (s_j, s_{j+1}, i_{j+1}) \in T, i_j \in I_I(t)$ *and* $v_j + 1 \in I(t)$.

I/O-IS may also be expanded to KS. The expansion works similar to IS, except for input sensitive transitions. These transitions are expanded as shown in Fig. 2.3. After the expansion, the efficient composition and model checking algorithms for ISs are applicable.

Since we want to examine reactive systems of connected I/O-ISs, we assume that for every xg-state and for every input evaluation there exists a successor xg-state. This means for edges with delay times greater one there has to be a fail state in which is visited if the actual input does not fulfill the input restriction. This implies that transitions either have no input restriction or, if a transition with delay time δ has an input restriction, there must exist a transition with interval $[1, \delta - 1]$ which connects the same starting state with the fail state. For unit-delay edges we have to ensure that for all input evaluations there exists a successor state. In general, the I/O-interval structures are not restricted to reactive systems, but the examples studied later should be reactive.

For an easier understanding of the following case studies, we introduce a graphical notation for I/O-IS in Fig. 2.4. The a_i represent the inputs of the modeled sub-system, and f denotes a function $f : P_I^n \to I\!B$.

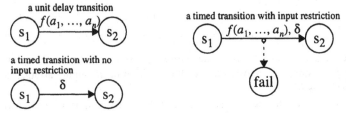

Fig. 2.4. Graphical notations

3 Specifying Real-Time Properties with CCTL

CCTL [11] is a temporal logic extending CTL with quantitative bounded temporal operators. Two new temporal operators are introduced to make the specification of timed properties easier. The syntax of CCTL is the following:

$$p \mid \neg\varphi \mid \varphi \wedge \varphi \mid \varphi \vee \varphi \mid \varphi \to \varphi \mid \varphi \leftrightarrow \varphi$$

$$\varphi := \mid EX_{[a]}\varphi \mid EF_{[a,b]}\varphi \mid EG_{[a,b]}\varphi \mid E(\varphi \, U_{[a,b]}\varphi) \mid E(\varphi \, C_{[a]}\varphi) \mid E(\varphi \, S_{[a]}\varphi) \qquad (7)$$

$$\mid AX_{[a]}\varphi \mid AF_{[a,b]}\varphi \mid AG_{[a,b]}\varphi \mid A(\varphi \, U_{[a,b]}\varphi) \mid A(\varphi \, C_{[a]}\varphi) \mid A(\varphi \, S_{[a]}\varphi)$$

where $p \in P$ is an atomic proposition and $a \in I\!N$ and $b \in I\!N \cup \{\infty\}$ are time-bounds. All interval operators can also be accompanied by a single time-bound only. In this case the lower bound is set to zero by default. If no interval is specified, the lower bound is implicitly set to zero and the upper bound is set to infinity. If the X-operator has no time bound, it is implicitly set to one. The semantics of the logic is given as a validation relation:

Definition 3.1. Given an IS $\mathfrak{S} = (P, S, T, L, I)$ and a configuration $g_0 = (s, v) \in G$.

$$g_0|=p \qquad :\Leftrightarrow p \in L(s)$$
$$g_0|=\neg\varphi \qquad :\Leftrightarrow \text{not } g_0|=\varphi \tag{8}$$
$$g_0|=(\varphi \wedge \psi) :\Leftrightarrow g_0|=\varphi \text{ and } g_0|=\psi$$

$$g_0 \models EG_{[a,b]}\varphi \qquad :\Leftrightarrow \text{there ex. a run } r = (g_0, g_1, \dots) \text{ s.t.}$$
$$\text{for all } a \leq i \leq b \text{ holds } g_i \models \varphi$$

$$g_0 \models E(\varphi \; U_{[a,b]}\psi) :\Leftrightarrow \text{there ex. a run } r = (g_0, g_1, \dots) \text{ and an } a \leq i \leq b \text{ s.t.}$$
$$g_i \models \psi \text{ and for all } j < i \text{ holds } g_j \models \varphi \tag{9}$$

$$g_0 \models E(\varphi \; C_{[a]}\psi) \quad :\Leftrightarrow \text{there ex. a run } r = (g_0, g_1, \dots) \text{ s.t.}$$
$$\text{if for all } i < a \text{ holds } g_i \models \varphi \text{ then } g_a \models \psi$$

$$g_0 \models E(\varphi \; S_{[a]}\psi) \quad :\Leftrightarrow \text{there ex. a run } r = (g_0, g_1, \dots) \text{ s.t.}$$
$$\text{for all } i < a \text{ holds } g_i \models \varphi \text{ and } g_a \models \psi$$

The other operators may be derived by the defined ones, e.g.:

$$EF_{[a,b]}\varphi \qquad \equiv E(true \; U_{[a,b]}\varphi)$$
$$A(\varphi \; C_{[a]}\psi) \equiv \neg E(\varphi \; S_{[a]}\neg\psi) \tag{10}$$
$$A(\varphi \; S_{[a]}\psi) \equiv \neg E(\varphi \; C_{[a]}\neg\psi)$$

Examples for the use of CCTL may be found in the case studies.

4 Composition and Model Checking

Model checking as described in [8] works on exactly one IS, but real-life systems are usually described modular by many interacting components. In order to make model checking applicable to networks of communicating components, it is necessary to compute the product structure of all submodules. This product-computation is called composition.

Due to space limitations we do not describe the composition algorithms in detail. The principal idea is to define a reduction operation (*reduce*) on KS which replaces adjacent stutter states by timed edges. With this operation we are able to define the composition by:

$$\Im_1 \| \Im_2 := reduce(expand(\Im_1) \| expand(\Im_2)) \tag{11}$$

The algorithms described in [11] works as follows:

1. expand all modules and substitute their input variables by the connected outputs
2. compose the expanded structures
3. reduce the composition KS

In order to avoid the computation of the complete unit-delay composition structure, Step 2 and Step 3 may be performed simultaneously by incrementally adding only the reachable transitions.

Since we adapt the expansion to I/O-ISs, the composition algorithms are also applicable if we work with closed systems, where every free variable is bounded by a module. After composition, the efficient model checking algorithms for IS may be used for verification. These algorithms are described in detail in [8]. Composition as well as the model checking algorithms use the symbolic representation of state sets and transition relations with MTBDDs during computation.

5 Case Studies

In this section, we show the flexibility of our approach to model real-time systems. We have chosen three systems out of very different areas and on different levels of abstraction. This section shows, that we do not necessarily need timed automata to represent complex real-time systems.

5.1 Modeling Hardware with Time Delays

A first class of real-time systems are digital hardware circuits with delay times. These circuits are composed by digital gates like AND gates or flipflops, but they have a specific timing behavior. Impulses on the inputs which are shorter than the delay time of a gate are suppressed.

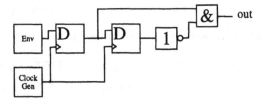

Fig. 5.1. The circuit of the single-pulser

Only if an impulse stays constant at least for the delay time, the gate may change its outputs. For modeling flipflops, we assume a setup and a hold time. If the input signals violate these time-constraints, the flipflop remains in its actual state. As an easy example we choose the single-pulser circuit [12] which is shown in Fig. 5.1. Figure 5.2 shows the basic gates (initial states are bold). For modeling the AND gate, input sensi-

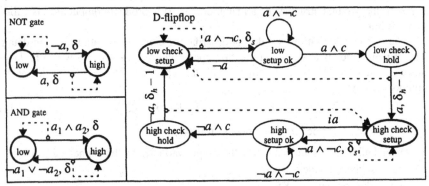

Fig. 5.2. Basic gates with input i, i_1, i_2, clock c and delay time δ resp. setup δ_s and hold δ_h

tive edges have been used: If the system starts in the state *high*, and the inputs fulfil $\neg a_1 \vee \neg a_2$ for δ time units then it changes to the state *low*. If the inputs fulfil $a_1 \wedge a_2$ before δ time units are passed, the structure remains in the state *high*. Here, I/

O-IS allow a very clear and compact modeling of timing constraints and communication behavior.

The clock generator is modeled by an I/O-IS with two states (*high* and *low*) which toggles between both states every cycle-time. The environment is either a human pressing the button (which should be single pulsed) or it's a bouncing pulse.

The specification checks that an output signal appears if the input stays high long enough:

$$spec1 := AG(\neg Env.out \rightarrow EX(A(Env.out\ C_{[2\delta_c + \delta_h]}AF_{[\delta_a - \delta_h]}And.out))) \quad (12)$$

The following specification verifies that the output stays high for one cycle period and then changes to low for a further cycle and afterwards rests low until the input becomes high:

$$spec2 := AG(\neg And.out \rightarrow EX(And.out \rightarrow A(And.out\ C_{[2\delta_c + \delta_n]}tmp1)))$$

$$tmp1 := AG_{[2\delta_c - 1]}\neg And.out \wedge A((\neg And.out)UEnv.out) \quad (13)$$

5.2 Modeling a Production Automation Systems

The production cell is a case study to evaluate different formal methods for verification [13]. A schematic deposition of the production cell is shown in Fig. 5.3. A feed belt moves work pieces to the elevation rotary table. This table lifts the pieces to the robot arm 1. The robot arm 1 moves the work pieces to the press, where robot arm 2 removes them after the work has

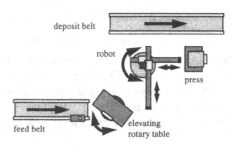

Fig. 5.3. Schematic disposition of the production cell

performed. Robot arm 2 drops the work pieces over the deposit belt. The robot has two arms to gain a maximal performance of the press.

Modeling the feed belt (FB, Fig. 5.4): The FB delivers the cell with work pieces (wp) every δ_{min} or δ_{max} seconds (state feed). If the wp is moved to the end of the belt (state work piece present) the belt stops rotating until the *next* signal raises. Then it shifts the wp to the elevating rotary table. If the table is not in the right position, the wp falls to the ground or blocks the table (state fail). This action takes δ_{reload} seconds for completion

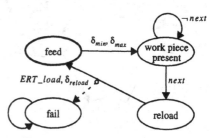

Fig. 5.4. I/O-IS modeling the feed belt

Modeling the elevating rotary table (ERT, Fig. 5.5): The ERT has a belt for loading wps, which starts rotating by a rising *load* signal. The *mov* signal makes the ERT moving and rotating from the low loading position to the high unloading position (and vice versa). The signal *pick* (*drop*) indicates that a wp is picked up (dropped down) by the robot arm 1. The signal *pick* is a combination of many signals, the right robot rotary table position, the out position of the arm, the empty signal of the arm and the magnet on signal of the controller.

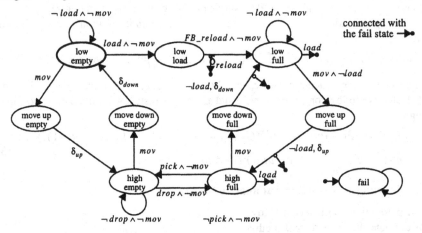

Fig. 5.5. I/O-IS modeling the elevation rotary table

The **robot** is modeled by three separate modules, the two **arms (RA, Fig. 5.6)** and the **robot rotary table (RRT, Fig. 5.7)**. A RA may be empty or full (a wp is hanging on the RA) and it may be extended (out) or retracted (in). The fail state indicates that the wp falls to the ground. The RRT has three positions: unload ERT (UE) unload press (UP) and drop wp (DW). Every movement between these positions is possible.

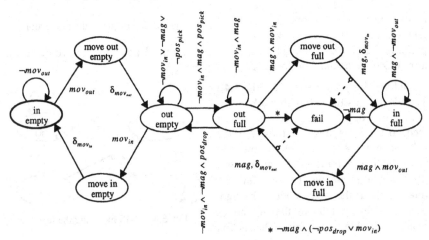

Fig. 5.6. I/O-IS modeling the robot arm

Modeling the press (P, Fig. 5.8): The P has three positions: low for unloading wps by RA2, mid for loading wps by RA1 and press. The signal *lift* (*lower*) makes the press

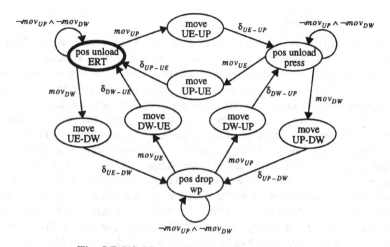

Fig. 5.7. I/O-IS modeling the robot rotary table

to move to the next higher (lower) position. The *drop* and *pick* signals are connected with the RAs and the RRT.

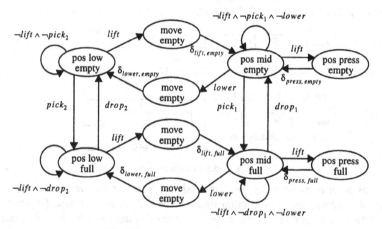

Fig. 5.8. I/O-IS modeling the press

The deposit belt is not modeled by a separate module, it is assumed that the belt is always running. If the RA 2 drops a wp over the deposit belt, it is transported.

The controlling unit is a synchronous FSM described with an output and a transition function. We assume a sampling rate of 5ms for the external signals. The controller manages the correct interaction between all modules.

The checked specifications check time bounds for the arrival of the first pressed wp, or the cycle time of pressed wps. It is also shown that 4 wps may stay in the production cell, but not five. Also some wait times for the wps are shown.

5.3 Modeling a Bus Protocol

The next example is the arbitration mechanism of a bus protocol. We modeled the J1850 protocol arbitration [14] which is used in on- and off-road vehicles. The protocol is a CSMA/CR protocol. Every node listens to the bus before sending (carrier sense, CS). If the bus is free for a certain amount of time, the node starts sending. It may happen that two or more nodes simultaneously start sending (multiple access, MA). Therefore, while sending, every node listens to the bus and compares the received signals to the send signals. If they divide, it looses arbitration (collision resolution, CR) and waits until the bus is free again. A sender distinguishes between two sending modes, a passive and an active mode. Active signals override passive signals on the bus. Succeeding bits are alternately send active and passive. The bits to be send are encoded by a variable pulse width: a passive zero has a pulse width of $64\mu sec$, a passive one bit takes $128\mu sec$, an active zero bit takes $128\mu sec$ and an active one bit takes $64\mu sec$. The bus is simply the union of all actively send signals. The arbitration is a bit-by-bit arbitration, since a (passive/active) zero shadows a one bit. Before sending the first bit, the nodes send an SOF (start of frame) signal, which is active and takes $200\mu sec$. In Fig. 5.9 some examples of arbitration are shown. We assume an exact frame length of 8 bits. After sending the last bit, the sender sends a passive signal of $280\mu sec$, the end of frame (EOF) signal.

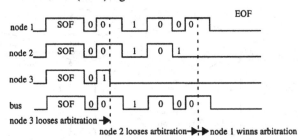

Fig. 5.9. Some examples of arbitration

One bus node is modeled by two sub-modules: a sender/receiver and a counter. Initially, all modules are in their initial states. If the node decides to send (indeterministically) the sender/receiver listens to the bus. If the bus stays low for δ_{CS} time units, the module changes to the SOF state. The counter is triggered by the continue high/low states of the sender. In the initial state, the counter module sends the *count* signal. After sending the SOF signal, the sender sends alternately passive and active one and zero bits. If the bus becomes active while sending a passive bit, the sender/receiver changes to the CS state and tries sending again later.

6 Experimental Results

In this section we compare the MTBDD model checking approach against an ROBDD approach which uses the complete expanded unit-delay transition relation for verification. Both algorithms prove the same formulas, only the representation and the model checking techniques are different. For the ROBDD approach we use our unit-delay model checker which uses the same BDD package as the MTBDD approach.

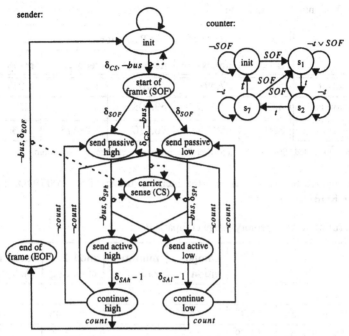

Fig. 5.10. Two submodules modeling one bus node

Both algorithms start with the same description of the networks of I/O-ISs and expand these descriptions to unit-delay structures. The MTBDD approach then performs the IS composition [11]. Afterwards it uses the algorithms presented in [8] for model checking. The ROBDD approach preforms standard product composition on Kripke structures and then checks the resulting Kripke structure against the same specifications.

Table 1 shows the number of ROBDD respective MTBDD nodes for the three case studies presented in section 5. The J1850 example contains three bus nodes.The max delay column indicates the maximal delay times used in the models. The ROBDD column shows the number of nodes necessary for the corresponding Kripke structure. The MTBDD column is divided into three sub columns, the first one shows the number of MTBDD nodes computed by the composition algorithm, the second and the third column shows the number of MTBDD nodes after applying a minimization algorithm (reencoding, minimal path) presented in [11]. The reencode algorithm reduces the number of MTBDD variables encoding the clock values (introduced by the expansion step) by substituting these encodings by new ones with a minimal number of encoding variables. The minpath algorithm adds unreachable states to the state space in the MTBDD representation to shrink the number of MTBDD nodes from the root to the leaves. Both algorithms do not affect the behavior of the system.

Table 1. BDD nodes comparison

	max delay	ROBDD nodes	MTBDD nodes		
			no min.	reenc.	minp.
single pulser	500	27,691	4,176	1,546	1,917
production cell	4,000	1,972,238	69,003	_a	18,999
J1850	300	350,358	76,610	389,478	50,301

a. memory exceeded 500MB due to a bad variable ordering required for this algorithm

Table 2 compares the run-times and the memory usage of the MTBDD approach against the ROBDD approach.

Table 2. run-time and memory usage comparison

	complete	composition	minimization	model checking	memory (MByte)
single-pulser					
MTBDD-approach	15.77	14.83	0.43	0.46	4.59
ROBDD-approach	11.59	8.84	-	2.56	5.23
production cell					
MTBDD-approach	39:38	38:01	00:18	01:17	19.22
ROBDD-approach	54:43	27:07	-	27:20	177.6
J1850 bus arbitration					
MTBDD-approach	05:06	00:37	00:09	04:19	22.2
ROBDD-approach	248:28	01:07	-	247:19	78.2

7 Conclusion

In this paper we have presented a new modeling formalism (I/O-interval structure) which is well suited for many real-time systems of different application areas and abstraction levels. The formalism extends interval structures by a proper communication method based on signal communication between modules. This technique overcomes the problem of splited edges appearing in interval structures while introducing free input variables.

On the other hand, this signal-based communication allows a natural modeling of hardware systems, and allows an easy combination of timed systems and synchronous controllers. This formalism leads to a compact model representation of real-time systems with MTBDDs. Moreover, after composing I/O-interval structures, the efficient model checking algorithms working on MTBDDs are applicable to the modeled systems.

References

[1] R. Alur and D. Dill. Automata for Modeling Real-Time Systems. In *International Colloquium on Automata, Languages and Programming, LNCS*, NY, 1990. Springer-Verlag.

[2] R. Alur, C. Courcoubetics, and D. Dill. Model Checking for Real-Time Systems. In *Symposium on Logic in Computer Science*, Washington, D.C., 1990. IEEE Computer Society Press.

[3] T. Henzinger, X. Nicollin, J. Sifakis, and S. Yovine. Symbolic Model Checking for Real-Time Systems. In *IEEE Symposium on Logic in Computer Science (LICS)*, pages 394–406, Santa-Cruz, California, June 1992. IEEE Computer Scienty Press.

[4] M. Bozga, O. Maler, A. Pnueli, and S. Yovine. Some progress in the symbolic verification of timed automata. In O. Grumberg, editor, *Conference on Computer Aided Verification (CAV)*, volume 1254 of *LNCS*, pages 179–190. Springer Verlag, June 1997.

[5] S. Campos and E. Clarke. Real-Time Symbolic Model Checking for Discrete Time Models. In T. Rus and C. Rattray, editors, *Theories and Experiences for Real-Time System Development*, AMAST Series in Computing. AMAST Series in Computing, May 1994.

[6] J. Frößl, J. Gerlach, and T. Kropf. An Efficient Algorithm for Real-Time Model Checking. In *European Design and Test Conference (EDTC)*, pages 15–21, Paris, France, March 1996. IEEE Computer Society Press (Los Alamitos, California).

[7] J. Ruf and T. Kropf. Using MTBDDs for discrete timed symbolic model checking. *Multiple-Valued Logic – An International Journal*, 1998. Special Issue on Decision Diagrams, Gordon and Breach.

[8] J. Ruf and T. Kropf. Symbolic model checking for a discrete clocked temporal logic with intervals. In *Conference on Correct Hardware Design and Verification Methods (CHARME)*, pages 146–166, Montreal, Canada, October 1997. Chapman and Hall.

[9] R. Bahar, E. Frohm, C. Gaona, G. Hachtel, E. Macii, A. Pardo, and F. Somenzi. Algebraic Decision Diagrams and Their Applications. In *IEEE/ACM International Conference on Computer Aided Design (ICCAD)*, pages 188–191, Santa Clara, California, November 1993. ACM/IEEE, IEEE Computer Society Press.

[10] E. Clarke, K. McMillian, X. Zhao, M. Fujita, and J.-Y. Yang. Spectral Transforms for large Boolean Functions with Application to Technologie Mapping. In *ACM/IEEE Design Automation Conference (DAC)*, pages 54–60, Dallas, TX, June 1993.

[11] J. Ruf and T. Kropf. Using MTBDDs for composition and model checking of real-time systems. In *FMCAD 1998*. Springer, November 1998.

[12] S. Johnson, P. Miner, and A. Camilleri. Studies of the single pulser in various reasoning systems. In *International Conference on Theorem Provers in Circuit Design (TPCD)*, volume 901 of *LNCS*, Bad Herrenalb, Germany, September 1994. Springer-Verlag, 1995.

[13] C. Lewerentz and T. Lindner, editors. *Formal Development of Reactive Systems - Case Study Production Cell*, number 891 in LNCS. Springer, 1995.

[14] SAE. J1850 class B data communication network interface. *The Engeneering Society For Advancing Mobility Land Sea Air and Space*, October 1995.

"Have I Written Enough Properties?" - A Method of Comparison between Specifiation and Implementation

Sagi Katz, Orna Grumberg

Computer Science Department, Technion, Haifa, Israel

{ksagi,orna}@cs.technion.ac.il

Danny Geist - IBM Haifa Research Lab., Haifa, Israel

geist@haifa.vnet.ibm.com

Abstract. This work presents a novel approach for evaluating the quality of the model checking process. Given a model of a design (or implementation) and a temporal logic formula that describes a specification, model checking determines whether the model satisfies the specification. Assume that all specification formulas were successfully checked for the implementation. Are we sure that the implementation is correct? If the specification is incomplete, we may fail to find an error in the implementation. On the other hand, if the specification is complete, then the model checking process can be stopped without adding more specification formulas. Thus, knowing whether the specification is complete may both avoid missed implementation errors and save precious verification time.

The completeness of a specification with respect to a given implementation is determined as follows. The specification formula is first transformed into a tableau. The simulation preorder is then used to compare the implementation model and the tableau model. We suggest four comparison criteria, each revealing a certain dissimilarity between the implementation and the specification. If all comparison criteria are empty, we conclude that the tableau is bisimilar to the implementation model and that the specification fully describes the implementation. We also conclude that there are no redundant states in the implementation.

The method is exemplified on a small hardware example. We implemented our method symbolically as an extension to SMV. The implementation involves efficient OBDD manipulations that reduce the number of OBDD variables from $4n$ to $2n$.

1 Introduction

This work presents a novel approach for evaluating the quality of the model checking process. Given a model of the design (or implementation) and a temporal logic formula that describes a specification, model checking [8, 2] determines whether the model satisfies the specification.

Assume that all specification formulas were successfully checked for the implementation. Are we sure that the implementation is correct? If the specification is incomplete, we may fail to find an error in the implementation. On the other hand, if the specification is complete, then the model checking process can be stopped without checking additional specification formulas. Thus, knowing whether the specification is complete may both avoid missed implementation errors and save precious verification time.

Below we describe our method to determine whether a specification is complete with respect to a given implementation. We restrict our attention to safety properties written in the universal branching-time logic ACTL [3]. This logic is relatively restricted, but can still express most of the specifications used in practice. Moreover, it can fully characterizes every deterministic implementation. We consider a single specification formula (the conjunction of all properties).

We first apply model checking to verify that the specification formula is true for the implementation model. The formula is then transformed into a *tableau* [3]. By definition, since the formula is true for the model, the tableau is greater by the *simulation preorder* [9] than the model.

We defined a *reduced tableau* for ACTL safety formulas. Our tableau is based on the *Particle tableau* for LTL, presented in [6]. We further reduce their tableau by removing redundant tableau states.

We next use the simulation preorder to find differences between the implementation and its specification. For example, if we find a reachable tableau state with no corresponding implementation state, then we argue that one of the two holds. Either the specification is not restrictive enough or the implementation fails to implement a meaningful state. Our method will not be able to determine which of the arguments is correct. However, the evidence for the dissimilarity (in this case a tableau state that none of the implementation states are mapped to) will assist the designer to make the decision.

We suggest four *comparison criteria*, each revealing a certain dissimilarity between the implementation and specification. If all comparison criteria are empty, we conclude that the tableau is bisimilar to the implementation model and that the specification fully describes the implementation. We also conclude that there are no redundant states in the implementation.

The practical aspects of this method are straightforward. Model checking activity in industry executes the following methodology: A verification engineer reads the specification, sets up a work environment and then proceeds to present the model checker with a sequence of properties in order to verify the design correctness [1]. The design (or implementation) on which this activity is executed can be quite large nowadays. As a result the set of properties written and verified becomes large as well, to the point that the engineer loses control over it.

A large property set makes it necessary to construct tools to evaluate its overall quality. The basic question to answer is: "Have I written enough properties?". The current solution is to manually review the property set. However, this solution is not scalable and furthermore since it is done manually it makes the description of model checking as "formal verification" imprecise. This inadequate solution indicates a growing need for tools that may be able to tell the engineer when the design is "bug-free" and therefore cut down development time.

Quality evaluation of verification activity is not new. Traditional verification methods have developed measurement criteria to measure the quality of test suites [11]. This area of verification is called *coverage*. The notion of coverage in model checking is to have a specification that covers the entire functionality required from the implementation. This can be divided into two questions:

1. Whether the environment is rich enough to provide all possible input sequences.
2. Whether the specification contains a sufficient set of properties.

The method we present addresses both problems as will be later shown by an example.

We compared a small hardware example with the reduced tableau. For the complete specification formula we received a reduced tableau with 20 states. The tableau presented in [3] would have a state space of 2^{15} states for this formula. It is interesting to note that in this example not all the implementation variables are observable.

We implemented our method symbolically as an extension to the symbolic model checker SMV [7]. Given a model with n state variables, a straightforward implementation of this method can create intermediate results that consists of $4n$ OBDD variables. However, our implementation reduces the required number of OBDD variables from $4n$ to $2n$.

The main contributions of our paper can be summarized as follows:

- We suggest for the first time a theoretical framework that provides quality evaluation for model checking. The suggested comparison criteria can assist the designer in finding errors in the design by indicating points in which the design and the specification disagree, and suggest criteria for terminating the verification effort.
- We implemented our method symbolically within SMV. Thus, it can be invoked automatically once the model checking terminates successfully. Of a special interest is the symbolic computation of the simulation relation for which no good symbolic algorithm is known.
- We defined a new reduced tableau for ACTL that is often significantly smaller in the number of states and transitions than known tableaux for ACTL.

In these days, another work on coverage of model checking has been independently developed [5]. The work computes the percentage of states in which a change in an observable proposition will not affect the correctness of the specification. Their evidence is closely related to our criterion of *Unimplemented State*. In their paper they list a number of limitations of their work. They are unable to give path evidence, cannot point out functionality missing in the model, and they have no indication that the specification is complete. In the conclusion we explain how our work solves these problems.

The analysis we perform compares the two models and tries to identify dissimilarities. It is therefore related to tautology checking of finite state machines as is done in [10]. However the method in [10] is suggested as an alternative to model checking and not as a complementary method.

The rest of this paper is organized as follows. Section 2 gives the necessary background. Section 3 describes the comparison criteria and the method for their use. Section 4 exemplifies the different criteria by applying the method to a small hardware circuit. Section 5 presents symbolic algorithms that implement our method. In Section 6 we discuss the reduced tableau for ACTL safety formulas. Finally, the last section describes future work and concludes the paper.

2 Preliminaries

Our specification language is the universal branching-time temporal logic ACTL [3], restricted to safety properties. Let AP be a set of atomic propositions. The set of ACTL *safety formulas* is defined inductively in negation normal form, where negations are applied only to atomic propositions. It consists of the temporal operators $\mathbf{X}$ ("next-state") and $\mathbf{W}$ ("weak until") and the path quantifier $\mathbf{A}$ ("for all paths").

- If $p \in AP$ then both p and $\neg p$ are ACTL safety formulas.
- If φ_1 and φ_2 are ACTL safety formulas then so are $\varphi_1 \wedge \varphi_2$, $\varphi_1 \vee \varphi_2$, $\mathbf{AX}\varphi_1$, and $\mathbf{A}[\varphi_1 \mathbf{W} \varphi_2]^1$.

We use Kripke structures to model our implementations. A *Kripke structure* is a tuple $M = (S, S_0, R, L)$ where S is a finite set of states; $S_0 \subseteq S$ is the set of initial states; $R \subseteq S \times S$ is the transition relation that must be total; and $L : S \rightarrow 2^{AP}$ is the labeling function that maps each state to the set of atomic propositions true at that state.

A *path* in M from a state s is a sequence $s_0, s_1, \ldots$ such that $s_0 = s$ and for every i, $(s_i, s_{i+1}) \in R$.

The logic ACTL is interpreted over a state s in a Kripke structure M. The formal definition is omitted here. Intuitively, $\mathbf{AX}\varphi_1$ is true in s if all its successors satisfy φ_1. $\mathbf{A}[\varphi_1 \mathbf{W} \varphi_2]$ is true in s if along every path from s, either φ_1 holds forever or φ_2 eventually holds and φ_1 holds up to that point. We say that a structure M satisfies a formula φ, denoted $M \models \varphi$, if every initial state of M satisfies φ.

Let $M = (S, S_0, R, L)$ and $M' = (S', S_0', R', L')$ be two Kripke structures over the same set of atomic propositions AP. A relation $SIM \subseteq S \times S'$ is a *simulation preorder* from M to M' [9] if for every initial state s_0 of M there is an initial state s_0' of M' such that $(s_0, s_0') \in SIM$. Moreover, if $(s, s') \in SIM$ then the following holds:

- $L(s) = L'(s')$, and
- $\forall s_1[(s, s_1) \in R \Longrightarrow \exists s_1'[(s', s_1') \in R' \wedge (s_1, s_1') \in SIM]]$.

---

[1] Full ACTL includes also formulas of the form $\mathbf{A}[\varphi_1 \mathbf{U} \varphi_2]$ ("strong until").

If there is a simulation preorder from M to M', we write $M \leq M'$ and say that M *simulates* M'.

It is well known [3] that if $M \leq M'$ then for every ACTL formula φ, if $M' \models \varphi$ then $M \models \varphi$. Furthermore, for every ACTL safety formula ψ it is possible to construct a Kripke structure $T(\psi)$, called a *tableau*[2] for ψ, that has the following *tableau properties* [3].

- $T(\psi) \models \psi$.
- For every structure M, $M \models \psi \iff M \leq T(\psi)$.

Intuitively, the simulation preorder relates two states if the computation tree starting from the state of the smaller model can be embedded in the computation tree starting from the state of the greater one. This, however, is not sufficient in order to determine how similar the two structures are. Instead, we use the *reachable simulation preorder* that relates two states if they are in the simulation preorder and are also reachable from initial states along corresponding paths.

Formally, let $SIM \subseteq S \times S'$ be the *greatest* simulation preorder from M to M'. The *reachable simulation preorder* for SIM, $ReachSIM \subseteq SIM$, is defined by: $(s, s') \in ReachSIM$ if and only if there is a path $\pi = s_0, s_1, \ldots, s_k$ in M with $s_0 \in S_0$ and $s_k = s$ and a path $\pi' = s'_0, s'_1, \ldots, s'_k$ in M' with $s'_0 \in S'_0$ and $s'_k = s'$ such that for all $0 \leq j \leq k$, $(s_j, s'_j) \in SIM$.
In this case, the paths π and π' are called *corresponding paths* leading to s and s'.

Lemma 1. *ReachSIM is a simulation preorder from M to M'.*

The proof of the lemma is postponed to Appendix A.

3 Comparison Criteria

Let $M = (S_i, S_{0i}, R_i, L_i)$ be an implementation structure and $T(\psi) = (S_t, S_{0t}, R_t, L_t)$ be a tableau structure over a common set of atomic propositions AP. For the two structures we consider only reachable states that are the start of an infinite path.

Assume $M \leq T(\psi)$. We define four criteria, each is associated with a set. A criterion is said to hold if the appropriate set is empty. For convenience we name each criterion the same as the appropriate set. The following sets define the criteria :

1. $UnImplementedStartState = \{s_t \in S_{0t} \mid \forall s_i \in S_{0i} [(s_i, s_t) \notin ReachSIM]\}$
   An Unimplemented Start State is an initial tableau state that has no corresponding initial state in the implementation structure. The existence of such a state may indicate that the specification does not properly constrain the set of start states. It may also indicate the lack of a required initial state in the implementation.

---

[2] The tableau for full ACTL is a *fair* Kripke structure (not defined here). It has the same properties except that $\models$ and $\leq$ are defined for fair structures.

2. $UnImplementedState = \{s_t \in S_t \mid \forall s_i \in S_i [(s_i, s_t) \notin ReachSIM]\}$
   An Unimplemented State is a state of the tableau that has no corresponding state in the implementation structure. This difference may suggest that
   the specification is not tight enough, or that a meaningful state was not
   implemented.

3. $UnImplementedTransition = \{(s_t, s'_t) \in R_t \mid \exists s_i, s'_i \in S_i,$
   $[(s_i, s_t) \in ReachSIM, (s'_i, s'_t) \in ReachSIM \text{ and } (s_i, s'_i) \notin R_i]\}$
   An Unimplemented Transition is a transition between two states of the
   tableau, for which a corresponding transition in the implementation does
   not exist. The existence of such a transition may suggest that the specification is not tight enough, or that a required transition (between reachable
   implementation states) was not implemented.

4. $ManyToOne = \{s_t \in S_t \mid \exists s_{1i}, s_{2i} \in S_i [(s_{1i}, s_t) \in ReachSIM, (s_{2i}, s_t) \in$
   $ReachSIM \text{ and } s_{1i} \neq s_{2i}]\}$
   A Many To One state is a tableau state to which multiple implementation
   states are mapped. The existence of such a state may indicate that the specification is not detailed enough. It may also suggest that the implementation
   contains redundancy.

Our criteria are defined for any tableau that has the tableau properties as defined
in Section 2. Any dissimilarity between the implementation and the specification
will result in a non empty criterion. Empty criteria indicate completeness, but
they are hard to obtain on traditional tableaux since such tableaux contain
redundancies. In the reduced tableau presented in Section 6, redundancies are
removed and therefore empty criteria are more likely to be achieved.

Given a structure M and a property ψ our method consists of the following
steps:

1. Apply model checking to verify that $M \models \psi$.
2. Build a (reduced) tableau $T(\psi)$ for ψ.
3. Compute SIM of $(M, T(\psi))$.
4. Compute $ReachSIM$ of $(M, T(\psi))$ from SIM of $(M, T(\psi))$.
5. For each of the comparison criteria, evaluate if its corresponding set is empty
   and if not present evidence for its failure.

Theorem 2. *Let M be an implementation model and ψ be an ACTL safety
formula such that $M \models \psi$. Let $T(\psi)$ be a tableau for ψ that has the tableau
properties. If the comparison criteria 1-3 hold then $T(\psi) \leq M$.*

The proof of this theorem is left to Appendix B. The proof implies that if criteria 1-3 hold then $T(\psi)$ and M are in fact *bisimilar*. The fourth criterion is
not necessary for completeness since whenever there are several non-bisimilar
implementation states that are mapped to the same tableau state, then there is
also an unimplemented state or transition. However, this criterion may reveal
redundancies in the implementation.

It is important to note that the goal is not to find a smaller set of criteria
that guarantees the specification completeness. The purpose of the criteria is to

assist the designer in the debugging process. Thus, we are looking for meaningful criteria that can distinguish among different types of problems and identify them. In Section 6 we define an additional criterion that can reveal redundancy in the specification.

4 Example

Consider a synchronous arbiter with two inputs, $req0, req1$ and two outputs $ack0, ack1$. The assertion of ack_i is a response to the assertion of req_i. Initially, both outputs of the arbiter are inactive. At any time, at most one acknowledge output may be active. The arbiter grants one of the active requests in the next cycle, and uses a round robin algorithm in case both request inputs are active. Furthermore in the case of *simultaneous assertion* (i.e. both requests are asserted and were not asserted in the previous cycle), request 0 has priority in the first *simultaneous assertion* occurrence. In any additional occurrence of *simultaneous assertion* the priority rotates with respect to the previous occurrence.

The implementation and the specification will share a common set of atomic propositions $AP = \{req0, req1, ack0, ack1\}$. An implementation of the arbiter M, written in the SMV language is presented below:

```
1)   var
2)     req0, req1, ack0, ack1, robin : boolean;
3)   assign
4)     init(ack0) := 0;
5)     init(ack1) := 0;
6)     init(robin) := 0;
7)   next(ack0) := case
8)     !req0              : 0;       – No request results no ack
9)     !req1              : 1;       – A single request
10)    !ack0 & !ack1      : !robin;  – Simultaneous requests assertions
11)    1                  : !ack0;   – Both requesting , toggle ack
12)   esac;
13)  next(ack1) := case
14)    !req1              : 0;       – No request results no ack
15)    !req0              : 1;       – A single request
16)    !ack0 & !ack1      : robin;   – simultaneous assertion
17)    1                  : !ack1;   – Both requesting , toggle ack
18)   esac;
19)  next(robin) := if req0 & req1 & !ack0 & !ack1 then !robin
20)                    else robin endif;    – Two simultaneous request assertions
```

From the verbal description given at the beginning of the section, one may derive a temporal formula that specifies the arbiter :

$\psi = \neg ack0 \wedge \neg ack1 \wedge$

$\quad \mathbf{A}[(\neg req0 \vee \neg req1 \vee ack0 \vee ack1)\mathbf{W}$

$\qquad (req0 \wedge req1 \wedge \neg ack0 \wedge \neg ack1 \wedge \mathbf{AX}ack0)] \qquad\qquad\qquad \wedge \qquad - \varphi_0$

$\quad \mathbf{AG}($

$\qquad (\neg ack0 \vee \neg ack1) \qquad\qquad\qquad\qquad\qquad\qquad\qquad\qquad \wedge \qquad - \varphi_1$

$\qquad (\neg req0 \wedge \neg req1 \to \mathbf{AX}(\neg ack0 \wedge \neg ack1)) \qquad\qquad \wedge \qquad - \varphi_2$

$\qquad (req0 \wedge \neg req1 \to \mathbf{AX}ack0) \qquad\qquad\qquad\qquad\quad \wedge \qquad - \varphi_3$

$\qquad (\neg req0 \wedge req1 \to \mathbf{AX}ack1) \qquad\qquad\qquad\qquad\quad \wedge \qquad - \varphi_4$

$\qquad (req1 \wedge ack0 \to \mathbf{AX}ack1) \qquad\qquad\qquad\qquad\qquad \wedge \qquad - \varphi_5$

$\qquad (req0 \wedge ack1 \to \mathbf{AX}ack0) \qquad\qquad\qquad\qquad\qquad \wedge \qquad - \varphi_6$

$\qquad (req0 \wedge req1 \wedge \neg ack0 \wedge \neg ack1 \to \mathbf{AX}(ack0 \to$

$\qquad \mathbf{A}[(\neg req0 \vee \neg req1 \vee ack0 \vee ack1)\mathbf{W}$

$\qquad\quad (req0 \wedge req1 \wedge \neg ack0 \wedge \neg ack1 \wedge \mathbf{AX}ack1)])) \qquad \wedge \qquad - \varphi_7$

$\qquad (req0 \wedge req1 \wedge \neg ack0 \wedge \neg ack1 \to \mathbf{AX}(ack1 \to$

$\qquad \mathbf{A}[(\neg req0 \vee \neg req1 \vee ack0 \vee ack1)\mathbf{W}$

$\qquad\quad (req0 \wedge req1 \wedge \neg ack0 \wedge \neg ack1 \wedge \mathbf{AX}ack0)])) \qquad) \qquad - \varphi_8$

where $\mathbf{AG}\varphi \equiv \mathbf{A}[\varphi \mathbf{W} false]$. We verified that $M \models \psi$ using the SMV model checker. We then applied our method. We found that all comparison criteria hold. We therefore concluded that ψ is a complete specification for M.

In order to exemplify the ability of our method we changed the implementation and the specification in different ways. In all cases the modified implementation satisfied the modified specification. However, our method reported the failure of some of the criteria. By examining the evidence supplied by the report, we could detect flaws in either the implementation or the specification.

4.1 Unimplemented Transition Evidence

Consider a modified version of the implementation M, named M_{trans} obtained by adding the line : $robin \& ack1 : \{0,1\}$;
between line (10) and line (11), and by adding the line :
$ack1 : !next(ack0)$; between line (16) and line (17).
Consider also the modified formula ψ_{trans} obtained from ψ by replacing φ_6 with:
$(req0 \wedge ack1 \to \mathbf{AX}(ack0 \vee ack1)) \wedge$
SMV shows that $M_{trans} \models \psi_{trans}$. However, applying the comparison method on M_{trans} and ψ_{trans}, reports an *Unimplemented Transition*. It supplies as an evidence a transition between tableau states s_t and s_t' such that $L_t(s_t) = L_t(s_t') = \{req0, req1, !ack0, ack1\}$. Such a transition is possible by ψ_{trans} but not possible in M_{trans} in case variable *robin* is not asserted.

If we check the reason for the incomplete specification we note that the evidence shows a cycle with $req0$ and $ack1$ asserted followed by a cycle were $ack1$ is asserted. This ill behavior violates the round robin requirement. The complete specification would detect that M_{trans} has a bug, since $M_{trans} \not\models \psi$.

4.2 Unimplemented State Evidence

Consider a modified version of the implementation M, named M_{unimp} obtained by adding line : $ack0 : \{0, 1\}$;
between lines (10) and line (11), and replacing line (2) with the following lines:
2.1) $req0\_temp, req1, ack0, ack1, robin : boolean$;
2.2) $define\ req0 := req0\_temp\ \&\ !(ack0\ \&\ ack1)$;
Here $req0\_temp$ is a free variable, and the input $req0$ is a restricted input such that if the state satisfies $ack0\&ack1$ then $req0$ is forced to be inactive.
Consider also the modified formula ψ_{unimp} obtained from ψ by deleting φ_1. SMV shows that $M_{unimp} \models \psi_{unimp}$. However, applying the comparison method on M_{unimp} and ψ_{unimp}, reports an *Unimplemented State*. It supplies as an evidence the state s_t such that $L_t(s_t) = \{req0, !req1, ack0, ack1\}$. This state is possible by ψ_{unimp} but not possible in M_{unimp}.
If we check the source of the incomplete specification we note that the evidence violates the mutual exclusion property. Both of the arbiter outputs $ack0$ and $ack1$ are active. The complete specification would detect that M_{unimp} has a bug, since $M_{unimp} \not\models \psi$.
Note that in this example we can also identify that $req0$ in M_{unimp} is a restricted input relative to the formula ψ_{unimp}. The state space of M_{unimp} does not include the states $\{req0, req1, ack0, ack1\}$ or $\{req0, !req1, ack0, ack1\}$. A restricted environment may hide bugs, so this is just as important as finding missing properties.

4.3 Many To One Evidence

A nonempty *Many To One* criterion may imply one of two cases. Redundant implementation, or incompleteness. The latter case is always accompanied with one of criteria 1-3. The former case where criteria 1-3 hold but we have a *Many To One* evidence implies that the implementation is complete with respect to the specification, but it is not efficient and contains redundancies. There is a smaller implementation that can preserve the completeness. This information may give insight on the efficiency of the implementation.
The following implementation M_{m2o} uses 5 implementation variables and two free inputs instead of 3 variables and two inputs of implementation M. Criteria 1-3 are met for M_{m2o} with respect to ψ.

1) var
2) $req0, req1, req0q, req1q, ack0q, ack1q, robin : boolean$;
3) $assign$
4) $init(req0q) := 0; init(req1q) := 0$;
5) $init(ack0q) := 0; init(ack1q) := 0$;
6) $init(robin) := 1$;
7) $define$
8) $ack0 := case$

```
9)   !req0q                : 0;        – No request results no ack
10)  !req1q                : 1;        – A single request
11)  !ack0q & !ack1q       : !robin;   – Simultaneous requests assertions
12)  1                     : !ack0q;   – Both requesting , toggle ack
13)  esac;
14)  ack1 := case
15)  !req1q                : 0;        – No request results no ack
16)  !req0q                : 1;        – A single request
17)  !ack0q & !ack1q       : robin;    – simultaneous assertion
18)  1                     : !ack1q;   – Both requesting , toggle ack
19)  esac;
20)  assign
21)  next(robin) := if req0 & req1 & !ack0 & !ack1 then !robin
22)      else robin endif;            – Two simultaneous request assertions
23)  next(req0q) := req0; next(req1q) := req1;
24)  next(ack0q) := ack0; next(ack1q) := ack1;
```

Applying model checking will show that $M_{m2o} \models \psi$.
In the above example we keep information of the current inputs $req0$ and $req1$, as well as their value in the previous cycle (i.e. $req0q$ and $req1q$). Intuitively, this duplicates each state in M to four states in the state space of M_{m2o}.

4.4 Unimplemented Start State Evidence

The *Unimplemented Start State* criterion does not hold when the specification is not restricted to the valid start states. Consider a specification formula obtained from ψ by removing the φ_0 subformula. Applying the comparison method on M and the modified formula would yield a *Unimplemented Start State* evidence of a tableau state s_{0t} such that $\{ack0, !ack1\} \subseteq L_t(s_{0t})$. Restricting the specification to the valid start states would cause the *Unimplemented Start State* criteria to hold.

4.5 Non-observable Implementation Variables

As can be seen in this example, a state of the implementation is not uniquely determined by $req0$, $req1$, $ack0$ and $ack1$. The variable *robin* effects the other variables, but it is a non-observable intermediate variable. This variable is not explicitly described in the specification, and does not appear in the common set of atomic propositions AP, referred to by the simulation preorder. Our criteria are defined with respect to observable variables only, but are not limited to systems where all the variables are observable.

5 Implementation of the Method

5.1 Symbolic Algorithms

In this section we present the symbolic algorithms that implement various parts of our method. In particular, we show how to compute symbolically the simulation relation. Our implementation will require less memory than the naive implementation since we reduce the number of OBDD variables. In Section 5.2 we show how this is achieved.

For conciseness, we use $R(s, s')$, $S(s)$ etc. instead of $(s, s') \in R$, $s \in S$.

Computing SIM: Let $M = (S_i, S_{0i}, R_i, L_i)$ be the implementation structure and let $T(\psi) = (S_t, S_{0t}, R_t, L_t)$ be a tableau structure. The following pseudo-code depicts the algorithm for computing SIM:

$Init$: $SIM_0(s_i, s_t) := \{ (s_i, s_t) \in S_i \times S_t \mid L_i(s_i) = L_t(s_t) \}$; $j := 0$
$Repeat$ {
$SIM_{j+1} := \{ (s_i, s_t) \mid \forall s'_i [R_i(s_i, s'_i) \rightarrow \exists s'_t [R_t(s_t, s'_t) \wedge SIM_j(s'_i, s'_t)]] \wedge SIM_j(s_i, s_t) \}$
$j := j + 1$ } $until$ $SIM_j = SIM_{j-1}$
$SIM := SIM_j$

Computing $ReachSIM$: Given the simulation relation SIM of the pair $(M, T(\psi))$ the following pseudo-code depicts the algorithm for computing $ReachSIM$:
$Init$: $ReachSIM_0 := (S_{0i} \times S_{0t}) \cap SIM$; $j := 0$
$Repeat$ {
$ReachSIM_{j+1} := ReachSIM_j \cup$
$\{ (s'_i, s'_t) \mid \exists s_i, s_t (ReachSIM_j(s_i, s_t) \wedge R_i(s_i, s'_i) \wedge R_t(s_t, s'_t) \wedge SIM(s'_i, s'_t)) \}$
$j := j + 1$ } $until$ $ReachSIM_j = ReachSIM_{j-1}$
$ReachSIM := ReachSIM_j$

5.2 Efficient OBDD Implementation

We now turn our attention to improving the performance of the algorithms described in the previous section. We assume that an implementation of such an algorithm will be done within a symbolic model checker such as SMV [7]. Since formal analysis always suffers from state explosion it is necessary to find methods to efficiently utilize computer memory. When working with OBDDs one possible way to do so is to try to minimize the number of OBDD variables that any OBDD created during the computation will have.

We can see from the algorithms presented before that some of the sets, constructed in intermediate computation steps, are defined over four sets of states: implementation states, specification states, tagged (next) implementation states, and tagged (next) specification states. For example, the computation of SIM_{j+1} is defined by means of the implementation states s_i, specification states s_t, tagged implementation states s'_i (representing implementation next states), and tagged specification states s'_t (representing specification next states).

Assume that we need at most n bits to encode each set of states. Then potentially some of the OBDDs created in the intermediate computations will have $4n$ OBDD variables. However, by breaking the algorithm operations to smaller ones and manipulating OBDDs in a nonstandard way we managed to bound the number of variables of the OBDDs created in intermediate computations by $2n$.

We define two operations, *compose* and *compose_odd*, that operate on two OBDDs a and b over a total number of $3n$ variables. As explained later, the main advantage of these operations is that they can be implemented using only $2n$ OBDD variables.

$$compose(\mathbf{y}, \mathbf{u}) \equiv \exists \mathbf{x}(a(\mathbf{x}, \mathbf{y}) \wedge b(\mathbf{x}, \mathbf{u})) \tag{1}$$

$$compose\_odd(\mathbf{y}, \mathbf{u}) \equiv \exists \mathbf{x}(a(\mathbf{y}, \mathbf{x}) \wedge b(\mathbf{u}, \mathbf{x})). \tag{2}$$

SIM and $ReachSIM$ can be implemented using *compose* and *compose_odd* as follows. Let $\mathbf{v_i}$, $\mathbf{v_i'}$ be the encoding of the states s_i, s_i' respectively. Similarly, let $\mathbf{v_t}$, $\mathbf{v_t'}$ be the encoding of s_t, s_t' respectively.

$SIM_{j+1}(\mathbf{v_i}, \mathbf{v_t}) := SIM_j(\mathbf{v_i}, \mathbf{v_t}) \wedge$
$\neg compose\_odd(R_i(\mathbf{v_i}, \mathbf{v_i'}), \neg compose\_odd(R_t(\mathbf{v_t}, \mathbf{v_t'}), SIM_j(\mathbf{v_i'}, \mathbf{v_t'})))$

$ReachSIM_{j+1}(\mathbf{v_i'}, \mathbf{v_t'}) := ReachSIM_j(\mathbf{v_i'}, \mathbf{v_t'}) \vee$
$(compose(compose(ReachSIM_j(\mathbf{v_i}, \mathbf{v_t}), R_i(\mathbf{v_i}, \mathbf{v_i'})), R_t(\mathbf{v_t}, \mathbf{v_t'})) \wedge SIM(\mathbf{v_i'}, \mathbf{v_t'}))$

The derivation of these expressions can be found in Appendix C. The algorithms above require that the implementation and specification "step" together along the transition relation. We break this to stepping along one, followed by stepping along the other. This is possible since transitions of the two structures are independent.

The comparison criteria *Unimplemented Transition* and *Many To One* can also be implemented with these operations. The two other criteria are defined over $2n$ variables and do not require such manipulation.

$UnimplementedTransition(\mathbf{v_t}, \mathbf{v_t'}) := R_t(\mathbf{v_t}, \mathbf{v_t'}) \wedge$
$compose(compose(\neg R_i(\mathbf{v_i}, \mathbf{v_i'}), ReachSIM(\mathbf{v_i}, \mathbf{v_t})), ReachSIM(\mathbf{v_i'}, \mathbf{v_t'}))$

$ManyToOne(\mathbf{v_t}) :=$
$\exists \mathbf{v_1}(ReachSIM(\mathbf{v_1}, \mathbf{v_t}) \wedge compose((\mathbf{v_1} \neq \mathbf{v_2}), ReachSIM(\mathbf{v_2}, \mathbf{v_t})))$

The details of these derivations can be found in Appendix C.

Up to now we showed how to reduce the number of OBDD variables from $4n$ to $3n$. We now show how to further reduce this number to $2n$. Our first step is to use the same OBDD variables to represent the implementation variables $\mathbf{v_i}$ and the specification variables $\mathbf{v_t}$. These OBDD variables will be referred to as *untagged*. Similarly, we use the same OBDD variables to represent $\mathbf{v_i'}$ and $\mathbf{v_t'}$. They will be referred to as *tagged* OBDD variables.

We also specify that whenever we have relations over both implementation variables and specification variables then the implementation variables are represented by untagged OBDD variables while the specification variables are represented by tagged OBDD variables. Note that now the relations R_i, R_t, SIM,

ReachSIM are all defined over the same sets of OBDD variables. Consequently, in all the derived expressions we apply *compose* and *compose_odd* to OBDDs that share variables, i.e. **y** and **u** are represented by the same OBDD variables. The implementation of *compose* and *compose_odd* uses non-standard OBDD operations in such a way that the resulting OBDDs are also defined over the same $2n$ variables.

Notice that this requires that the OBDD variable change semantics in the result (e.g., in Equation 1 **y** is represented by tagged OBDD variables in the input parameters and by untagged variables in the result). OBDD packages can easily be extended with these operations.

6 Reduced Tableau and Redundancies in Specification

6.1 Smaller Tableau Structure

When striving for completeness, the size of tableau structures as defined in [3] is usually too large to be practical, and may be much larger than the state space of the given implementation. This is because the state space of such tableaux contain all combinations of subformulas of the specification formula. Such tableaux usually contain many redundant states, that can be removed while preserving the tableau properties. If not removed, these states may introduce evidences which are not of interest.

Much of the redundancies can be eliminated if each state contains *exactly* the set of formulas required for satisfying the specification formula. Consider for example the ACTL formula **AXAX**p. Its set of subformulas is $\{\mathbf{AXAX}p, \mathbf{AX}p, p\}$. We desire a tableau structure in which each state contains only the set of subformulas required to satisfy the formula. In this case, the initial state should satisfy **AXAX**p, its successor should satisfy **AX**p and its successor should satisfy p. In each of these states all unmentioned subformulas have a "don't care" value. Thus, one state of the reduced tableau represents many states. For instance, the initial state $\{\mathbf{AXAX}p\}$ represents four initial states in the traditional tableau [3]. In such examples we may get a linear size tableau instead of an exponential one.

Following the above motivation, the reduced tableau will be defined over a *3-value labeling* for atomic propositions, i.e., for an atomic proposition p, a state may be labeled by either p, $\neg p$ or neither of them. Also, only the reachable portion of the structure will be constructed.

Further reduction may be obtained if the set of successors for each state is constructed more carefully. If a state s has two successors s' and s'', such that the set of formulas of s'' is contained in the set of formulas of s', then s' is not constructed. Any tableau behavior starting at s' has a corresponding behavior from s''. Thus, it is unnecessary to include both.

Given an ACTL safety formulas, the definition of the reduced tableau is derived from the Particle tableau for LTL, presented in [6] by replacing the use of the **X** temporal operator by **AX**. Since the only difference between LTL and ACTL is that temporal operators are always preceded by the universal path quantifier, this change is sufficient. In general, we will obtain a smaller tableau since we also avoid the construction of redundant successors.

Since the reduced tableau is based on the 3-value labeling, the definition of satisfaction and simulation preorder are changed accordingly. Our reduced tableau $T(\psi)$ for ACTL then has the same properties as the one in [3]:

- $T(\psi) \models \psi$.
- For every Kripke structure M, $M \leq T(\psi)$ if and only if $M \models \psi$.

Note that adopting the reduced tableau also requires modifications to our criteria due to the 3-value labeling semantics.

6.2 Reduced Tableau Results

We have defined the reduced tableau and proved its tableau properties. In addition we have adapted the comparison criteria to comply with the *3-value labeling*. We also coded the reduced tableau construction and the comparison criteria into the SMV model checker, performing the structure comparison in a symbolic manner.

We have run the arbiter example of Section 4 with the reduced tableau. For the complete specification formula ψ presented there we received a structure with 20 states. A traditional tableau structure would have a state space of 2^{15} states for ψ.

6.3 Identifying Redundancies in the Specification

Section 3 defines criteria that characterize when a specification is rich enough (i.e., complete). We would like also to determine whether a complete specification contains redundancies, i.e., subformulas that can be removed or be rewritten without destroying the completeness of the specification.

Given the reduced tableau, we suggest a new criterion, called *One To Many*, that identifies implementation states that are mapped (by *ReachSIM*) to multiple tableau states. Finding such states means that there is a smaller structure that corresponds to an equivalent specification formula. The criterion $OneToMany$ is defined by:

$OneToMany = \{s_i \in S_i \mid \exists s_{1t}, s_{2t} \in S_t[(s_i, s_{1t}) \in ReachSIM \land (s_i, s_{2t}) \in ReachSIM \land s_{1t} \neq s_{2t}]\}$.

6.4 One To Many Example

The following example demonstrates the One to Many criterion. It identifies a redundant sub formula, which does not add to the completeness of the specification formula. Consider the following specification formula :

$\psi_{One2Many} =$
$\quad \neg ack0 \wedge \neg ack1 \qquad\qquad\qquad\qquad\qquad \wedge$
$\quad \mathbf{A}[(\neg req0 \vee \neg req1 \vee ack0 \vee ack1)\mathbf{W}$
$\qquad (req0 \wedge req1 \wedge \neg ack0 \wedge \neg ack1 \wedge \mathbf{AX}ack0)] \wedge \qquad\qquad - \varphi_0$
$\quad \mathbf{AG}($
$\qquad (\neg ack0 \vee \neg ack1) \qquad\qquad\qquad\qquad \wedge \qquad\qquad\qquad - \varphi_1$
$\qquad (\neg req0 \wedge \neg req1 \wedge \mathbf{AX}(\neg ack0 \wedge \neg ack1) \qquad \vee \qquad\quad - \varphi_2$
$\qquad req0 \wedge \neg req1 \wedge \mathbf{AX}ack0 \qquad\qquad\qquad \vee \qquad\quad - \varphi_3$
$\qquad \neg req0 \wedge req1 \wedge ack1 \wedge \mathbf{AX}ack1 \qquad\qquad \vee \qquad\quad - \varphi_4$
$\qquad req0 \wedge req1 \wedge ack0 \wedge \mathbf{AX}ack1 \qquad\qquad \vee \qquad\quad - \varphi_5$
$\qquad req0 \wedge req1 \wedge ack1 \wedge \mathbf{AX}ack0 \qquad\qquad \vee \qquad\quad - \varphi_6$
$\qquad req1 \wedge ack1 \wedge \mathbf{AX}(ack0 \wedge \neg ack1) \qquad\qquad \vee \qquad\quad - \varphi_{redundant}$
$\qquad req0 \wedge req1 \wedge \neg ack0 \wedge \neg ack1 \wedge \mathbf{AX}($
$\qquad ack0 \wedge \mathbf{A}[(\neg req0 \vee \neg req1 \vee ack0 \vee ack1)\mathbf{W}$
$\qquad (req0 \wedge req1 \wedge \neg ack0 \wedge \neg ack1 \wedge \mathbf{AX}ack1)] \vee$
$\qquad ack1 \wedge \mathbf{A}[(\neg req0 \vee \neg req1 \vee ack0 \vee ack1)\mathbf{W}$
$\qquad (req0 \wedge req1 \wedge \neg ack0 \wedge \neg ack1 \wedge \mathbf{AX}ack0)])))) \qquad\quad - \varphi_7$

Our method reported that for $\psi_{One2Many}$ criteria 1-4 are met. In addition, it reported that the *One To Many* criterion is not met. As an evidence it provides the implementation state s_i such that $L_i(s_i) = \{req0, req1, \neg ack0, ack1\}$. This state is mapped to s_{1t} and s_{2t} of the reduced tableau for which $L_t(s_{1t}) = \{req0, req1, \neg ack0, ack1\}$ and $L_t(s_{2t}) = \{req1, \neg ack0, ack1\}$.

We may note that $\varphi_{redundant}$ sub formulas agrees with φ_6 for states labeled with $\{req0, req1\}$, and does not agree with φ_4 for states labeled with $\{\neg req0, req1\}$. Since it comes as a disjunct, it does not limit the reachable simulation, and does not add allowed behavior. Deleting sub formula $\varphi_{redundant}$ leaves a specification formula such that criteria 1-4 are met and the *One to Many* criterion is also met.

7 Future Work

In this paper we presented a novel approach for evaluating the quality of the model checking process. The method we described can give an engineer the confidence that the model is indeed "bug-free" and reduce the development time.

We are aware that the work we have done is not complete. There are a few technical issues that will have to be addressed:

1. **State explosion:** The state explosion problem is even more acute than with model checking because we have to perform symbolic computations while M and $T(\psi)$ are both in memory. This implies that at present the circuits that we can apply this method to are smaller than those that we can model check. Therefore we currently cannot provide a solution for large models. However we believe that over time optimizations in this area will be introduced as

was done for model checking. We are investigating the possibility of running this method separately on small properties and then combining the results. Another solution to the state explosion is to compute the criteria "on-the-fly" together with the computation of *ReachSIM* and to discover violations before *ReachSIM* is fully computed.

A third solution is to use the algorithm in [5] as a preliminary step, and try to expand it to fully support our methodology. The definition of Unimplemented State is closely related to the evidences in [5]. On the other hand, our Unimplemented Transition criterion provides path evidences, while path coverage is not addressed by the methodology of [5]. Furthermore, our method can indicate that the specification and the implementation totally agree. This may serve as an indication that the verification process can be stopped.

2. **Irrelevant information:** Similar to the area of traditional simulation coverage, measurement of quality produces a lot of information which is often irrelevant. A major problem is that specifications tend to be incomplete by nature and therefore we do not necessarily want to achieve a bisimulation relation between the specification and implementation. Therefore, it will eventually be necessary to devise techniques to filter the results such that only the interesting evidences are reported.

   We are also investigating whether the reduced tableau described in Section 6 is optimal in the sense that it does not contain any redundancies.

3. **Expressivity:** Our specification language is currently restricted to ACTL safety formulas. It is straight forward to extend our method to full ACTL. This will require, however, to add fairness constraints to the tableau structure and to use the *fair simulation preorder* [3]. Unfortunately, there is no efficient algorithm to implement fair simulation [4]. Thus, it is currently impractical to use full ACTL. There is a need to find logics that are both reasonable in terms of their expressivity and practical in terms of tableau construction and comparison criteria.

Acknowledgment: We thank Ilan Beer for suggesting to look into the problem of coverage in model checking. The first author thanks Galileo Technology for the opportunity to work on the subject.

References

1. I. Beer, S. Ben-David, C. Eisner, and A. Landver. Rulebase - an industry oriented formal verification tool. In *33th Design Automation Conference*, 1996. DAC.
2. E.M. Clarke, O. Grumberg, and D. Peled. *Model Checking*. MIT press, 1999. To appear.
3. O. Grumberg and D.E. Long. Model checking and modular verification. *ACM Trans. on Programming Languages and Systems*, 16(3):843–871, 1994.

4. T. A. Henzinger, O. Kupferman, and S. K. Rajamani. Fair simulation. In *Proc. of the 7th Conference on Concurrency Theory (CONCUR'97)*, volume 1243 of *LNCS*, Warsaw, July 1997.

5. Hoskote, Kam, Ho, and Zhao. Coverage estimation for symbolic model checking. In *proceedings of the 36rd Design Automation Conference (DAC'99)*. IEEE Computer Society Press, June 1999.

6. Z. Manna and A. Pnueli. *Temporal verifications of Reactive Systems - Safety*. Springer-Verlag, 1995.

7. K. L. McMillan. *The SMV System DRAFT*. Carnegie Mellon University, Pittsburgh, PA, 1992.

8. K. L. McMillan. *Symbolic Model Checking*. Kluwer Academic Press, Norwell, MA, 1993.

9. R. Milner. An algebraic definition of simulation between programs. In *In proceedings of the 2nd International Joint Conference on Artificial Intelligence*, pages 481–489, September 1971.

10. T. Filkorn. A method for symbolic verification of synchronous circuits. In D. Borrione and R. Waxman, editors, *Proceedings of The Tenth International Symposium on Computer Hardware Description Languages and their Applications*, IFIP WG 10.2, pages 249–259, Marseille, April 1991. North-Holland.

11. Elaine J. Weyuker and Bingchiang Jeng. Analyzing partition testing strategies. *IEEE Transactions on Software Engineering*, 2(17), July 1991.

A   Proof of Lemma 1

Lemma 1 *ReachSIM is a simulation preorder from M to M'.*

Proof. Clearly, for initial states, $(s_0, s_0') \in SIM$ if and only if $(s_0, s_0') \in ReachSIM$. Thus, for every initial state of M there is a $ReachSIM$-related initial state of M'. Let $(s, s') \in ReachSIM$. First we note that since $ReachSIM \subseteq SIM$, $(s, s') \in SIM$ and therefore $L(s) = L'(s')$.

Now let $(s, s_1) \in R$. Then there is s_1' such that $(s', s_1') \in R'$ and $(s_1, s_1') \in SIM$. Since $(s, s') \in ReachSIM$, there are corresponding paths π and π' leading to s and s'. These paths can be extended to corresponding paths leading to s_1 and s_1'. Thus, $(s_1, s_1') \in ReachSIM$. $\square$

B   Proof of Theorem 2

Theorem 1 *Let M be an implementation model and ψ be an ACTL safety formula such that $M \models \psi$. Let $T(\psi)$ be a tableau for ψ that satisfy the tableau properties. If the comparison criteria 1-3 hold then $T(\psi) \leq M$.*

Proof. Since $M \models \psi$, $M \leq T(\psi)$. Thus, there is a simulation preorder $SIM \subseteq S_i \times S_t$. Let $ReachSIM$ be the reachable simulation preorder for SIM. Then $ReachSIM^{-1} \subseteq S_t \times S_i$ is defined by $(s_t, s_i) \in ReachSIM^{-1}$ if and only if $(s_i, s_t) \in ReachSIM$. We show that $ReachSIM^{-1}$ is a simulation preorder from $T(\psi)$ to M.

Let s_{0t} be an initial state of $T(\psi)$. Since $UnImplementedStartState$ is empty, there must be an initial state s_{0i} of M such that $(s_{0i}, s_{0t}) \in ReachSIM$. Thus, $(s_{0t}, s_{0i}) \in ReachSIM^{-1}$.

Now let $(s_t, s_i) \in ReachSIM^{-1}$. Since $(s_i, s_t) \in ReachSIM$, $L_t(s_t) = L_i(s_i)$. Let $(s_t, s'_t) \in R_t$. Since $UnimplementedState$ is empty, there must be a state $s'_i \in S_i$ such that $(s'_i, s'_t) \in ReachSIM$. Since $UnImplementedTransition$ is empty we get $(s_i, s'_i) \in R_i$. Thus, s'_i is a successor of s_i and $(s'_t, s'_i) \in ReachSIM^{-1}$. We conclude that $ReachSIM^{-1}$ is a simulation preorder and therefore $T(\psi) \leq M$. □

Note that, since $ReachSIM$ and $ReachSIM^{-1}$ are both simulation preorders, $ReachSIM$ is actually a bisimulation relation.

C Derivation of the *Compose* Formulas

Following is the algebraic derivation that enables the use of the *compose* and *compose_odd* operations described in Section 5.

- Derivation of SIM_j:

  $SIM_{j+1}(v_i, v_t) :=$
  $\forall v'_i [R_i(v_i, v'_i) \rightarrow \exists v'_t [R_t(v_t, v'_t) \wedge SIM_j(v'_i, v'_t)]] \wedge SIM_j(v_i, v_t) =$
  $\forall v'_i [\neg R_i(v_i, v'_i) \vee \exists v'_t [R_t(v_t, v'_t) \wedge SIM_j(v'_i, v'_t)]] \wedge SIM_j(v_i, v_t) =$
  $\neg \exists v'_i [R_i(v_i, v'_i) \wedge \neg \exists v'_t [R_t(v_t, v'_t) \wedge SIM_j(v'_i, v'_t)]] \wedge SIM_j(v_i, v_t) =$
  $\neg compose\_odd(R_i(v_i, v'_i), \neg compose\_odd(R_t(v_t, v'_t), SIM_j(v'_i, v'_t))) \wedge SIM_j(v_i, v_t)$

- Derivation of $ReachSIM_j$:

  $f_{j+1}(v_t, v'_i) :=$
  $\exists v_i (ReachSIM_j(v_i, v_t) \wedge R_i(v_i, v'_i)) = compose(ReachSIM_j(v_i, v_t), R_i(v_i, v'_i))$
  $g_{j+1}(v'_i, v'_t) :=$
  $\exists v_t (f_{j+1}(v_t, v'_i) \wedge R_t(v_t, v'_t)) = compose(f_{j+1}(v_t, v'_i), R_t(v_t, v'_t))$
  $g_{j+1}(v_i, v_t) := g_{j+1}(v'_i, v'_t)$
  $ReachSIM_{j+1}(v_i, v_t) := (g_{j+1}(v_i, v_t) \wedge SIM(v_i, v_t)) \vee ReachSIM_j(v_i, v_t)$

- Derivation of $ManyToOne$:

  $ManyToOne(v_t) :=$
  $\exists v_1, v_2 (ReachSIM(v_1, v_t) \wedge ReachSIM(v_2, v_t) \wedge (v_1 \neq v_2)) =$
  $\exists v_1 (ReachSIM(v_1, v_t) \wedge \exists v_2 ((v_2 \neq v_1) \wedge ReachSIM(v_2, v_t))) =$
  $\exists v_1 (ReachSIM(v_1, v_t) \wedge compose((v_1 \neq v_2), ReachSIM(v_2, v_t)))$

- Derivation of $UnimplementedTransition$:

  $f(v'_i, v_t) :=$
  $\exists v_i (\neg R_i(v_i, v'_i) \wedge ReachSIM(v_i, v_t)) = compose(\neg R_i(v_i, v'_i), ReachSIM(v_i, v_t))$
  $g(v_t, v'_t) :=$
  $\exists v'_i (f(v'_i, v_t) \wedge ReachSIM(v'_i, v'_t)) = compose(f(v'_i, v_t), ReachSIM(v'_i, v'_t))$
  $UnimplementedTransition(v_t, v'_t) := g(v_t, v'_t) \wedge R_t(v_t, v'_t)$

Program Slicing of Hardware Description Languages*

E. M. Clarke[1,6], M. Fujita[3], S. P. Rajan[3], T. Reps[4,7], S. Shankar[1,5,6], and T. Teitelbaum[2,4]

[1] Carnegie Mellon University, Pittsburgh, PA, USA (emc+@cs.cmu.edu)
[2] Cornell University, Ithaca, NY, USA
[3] Fujitsu Labs of America, Sunnyvale, CA, USA ({fujita,sree}@fla.fujitsu.com)
[4] Grammatech, Inc., Ithaca, NY, USA ({reps, tt}@grammatech.com)
[5] Hunter College and the Graduate School, The City University of New York, New York, NY, USA (sshankar@roz.hunter.cuny.edu)
[6] Verysys Design Automation, Inc., Fremont, CA, USA
[7] University of Wisconsin, Madison, WI, USA

Abstract. Hardware description languages (HDLs) are used today to describe circuits at all levels. In large HDL programs, there is a need for source code reduction techniques to address a myriad of problems in formal verification, design, simulation, and testing. Program slicing is a static program analysis technique that allows an analyst to automatically extract portions of programs relevant to the aspects being analyzed. We extend program slicing to HDLs, thus allowing for automatic program reduction to allow the user to focus on relevant code portions. We have implemented a VHDL slicing tool composed of a general inter-procedural slicer and a front-end that captures VHDL execution semantics. This paper provides an overview of program slicing, a discussion of how to slice VHDL programs, a description of the resulting tool, and a brief overview of some applications and experimental results.

1 Introduction

Hardware description languages (HDLs) are used today to describe circuits at all levels from conceptual system architecture to low-level circuit implementations suitable for synthesis. There are also several tools that apply model checking [1] to formally verify correctness of HDL designs (one such system for VHDL is described in [2]). The fundamental problem in model checking is state explosion,

* This research is supported in part by the Semiconductor Research Corporation (SRC) (Contract 97-DJ-294), the National Science Foundation (NSF) (Grants CCR-9505472, CCR-9625667, CCR-9619219), the Defense Advanced Research Projects Agency (DARPA) (Contract DABT63-96-C-0071), the United States-Israel Binational Science Foundation (Grant 96-00337), IBM, and the University of Wisconsin (Vilas Associate Award). Any opinions, findings and conclusions or recommendations expressed in this material are those of the authors and do not necessarily reflect the views of the supporting agencies.

and there is consequently a need to reduce the size of HDL descriptions so that their corresponding models have fewer states. For many designs, it is not even possible to build the state transition relation, and the need for HDL program reduction techniques is even more critical in these cases.

HDL reduction is also useful for other design, simulation, and testing tasks since a major lack in current simulation methodologies is the need for structured design and analysis techniques. The use of automatic reduction techniques allows an analyst to focus on relevant code portions for further study. Moreover, with the increasing use of reusable libraries of existing code, reduction techniques are also useful to simplify the usage and/or modification of unstructured libraries.

Several of these desiderata have close parallels in the software-engineering domain, where it is desirable to understand and manipulate large programs. This is difficult to do, partly because of the presence of large quantities of irrelevant code. *Program slicing* was defined by Weiser [3] to cope with these problems by performing automatic decomposition of programs based on data- and control-flow analysis. A *program slice* consists of those parts of a program that can potentially affect (or be affected by) a *slicing criterion* (i.e., a set of program points of interest to the user). The identification of program slices with respect to a slicing criterion allows the user to reduce the original program to one that is simpler but functionally equivalent with respect to the slicing criterion.

Program slicing results in the software engineering world suggest that the techniques can also be applied to HDLs to solve many of the problems mentioned above. However, most traditional slicing techniques are designed for sequential procedural programming languages, and the techniques are not directly applicable to HDLs, which have a fundamentally different computation paradigm: An HDL program is a non-halting reactive system composed of a set of concurrent processes, and many HDL constructs have no direct analogue in more traditional programming languages. In this paper, we present an approach for slicing VHDL based on its operational semantics. Our approach is based on a mapping of VHDL constructs onto traditional programming language constructs, in a way that ensures that all traces of the VHDL program will also be valid traces in the corresponding sequential program. Corresponding to this approach, we have implemented a VHDL slicing tool consisting of a VHDL front-end coupled with a language-independent toolset intended for inter-procedural slicing of sequential languages such as C. We have applied the tool to some formal verification problems, and have achieved substantial state space reductions.

The remainder of the paper is organized as follows: Section 2 presents requisite background material while Section 3 presents our techniques for performing language-independent interprocedural slicing. Section 4 shows how we capture VHDL semantics for slicing. Section 5 describes the architecture and implementation of the VHDL slicing tool, and provides a walkthrough of a simple VHDL example. Section 6 lists some applications of slicing, and provides experimental results that concretely illustrate the benefits of slicing in reducing state space size for model checking. We compare our work with other approaches in Section 7. Finally, Section 8 summarizes our conclusions and briefly discusses our future plans in this area.

2 Background

Slicing is an operation that identifies semantically meaningful decompositions of programs, where the decompositions consist of elements that are not necessarily textually contiguous [4, 3, 5–8]. (See [9, 10] for surveys on slicing.) Slicing, and subsequent manipulation of slices, has applications in many software-engineering tasks, including program understanding, maintenance [11], debugging [12], testing [13, 14], differencing [15, 16], specialization [17], reuse [18], and merging [15].

There are two kinds of slices: a *backward slice* of a program with respect to a slicing criterion C is the set of all program elements that might affect (either directly or transitively) the values of the variables used at members of C; a *forward slice* with respect to C is the set of all program elements that might be affected by the computations performed at members of C.

A related operation is program *chopping* [19, 20]. A chop answers questions of the form "Which program elements serve to transmit effects from a given source element s to a given target element t?". Given sets of source and target program points, S and T, the *chop* consists of all program points that could transmit the effect of executing a member of S to a member of T.

It is important to understand the distinction between two different but related "slicing problems": the *closure slice* of a program P with respect to program point p and variable x identifies all statements and predicates of P that might affect the value of x at point p; the *executable slice* of P with respect to p and x is a reduced program that computes the same sequence of values for x at p (i.e., at point p the behavior of the reduced program with respect to x is indistinguishable from that of P). In *intra*procedural slicing, an executable slice can be obtained from the closure slice; however, in *inter*procedural slicing, where a slice can cross the boundaries of procedure calls, an executable slice is harder to obtain since the closure slice might contain different subsets of a procedure's parameters for different calls to the same procedure. However a closure slice can always be extended to an executable slice [21]. Our system does closure slicing, with partial support for executable slicing.

A second major design issue is the type of interprocedural slicing. Some slicing and chopping algorithms are *precise* in the sense that they track dependences transmitted through the program only along paths that reflect the fact that when a procedure call finishes, control returns to the site of the most recently invoked call [7, 8, 20]. In contrast, other algorithms are *imprecise* in that they safely, but pessimistically, track dependences along paths that enter a procedure at one call site, but return to a different call site [22, 19]. Precise algorithms are preferable because they return smaller slices. Precise slicing and chopping can be performed in polynomial time [7, 8, 20]. Our VHDL slicing tool supports precise interprocedural slicing and chopping.

3 Inter-procedural Slicing

The value of a variable x *defined* at p is directly affected by the values of the variables used at p and by the predicates that control how many times p is executed. Similarly, the value of a variable y *used* at p is directly affected by assignments to y that reach p and by the predicates that control how many times p is executed. Consequently, a slice can be obtained by following chains of dependences in the directly-affects relation. This observation is due to Ottenstein and Ottenstein [5], who noted that *procedure dependence graphs* (PDGs), which were originally devised for use in parallelizing and vectorizing compilers, are a convenient data structure for slicing.

The PDG for a procedure is a directed graph whose vertices represent the individual statements and predicates of the procedure. Vertices are included for each of the following constructs:

- Each procedure has an entry vertex.
- Each formal parameter has a vertex representing its initialization from the corresponding actual parameter.
- Each assignment statement has a vertex.
- Each control-structure condition (e.g., *if*) has a vertex.
- Each procedure call has a vertex.
- Each actual parameter to a procedure has a vertex representing the assignment of the argument expression to some implicit (generated) variable.
- Each procedure with a return value has a vertex representing the assignment of the return value to some generated name.
- Each formal parameter and local variable has a vertex representing its declaration.

A procedure's parameters may sometimes be implicit. If a procedure assigns to or uses a global variable x (either directly or transitively via a procedure call), x is treated as an "hidden" input parameter, thus giving rise to additional actual-in and formal-in vertices. Similarly, if a procedure assigns to a global variable x (either directly or transitively), x is treated as a "hidden" output parameter, thus giving rise to additional actual-out and formal-out vertices.

Denote the program code corresponding to a vertex V as $\#V$. PDG vertices are connected through the following types of edges:

- There is a *flow dependence edge* between two vertices v_1 and v_2 if there exists a program variable x such that v_1 can assign a value to x, v_2 can use the value in x, and there is an execution path in the program from $\#v_1$ to $\#v_2$ along which there is no assignment to x.
- There is a *control dependence edge* between a condition vertex v_c and a second vertex v if the truth of the condition of $\#v_c$ controls whether or not $\#v$ is executed.
- There is a *declaration edge* from the declaration vertex for a program variable, x, to each vertex that can reference x.

- There is a *summary edge* corresponding to each indirect dependence from a procedure call's actual parameters and its output(s). These edges are used to avoid recomputing these summary relationships, for efficiency reasons. They are actually computed after PDG construction.

Given PDGs for each procedure, a *system dependence graph* (SDG) is then constructed by connecting the PDGs appropriately using the following additional types of edges:

- There is a *call edge* from a procedure call vertex to the corresponding procedure entry vertex.
- There is a *parameter-in edge* between each actual parameter and the corresponding formal parameter.
- There is a *parameter-out edge* between each procedure output value vertex and the vertex for an implicit (generated) variable on the caller side designated to receive it.

Figure 1 illustrates a SDG for a small pseudocode program.

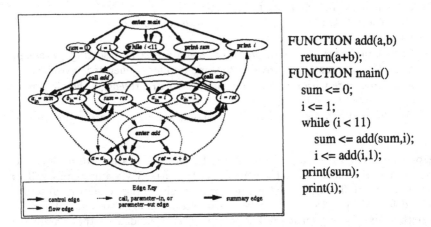

```
FUNCTION add(a,b)
    return(a+b);
FUNCTION main()
    sum <= 0;
    i <= 1;
    while (i < 11)
        sum <= add(sum,i);
        i <= add(i,1);
    print(sum);
    print(i);
```

Fig. 1. Sample SDG

The complete algorithm for building a SDG from a program is:

1. Build a Control Flow Graph (CFG) for each procedure in the program.
2. Build the call graph for the program.
3. Perform global variable analysis, turning global variables into hidden parameters of the procedures that reference or modify them.
4. Construct PDGs by doing control-dependence and flow-dependence analysis.
5. Optionally compress the PDG so that each strongly connected region is represented by one node.
6. Bring together the PDGs and the call graph to form the SDG.

7. Compute summary edges for procedures that describe dependences between the inputs and the outputs of each procedure.

Then, slices and chops are computed by following the chains of dependences represented in the edges of the SDG.

4 VHDL Slicing

Rather than creating an independent slicer built specifically for VHDL, our approach is to map VHDL constructs onto constructs for more traditional procedural languages (e.g., C, Ada), utilizing the operational semantics provided by the VHDL LRM [23]. Figure 2 lists the mapping between VHDL and traditional constructs that we use:

VHDL Construct	Traditional Construct
Process, Concurrent Assignment Function, Procedure	Procedure
Architecture variable	Local variable
Signal, Port	Global variable
Sequential Statement	Statement

Fig. 2. Mapping of VHDL Constructs

While many of these mappings may seem obvious, there are several major differences between VHDL and traditional programming languages which complicate the generation of the SDG. A VHDL program executes as a series of simulation cycles, as illustrated in Figure 3.

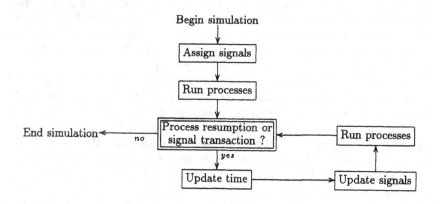

Fig. 3. Simplified VHDL simulation cycle

The VHDL computational paradigm differs fundamentally from traditional languages in three ways:

1. A VHDL program is a non-halting reactive system, rather than a collection of halting procedures.
2. A VHDL program is a concurrent composition of processes, without any explicit means for these processes to be invoked (in the manner of traditional procedures).
3. VHDL processes communicate through multiple-reader signals to which they are sensitive, instead of through parameters defined at a single procedure entry point.

VHDL procedures and functions are modeled in the traditional way. However, VHDL process models must capture the above differences, and we do this through three types of modifications:

- The CFGs model the non-halting reactive nature of VHDL processes.
- The PDGs capture an additional dependence corresponding to VHDL signal communication.
- An implicit generated master "main" procedure controls process invocation, analogous to the event queue that controls VHDL simulator execution.

These mechanisms are described below. Although the discussion only mentions processes, concurrent statements are to be treated analogously.

4.1 Constructing the CFG

CFG construction for traditional languages is well understood, and the identical technique is used for VHDL procedures and functions. VHDL processes require some CFG modifications. We first consider processes with an explicit sensitivity list or a single wait statement. The non-halting nature of processes is modeled simply by passing control from the end of the process back to its beginning. The wait statement provides the only complication. As suggested by Figure 3, from a wait statement, either control passes to the next statement or the simulation exits (in case the wait condition is never satisfied). This is simple to capture in the CFG by creating two corresponding child control-flow arcs from the wait statement. Figure 4 illustrates a CFG for a simple process.

The situation is substantially more complicated when there are multiple wait statements in the process. Although the above procedure still works, the resulting slice may be substantially larger than needed. Since each wait statement corresponds to a point where a region of the process may be invoked, a forward slice that affects a wait statement needs to include only the portion of the process between the wait statement and the next wait statement (and similarly for backward slices). To model this, we partition each process into regions corresponding to portions of processes between successive wait statements. More precisely, for each wait statement w in a process p, let T be the set of statements in some process execution trace starting at w and proceeding until another wait

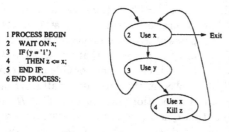

```
1 PROCESS BEGIN
2   WAIT ON x;
3   IF (y = '1')
4     THEN z <= x;
5   END IF;
6 END PROCESS;
```

Fig. 4. Sample CFG

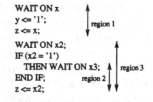

```
WAIT ON x
y <= '1';                    region 1
z <= x;

WAIT ON x2;
IF (x2 = '1')
  THEN WAIT ON x3;     region 3
END IF;         region 2
z <= x2;
```

Fig. 5. Process regions in the presence of multiple wait statements

statement w' with no intervening wait statements ($w' \notin T$). Let T be the set of all such traces. Then, there is a *process region* corresponding to w which includes all the statements in T. Figure 5 illustrates a simple example.

Note that we only require that each end node of a region precedes a wait statement; there may be multiple end nodes, and regions may overlap in the presence of wait statements within branching control structures (though very few VHDL programs have such control structures in practice). Then, a procedure for each process region is created, and a CFG for each of the resulting procedures is created as usual. To capture context information between process regions within the same process, all objects local to the process (e.g., variables) are treated as global variables after renaming to avoid conflicts with other processes (recall that the SDG build algorithms treat global variables as hidden parameters). Thus, for a process with W wait statements, W+1 procedures are created, one starting at each of the wait statements and one starting at the beginning of the process.

4.2 PDG Modifications

In traditional languages, inter-procedure communication occurs through global variables and parameters explicitly passed from the calling procedure to a called procedure. In contrast, VHDL process communication occurs through signals, and a process (or region) is invoked when it is at a wait statement w, and there is an event on a signal that w is sensitive to. This communication is captured through the notion of signal dependence (in addition to the dependence types listed in Section 3): A process region p is said to be *signal dependent* on statement s if s assigns a value to a signal that p is sensitive to. Rather than modeling this signal dependence explicitly in the PDG, we generate implicit procedure calls in the CFG every time a signal is potentially assigned. For example, every assignment to signal s is followed by implicit calls to every procedure (e.g., VHDL process region, concurrent assignment) that is sensitive to s.

4.3 The Master Process

The above changes do not handle the reactive nature of VHDL, since processes may also be invoked by events on input ports. For simplicity, the following discussion deals with processes, though the same arguments are also applicable to process regions. Consider a VHDL program $\Pi = \|_{i=1}^{n} P_i$, where the P_i's are the processes comprising the program (as before, other concurrent statements are treated as one-line processes for the purposes of this discussion). Partition Π into two disjoint sets Π_1, Π_2, where Π_1 is the set of processes that are sensitive to at least one input port (hence, $\Pi_2 = \Pi \setminus \Pi_1$). It is clearly not possible to determine *a priori* whether a process $P \in \Pi_1$ is invoked in the simulation (after its initial invocation). In contrast, any non-initial invocations of a process $Q \in \Pi_2$ must occur after an assignment to a signal that Q is sensitive to, and such invocations are handled using the signal dependences discussed above. Given these two observations, a CFG for the master process comprising the following (pseudocode) steps can be constructed:

> for $Q \in \Pi$ – initial invocations of each process
>   call Q
> while (true)   – subsequent invocations of Π_1 processes
>   for $P \in \Pi_1$
>   call P

Given PDGs for the master process and each procedure in the VHDL program, the SDG is constructed as usual.

4.4 Correctness

Our motivation for the VHDL mapping discussed above is captured in the following theorem: First, define a VHDL *process invocation trace* to be a sequence $T = \langle T_1, T_2, \ldots, T_i, \ldots \rangle$, where $T_i \in 2^{\Pi}$, and T_i is the set of processes that are invoked on simulation cycle i of the trace.

Theorem 1 *Let the VHDL program Π have a process invocation trace $T = \langle T_1, T_2, \ldots, T_i, \ldots \rangle$. Then, for any $P_j \in T_i$, either 1) $P_j \in \Pi_1$ or 2) there is a signal dependence from some statement in $P_k \in T_\ell$ to P_j for some $\ell < i$.*

The theorem can be seen to follow from VHDL operational semantics.

Since the two cases of the theorem are captured using the master process and our notion of signal dependence, any inter-process dependences will have corresponding call edges in the SDG, by construction, and correctness of our VHDL slicing semantics thus follows from the theorem.

5 The Slicer

Figure 6 illustrates the architecture of the VHDL slicer. The tool implements all the issues discussed above except process regions. The CFG Extractor and SDG Builder perform the algorithm described in Section 3 and output the SDG, as well as a map from PDG nodes to source-text references. The slicing and chopping algorithms are embedded in the slicer core.

The slicer user interface is through the source code: By maintaining appropriate maps between the underlying SDG on which slicing is performed and the

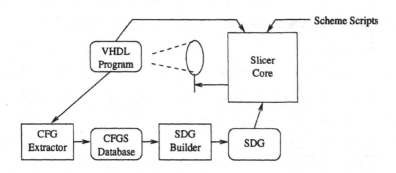

Fig. 6. VHDL Slicer Architecture

source code, the user specifies the slicing criterion by selecting program elements in the display of the source code, and the slice itself is displayed by highlighting the appropriate parts of the program. Slices may be forward or backward, and unions of slices may be computed using the GUI. The toolset GUI also supports browsing of projects and project files, as well as navigation through dependence graphs, slices, and chops.

5.1 Tool Walkthrough

To give a feel for the interface and some capabilities of our tool, we use a simple VHDL program, consisting of 1 D flip-flop and 2 logic functions (Figures 7(a),(b))[1]. The project view provides hierarchical summary information that is interactively viewable (partially shown in the figure), while the file view provides the actual text comprising the program.

Figure 7(c) shows a file view of the executable statements in the forward slice on the program point t1 <= t0 AND not(a);. As expected, the slice includes the flip-flop but not any input circuitry. Figure 7(d) shows a project view of the backward slice on the same program point as above. This time, the slice excludes the flip-flop. In large files, the colorbars to the right of the scrollbar allow the user to quickly scroll to the slice.

6 Applications

HDL slicing is useful in model checking to prove circuit correctness. The major problem in model checking is state space explosion. A backward slice on the set of statements assigning values to variables in the temporal specification results in a program subset consisting of only the statements that can potentially affect the correctness of the specification. Figure 8 illustrates the state space reduction

---

[1] The screenshots reproduced here are dithered monochrome versions of the color tool output, and thus suffer from some loss of clarity here

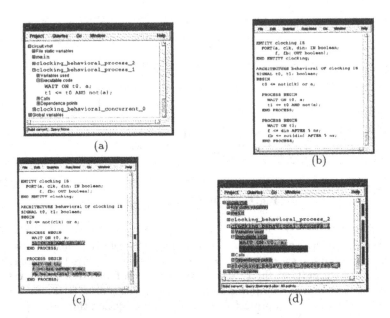

Fig. 7. Example (a) Project View, (b) File View, (c) Forward Slice, (d) Backward Slice

that was achieved in the verification of the controller logic for a RISC processor, using the model checker described in [2] and sliced with respect to two different CTL specifications (all builds and slicing operations needed negligible time).

	Processes	Concurrent Statements	Total States	Reachable States
Original	7	18	$1.8X10^{47}$	$2.5X10^{37}$
Sliced (safety)	4	3	$2.0X10^{31}$	$2.8X10^{21}$
Original	7	18	$1.8X10^{47}$	$6.5X10^{39}$
Sliced (liveness)	4	1	$3.1X10^{29}$	$1.1X10^{22}$

Fig. 8. Benefits of Slicing for Formal Verification

The benefits of slicing vary widely depending on the nature of the circuit and the property being verified, and the above circuit was selected to be a typical one. However, an interesting aspect of slicing-based program reduction is a reverse-scalability effect. Since smaller VHDL programs tend to have fewer irrelevant components, we have observed the benefits of slicing to improve (percentage-wise) as programs grow in size.

There are numerous other applications of slicing in hardware design, simulation, and testing. The reader is referred to [24] for a more detailed description of the applications of slicing, but we briefly list some of them in this section. Sample questions that a slicer can assist an engineer in answering include:

- How do you specialize (or modify) an existing IP design for reuse?
- What part of the design is relevant to the actual function (and not the design-for-test and debug circuitry)?
- What part of the circuit is in the control path (and not the datapath)?
- What code portions can potentially cause an unexpected signal value found in simulation?
- What portions of the circuit can be affected by changing a particular code segment?
- What potentially harmful interactions with other modules can result from changing a particular code segment?
- What execution paths are covered by a given test vector?
- What part of the circuit must be retested if a certain code segment is changed (i.e., regression testing)?
- What portion of a circuit is controllable from a given set of input ports?
- What portion of a circuit is observable from a given set of output ports?
- What portion of a circuit is testable (i.e., both controllable and observable) from a given set of input and output ports?

7 Related Work

The only other application of program slicing to HDLs that we are aware of is by [25], which discusses a number of issues and applications related to VHDL slicing (a resulting system that implements some of these is discussed in [26]). Our approach differs since it captures VHDL operational semantics within existing procedural frameworks so that the benefits of existing precise slicing technology can better be exploited. We believe that using such an approach will enable us to cover a larger subset of VHDL, and get finer slices. Also, to the best of our knowledge, we are the first to have used HDL slicing for formal verification.

A tool for slicing Promela, the input language for the Spin model checker, is currently being constructed ([27]). However, the concurrency issues dealt with there are different from those in VHDL.

SDG-like structures form the basis of many gate-level test-generation algorithms. However, our approach works at the VHDL source level, thus avoiding the heavy complexity of synthesis. Moreover, many applications of slicing described in Section 6 make sense only at the VHDL source level.

Another area of related work occurs in the model checking domain, where state space size is reduced using the *cone of influence* reduction (COI) or *localization* reduction ([28]).

COI can be expressed as a fixpoint computation that constructs the set of state variables that can potentially affect the value of a variable in the CTL specification (i.e., the set of variables in the *cone of influence* of the variable of interest). Alternatively, COI can be thought of as building a dependence graph for the program, and then using graph reachability to determine what parts of the specification are relevant to the variable of interest. The actual dependence graph may be either on the VHDL source-code (*pre-encoding*) or on the set of equations that represent the transition function (*post-encoding*), though the former is difficult.

The localization reduction performs a related function. Intuitively, it works by conservatively abstracting system components and verifying a localized version of the specification. If the localized version is not verifiable, the abstractions are iteratively relaxed by adding more components, until the specification is eventually provable. Added components are in the specification's COI.

Several differences between these two reductions and slicing are worth noting (the first 3 apply only if the reductions are done as post-encoding operations):

- In HDL formal verification, the difficulty often lies in model generation rather than model checking, and it is sometimes not even possible to build the model. Any post-encoding method obviously does not help in such cases.
- The model generation process often does some translation of the VHDL program into a restricted VHDL subset, and it is thus difficult or impossible to trace back to statements in the original program. Most of the design, simulation, and testing applications mentioned in this paper are consequently not possible using a post-encoding technique.
- One of the variables that the model size is a function of is the size of the input program (e.g., the bits needed to represent the program counters). Post-encoding reductions cannot reduce this overhead in general.
- Slicing permits more complex reductions of programs to be specified than is possible using COI. For example, suppose the specification is of the form "Signal x is always false". In verification, we are primarily interested in ensuring that counterexamples in the original program are also in the slice. Thus, we can select the set of all statements that potentially assign non-false values to x as the slicing criterion, and perform a backward slice with respect to these statements to produce the desired reduced program. In the most general case, the structure of the specification can be analyzed to determine the appropriate combination of forward and backward slices that result in an equivalent program.
- The precise interprocedural slicing technique used is based on "matched-parenthesis reachability" [7, 8, 29], which is more involved than the ordinary graph reachability used by pre-encoding COI. As mentioned in Section 2, not all SDG paths are possible execution paths, since paths in which calls and returns are mismatched can be excluded. Although the problem of determining the feasibility of a given SDG path is in general undecidable, such mismatched paths can be excluded using a balanced parenthesis language. Ordinary reachability is an example of CFL-reachability in which the CFL is the regular language e^* (where all edges are labeled with e), while this balanced parenthesis condition can not be expressed either with a regular language or with a linear-CFL. Thus, a pre-encoding COI requires a complete elaboration and unfolding of function calls to achieve the same effect, which results in greater design size.

8 Conclusions

In this paper, we have shown how to extend traditional slicing techniques to VHDL, using an approach based on capturing VHDL operational semantics with traditional constructs. We have implemented a tool for automatic slicing, and the paper listed many applications of the tool along with some experimental results showing the state space reduction achievable in model checking. We are currently pursuing further research along four lines. First, we are enhancing the class of supportable HDL (both VHDL and Verilog) constructs. Second, we are investigating techniques to achieve more precise slices, by capturing VHDL semantics more accurately in the SDGs. The current SDGs are conservative in allowing for more dependences than actually exist, and more inter-cycle analysis of VHDL can remove some of these dependences. Third, we are working on developing slicing techniques for general concurrent languages, since the techniques described here extend readily to other concurrent languages. Finally, we are developing a theoretical basis to generate slicing criteria from CTL specifications for use in formal verification.

References

1. E.M. Clarke, E.A. Emerson, and A.P. Sistla. Automatic verification of finite-state concurrent systems using temporal logic specifications. *ACM Transactions on Programming Languages and Systems*, 8(2):244–263, April 1986.
2. D. Déharbe, S. Shankar, and E.M. Clarke. Model checking VHDL with CV. In *Formal Methods in Computer Aided Design (FMCAD)*, page to appear, 1997.
3. M. Weiser. Program slicing. *IEEE Transactions on Software Engineering*, 10(4):352–357, 1984.
4. M. Weiser. *Program slices: Formal, psychological, and practical investigations of an automatic program abstraction method*. PhD thesis, University of Michigan, 1979.
5. K.J. Ottenstein and L.M. Ottenstein. The program dependence graph in a software development environment. In *Proceedings of the ACM SIGSOFT/SIGPLAN Software Engineering Symposium on Practical Software Development Environments*, pages 177–184, New York, NY, 1984. ACM Press.
6. J. Ferrante, K. Ottenstein, and J. Warren. The program dependence graph and its use in optimization. *ACM Transactions on Programming Languages and Systems*, 3(9):319–349, 1987.
7. S. Horwitz, T. Reps, and D. Binkley. Interprocedural slicing using dependence graphs. *ACM Transactions on Programming Languages and Systems*, 12(1):26–60, January 1990.
8. S. Horwitz, T. Reps, M. Sagiv, and G. Rosay. Speeding up slicing. In *Proceedings of the Third ACM SIGSOFT Symposium on the Foundations of Software Engineering*, pages 11–20, New York, NY, December 1994. ACM Press.
9. F. Tip. A survey of program slicing techniques. Technical Report CS-R9438, Centrum voor Wiskunde en Informatica, 1994.
10. D. Binkley and K. Gallagher. Program slicing. In M. Zelkowitz, editor, *Advances in Computers, Vol. 43*. Academic Press, San Diego, CA, 1996.

11. K.B. Gallagher and J.R. Lyle. Using program slicing in software maintenance. *IEEE Transactions on Software Engineering*, SE-17(8):751–761, August 1991.

12. J. Lyle and M. Weiser. Experiments on slicing-based debugging tools. In *Proceedings of the First Conference on Empirical Studies of Programming*, June 1986.

13. D. Binkley. Using semantic differencing to reduce the cost of regression testing. In *Proceedings of the 1992 Conference on Software Maintenance* (Orlando, FL, November 9-12, 1992), pages 41–50, 1992.

14. S. Bates and S. Horwitz. Incremental program testing using program dependence graphs. In *ACM Symposium on Principles of Programming Languages*, pages 384–396, 1993.

15. S. Horwitz, J. Prins, and T. Reps. Integrating non-interfering versions of programs. *ACM Transactions on Programming Languages and Systems*, 11(3):345–387, July 1989.

16. S. Horwitz. Identifying the semantic and textual differences between two versions of a program. In *SIGPLAN Conference on Programming Languages Design and Implementation*, pages 234–245, 1990.

17. T. Reps and T. Turnidge. Program specialization via program slicing. In O. Danvy, R. Glueck, and P. Thiemann, editors, *Proc. of the Dagstuhl Seminar on Partial Evaluation*, volume 1110 of *Lecture Notes in Computer Science*, pages 409–429, Schloss Dagstuhl, Wadern, Germany, February 1996. Springer-Verlag.

18. J.Q. Ning, A. Engberts, and W. Kozaczynski. Automated support for legacy code understanding. *Communications of the ACM*, 37(5):50–57, May 1994.

19. D. Jackson and E.J. Rollins. A new model of program dependences for reverse engineering. *SIGSOFT 94: Proceedings of the Second ACM SIGSOFT Symposium on the Foundations of Software Engineering*, (New Orleans, LA, December 7-9, 1994), *ACM SIGSOFT Software Engineering Notes*, 19, December 1994.

20. T. Reps and G. Rosay. Precise interprocedural chopping. *SIGSOFT 95: Proceedings of the Third ACM SIGSOFT Symposium on the Foundations of Software Engineering*, (Washington, DC, October 10-13, 1995), *ACM SIGSOFT Software Engineering Notes*, 20(4), 1995.

21. D. Binkley. Precise executable interprocedural slices. *ACM Letters on Programming Languages and Systems*, 2:31–45, 1993.

22. M. Weiser. Program slicing. *IEEE Transactions on Software Engineering*, SE-10(4):352–357, July 1984.

23. IEEE. *IEEE Standard VHDL Language Reference Manual*, 1987. Std 1076-1987.

24. E.M. Clarke, M. Fujita, S.P. Rajan, T. Reps, S. Shankar, and T. Teitelbaum. Program slicing of hardware description languages. Technical Report CMU-CS-99-103, Carnegie Mellon University, 1999.

25. M. Iwaihara, M. Nomura, S. Ichinose, and H. Yasuura. Program slicing on VHDL descriptions and its applications. In *Asian Pacific Conference on Hardware Description Languages (APCHDL)*, pages 132–139, 1996.

26. S. Ichinose, M. Iwaihara, and H. Yasuura. Program slicing on VHDL descriptions and its evaluation. Technical report, Kyushu University, 1998.

27. L. Millett and T. Teitelbaum. Slicing promela and its applications to protocol understanding and analysis. In *4th International SPIN Workshop*, pages 75–83, 1998.

28. Robert P. Kurshan. *"Computer-Aided Verification of Coordinating Processes"*. Princeton University Press, 1994.

29. T. Reps. Program analysis via graph reachability. In *Proc. of ILPS '97: Int. Logic Programming Symposium*, pages 5–19, Cambridge, MA, 1997. M.I.T.

Results of the Verification of a Complex Pipelined Machine Model

Jun Sawada[1] and Warren A. Hunt, Jr.[2]

[1] Department of Computer Sciences, University of Texas at Austin
[2] IBM Austin Research Laboratory

Abstract. Using a theorem prover, we have verified a microprocessor design, FM9801. We define our correctness criterion for processors with speculative execution and interrupts. Our verification approach defines an invariant on an intermediate abstraction that records the history of instructions. We verified the invariant first, and then proved the correctness criterion. We found several bugs during the verification process.

1 FM9801 and Correctness Criterion

We argue that even complex microprocessor design can be formally verified. As an evidence of our claim, we have mechanically verified our FM9801 microprocessor design. It has various features such as out-of-order issue and completion of instructions with Tomasulo's algorithm, speculative execution with branch prediction, precise handling of internal exceptions and external interrupts, and supervisor/user modes.

The FM9801 is formally specified in the ACL2 logic[KM96] at the *instruction-set architecture* (ISA) level and the *microarchitecture* (MA) level. These formal definitions are publicly available along with the FM9801 verification scripts[Saw]. The ISA sequentially executes instructions. Its behavior is specified with function ISA-step($ISA, intr$), which returns the ISA state after executing one instruction from state ISA, with interrupt signal $intr$. The MA is a clock cycle accurate model of the pipelined hardware design. Its behavioral function MA-step($MA, sigs$) returns the MA state after one clock cycle of execution with external signals $sigs$. We define ISA-stepn($ISA, intr\text{-}list, m$) as the recursive function that repeatedly applies the next state function ISA-step to state ISA m times, where $intr\text{-}list$ is a list of interrupt signals for each execution step. Similarly, we define MA-stepn($MA, sig\text{-}list, n$) as n applications of MA-step with a list of signals $sig\text{-}list$. Projection function proj(MA) returns the ISA state consisting of the program counter, the register file, and the memory in MA.

Our correctness criterion is whether our machine designs satisfy the commutative diagram shown in Fig. 1. For an arbitrary initial MA state MA_0, a list of signals $sig\text{-}list$, and a natural number n, if the initial state MA_0 and the final state $MA_n = $ MA-step($MA_0, sig\text{-}list, n$) are both pipeline flushed states, then

$$\text{proj}(\text{MA-stepn}(MA_0, sig\text{-}list, n)) = \text{ISA-step}(\text{proj}(MA_0), intr\text{-}list, m)$$

should hold for an appropriate list of interrupt signals $intr\text{-}list$ and a natural number m. We additionally assume that the executed program does not modify itself.

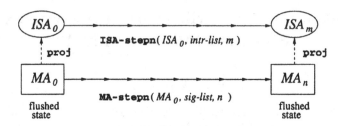

Fig. 1. Correctness Diagram

2 Invariant and Correctness Proof

We use an intermediate model of the MA state that mimics the behavior of speculative execution, exceptions, and external interrupts. This abstraction, which is called a *MAETT*, records executed instructions, each of which is represented with a data-structure holding the values related to the instruction[SH98].

Table 1 gives a list of properties we defined during our verification. Let Π be the set of properties in the table. Additionally, we define predicate CMI-p that holds iff the MA has committed any self-modifying code. Then $\bigwedge_{P \in \Pi} P$ is an *invariant under the constraint* ¬CMI-p [LL90], that is, every property in Π is preserved as long as no self-modifying program is committed. Since $\bigwedge_{P \in \Pi} P$ holds for any flushed pipeline states, $\bigwedge_{P \in \Pi} P$ is true for any reachable state from a flushed state, as long as no self-modifying code is committed.

Given that the final MA state MA_n in Fig. 1 satisfies $\bigwedge_{P \in \Pi} P$, we can show the commutative diagram holds. In fact, all we need to know is that the properties labeled 1 through 6 hold for MA_n. The rest of the properties are necessary for our inductive proofs.

The properties in Table 1 were obtained interactively during the verification process. Initially we started invariant verification by only considering the conjunction of properties labeled 1 through 6. Naturally, our first proof attempt failed. Then, we analyzed the failed proof, and added more properties to the conjunction. Eventually we identified all properties in Π. This was the most time consuming part of the verification.

The proof of the correctness criterion must bridge the complex time abstraction between the ISA level and the MA level. The state of each programmer visible component in the MA is related to different ISA states. These relations are expressed with properties labeled 1 through 4 in Table 1. The proof of the criterion can be found in our report.[SH].

3 Verification Summary

The FM9801 verification was carried out with the ACL2 theorem prover. First, we simulated our FM9801 specification using ACL2's execution capability. This eliminated most of the bugs in our original design before we started the formal verification process.

Table 1. List of the properties used to define our invariant.

#	Property Name	Brief Description
0	weak-invariants:	A well-formedness predicate for a MAETT.
1	pc-match-p:	Correct state of the program counter.
2	sregs-match-p:	Correct state of the special register file.
3	regs-match-p:	Correct state of the general register file.
4	mem-match-p:	Correct state of the memory.
5	no-speculative-commit-p:	No speculatively executed instruction commits.
6	MT-inst-invariants:	Valid intermediate data values in the pipeline.
7	correct-speculation-p:	Instructions following a mis-predicted branch are speculatively executed.
8	correct-exintr-p:	Externally interrupted instructions retires immediately.
9	in-order-dispatch-commit-p:	Instructions dispatch and commit in program order.
10	in-order-DQ-p:	The dispatch queue is a FIFO queue.
11	in-order-ROB-p:	The re-order buffer is a FIFO queue.
12	no-stage-conflict:	No structural conflict at pipeline stages.
13	no-robe-conflict:	No structural conflict in the re-order buffer.
14	in-order-LSU-inst-p:	Certain orders are preserved for instructions in the load-store unit.
15	consistent-RS-p:	Reservation stations keep track of instruction dependencies.
16	consistent-reg-tbl-p:	The register reference table keeps track of the newest instruction that updates each general register.
17	consistent-sreg-tbl-p:	The register reference table keeps track of the newest instruction that updates each special register.
18	consistent-MA-p:	The conjunction of miscellaneous conditions.
19	misc-invariants:	The conjunction of miscellaneous conditions.

The size of the ACL2 verification scripts and the time to certify the proofs for each stage of the verification are given in Table 2. The whole verification project took about 15 months. The verification of invariant $\bigwedge_{P \in \Pi} P$ occupied the largest portion in the ACL2 proof scripts and in our verification effort. This is not surprising because the proof of the invariant is the core of our verification process. We found several bugs that were not detected in the simulation, and all these bugs were detected during the verification of the invariant.

Table 2. ACL2 script size and CPU time with Pentium Pro 200MHz.

Type of ACL2 Script	ACL2 Script Size	CPU Time to Certify
Definitions of ISA and MA	140 KBytes	14 minutes
MAETT modeling	55 KBytes	6 minutes
Definitions of Our Invariant	89 KBytes	7 minutes
Proof of Shared Lemmas	481 KBytes	58 minutes
Proof of Our Invariant	1034 KBytes	211 minutes
Proof of Criterion	37 KBytes	11 minutes

We found 14 design faults in our machine design during the verification process. For instance, one bug was found in the control logic for the speculative execution and the branch prediction. A prediction is usually made for the branch instruction at the instruction fetch unit (IFU). If the branch instruction stalls in the IFU, then more than one branch predictions are made for a single branch instruction. In the original design, if the branch prediction outcomes differed, the machine did not correctly execute instructions from the branching point.

Table 3. Sizes of ACL2 proof scripts for different machines.

Verified Machine	Machine Spec	Total Verification
Small Example Machine	13 KBytes	169 KBytes
Pipelined Design presented in CAV '97	78 KBytes	757 KBytes
FM9801	140 KBytes	1909 KBytes

Although this verification was labor intensive, our technique seems to scale well with the size of machine. In Table 3, we compare the size of our machine specification and verification scripts with two other proof efforts, each with different machine sizes, where we employed a similar approach. The ratio of the size of the verified machine design and its verification script does not change much. We also note that the CPU time in Table 2 is relatively small. This is because we decompose a complex verification problem into small lemmas to avoid case explosions. Typically, the ACL2 theorem prover proves single lemmas in less than a minutes during our verification.

We have demonstrated that the pipelined machine with complex control features can be mechanically verified. Although the verification cost was high, we do not see any major difficulty in scaling the verification process for a more complex design. Until now, we have only used an theorem prover, but the combination of algorithmic approach could improve the verification efficiency. Improving invariant verification processes will make our technique more practical.

References

[KM96] Matt Kaufmann and J Strother Moore. ACL2: An industrial strength version of nqthm. In *Eleventh Annual Conference on Computer Assurance (COMPASS-96)*, pages 23–34. IEEE computer Society Press, June 1996.

[LL90] Leslie Lamport and Nancy Lynch. Distributed computing models and methods. In *Handbook of Theoretical Computer Science*, volume B, pages 1159–1199. The MIT Press, Cambridge, Ma., 1990.

[Saw] Jun Sawada. Verification scripts for FM9801 pipelined microprocessor design. Web page http://www.cs.utexas.edu/users/sawada/FM9801/.

[SH] Jun Sawada and Warren A. Hunt, Jr. Verification of FM9801: Out-of-order processor with speculative execution and exceptions that may execute self-modifying code. Unpublished Report. Personal contact: sawada@cs.utexas.edu.

[SH98] Jun Sawada and Warren A. Hunt, Jr. Processor verification with precise exceptions and speculative execution. In Alan J. Hu and Moshe Y. Vardi, editors, *computer Aided Verification (CAV '98)*, volume 1427 of *LNCS*, pages 135–146. Springer Verlag, 1998.

Hazard–Freedom Checking in Speed–Independent Systems*

Husnu Yenigun[1], Vladimir Levin[1], Doron Peled[1], and Peter A. Beerel[2]

[1] Bell Laboratories, 600 Mountain Av., Murray Hill, NJ 07974, USA
{husnu,levin,doron}@research.bell-labs.com
[2] University of Southern California, Los Angeles, CA 90089, USA
pabeerel@eiger.usc.edu

Abstract. We describe two approaches to use the model checking tool COSPAN to check the hazard freedom in speed–independent circuits. First, we propose a straight forward approach to implement a speed–independent circuit in S/R. Second, we propose a reduction technique over the first approach by restricting the original system with certain constraints. This reduction is implemented on the top of COSPAN which also applies its own reductions, including symbolic representation (BDD).

1 Introduction

Speed–Independent systems are a special subclass of asynchronous systems in which gates are modeled as instantaneous functional elements followed by arbitrarily long delay components while assuming zero delay in wires. The advantage of this approach is that the design works regardless of the delay of the individual gates thus eliminating the need for any timing assumptions in the circuit. There are several techniques developed to help verify speed–independent systems (e.g., [1–4]).

The design of speed–independent systems is complicated since one has to make sure that, the unwanted signals, *hazards*, which cause the circuit malfunction, do not appear in the design. In this paper we propose two approaches to use the model checking engine COSPAN [5] to check the hazard freedom of speed–independent circuits. In the first approach, a speed–independent system is completely specified in S/R (the input language of COSPAN), and COSPAN is used to check the states of the system exhaustively to search for a hazard state. This approach, however, suffers from the state explosion problem. Therefore, we propose a reduction method, similar to the partial order reduction [6, 7], to force COSPAN to search only a subset of the reachable states of the system. Our technique is based on the static partial order reduction [8].

2 Definitions

A speed–independent system SI is given as a tuple $SI = (G, I, F)$. G is the set of gates in the circuit. $I : G \to 2^G$ is a function giving the interconnection of

---

* This work was supported in part by SRC Contract no. 98-DJ-486.

the circuit. For two gates g_1, g_2, if $g_1 \in I(g_2)$, then the output of the gate g_1 is an input to the gate g_2. We can also have $g \in I(g)$ for a gate g whose output is also an input to itself. And finally, F is a function mapping each gate $g \in G$ to a boolean function, so $F(g)$ is a boolean function of arity $|I(g)|$.

A state s of the system is a function $s : G \to \{0, 1\}$ where for a gate $g \in G$, $s(g)$ is the current output value of gate g at s. Let S be the set of all such functions and $s_0 \in S$ be the initial state of the system.

Given a gate g, and a state s, $I_s(g)$ is the bit vector $[v_1, v_2, \ldots, v_{|I(g)|}]$ representing the current input vector to g at s. $F(g)(I_s(g))$, or shortly $F_s(g)$, is *the desired value* which the current inputs of gate g tend to derive as the new output of g. A gate g is said to be *enabled at state s* if $s(g) \neq F_s(g)$, i.e., the current value of the gate is different from the desired value of the gate. The set of enabled gates at s is given by $enabled(s) = \{g \in G | s(g) \neq F_s(g)\}$.

The interleaving semantics of speed–independent systems requires that only one of the enabled gates may change its value at a time. The transitions of a speed–independent system are given by the relation $T \subseteq S \times S$, such that, $(s, (s/g)) \in T \Leftrightarrow g \in enabled(s)$, where s/g is the state obtained from s by inverting the current output value of g.

A *run from s_1* is a finite sequence of states, $r = (s_1, s_2, \ldots, s_k)$ such that for $1 \leq i < k$, $(s_i, s_{i+1}) \in T$. The *full state space* (or reachable state space) $S_F \subseteq S$ of the system is the set of states such that $s \in S_F$ iff there exists a run $r = (s_1, s_2, \ldots, s_k)$ such that $s_1 = s_0$ and $s_k = s$.

A transition $(s, s/g)$ is called a *hazard–transition* if $\exists g' \neq g$ and $g' \in enabled(s)$ and $g' \notin enabled(s')$. A state s is called a *hazard–state* if there exists two gates g and g' such that $g, g' \in enabled(s)$ and $g' \notin enabled(s/g)$. A *hazard–run from s_1* is a run from s_1, $r = (s_1, s_2, \ldots, s_k)$ such that for all $1 \leq i \leq k - 1$, s_i is not a hazard–state, and s_k is a hazard–state. The system SI is said to be *hazardous* if there exists a hazard–run from s_0, the initial state of the system.

3 Reduced Hazard Check

A sequence of gates $c = g_1, g_2, \ldots, g_n$ is called a *gate–cycle* if $\forall i < n, j \leq n :$ $g_i \in I(g_{i+1})$, $g_n \in I(g_1)$ and $i \neq j$ implies $g_i \neq g_j$. A gate–cycle is a simple cycle in the interconnection structure of the gates. Let $\bar{c}$ denote the *set* of gates in the sequence c. Let $C = \{c_1, c_2, \ldots, c_m\}$ be the set of all gate–cycles in the given circuit and $G_{sticky} \subseteq G$ be a set of gates such that, $\forall c \in C$, $\exists g \in \bar{c}$ such that $g \in G_{sticky}$.

Given two gates g and g' such that $g \in I(g')$, and a state s, g is called *disabling–input for g' at s* if $g' \notin enabled(s)$ and $\forall s'$, $(s(g) = s'(g)) \wedge (s(g') = s'(g'))$ implies $g' \notin enabled(s')$. Intuitively, if g is disabling–input for g' at s, then g' cannot be enabled as long as g stays at its current value. g is called *enabling–input for g' at s* if $g' \notin enabled(s)$ and $g' \in enabled(s/g)$. Let $fanout(g) = \{g' | g \in I(g')\}$ denote the fanout gates of gate g.

A gate g is called *ample at s*, if (1) $g \in enabled(s)$; (2) $s(g) = 1$ or $g \notin G_{sticky}$; (3) $\forall g' \in fanout(g)$, g is a disabling–input for g' at s; and (4) $\exists g' \in fanout(g)$,

g is an enabling–input for g' at s. State s is an *ample state* if there exists an ample gate g at s.

The defining feature of our reduced algorithm over that of the original algorithm is that it explores only one transition associated with an ample gate out of every ample state whereas it explores all transitions from non-ample states. The reduced algorithm is still guaranteed to catch a hazard if the system is not hazard free.

4 Implementation

To implement our algorithms within COSPAN, we developed two different S/R process–type libraries for basic gates. Each gate is represented by a S/R process instantiated from the appropriate S/R process type.

In the first library, a process type has a state variable named *out* which keeps the current output value of the gate and a selection variable named *output*, assigned to the value of the state variable *out*, to inform the current output value to the fanout gates using appropriate instantiation connections. It also has another selection variable, *enabled*, which is set to true whenever the gate is enabled and a selection variable, named *hazard_flag*, which is set to true whenever the current state is a hazard state on this gate. If the process type implements an n input basic gate, then it has $2n + 1$ formal parameters. Specifically, it imports the *output* and *enabled* selection variables from each fanin gate g and a last formal parameter from a process called *Asynchrony_Manager*, or shortly *AM*. Since an S/R system is a synchronous system, whereas a speed–independent system is asynchronous, we mimic the asynchrony by using AM. Every gate informs AM whether it is enabled or not (via its *enabled* selection variable). AM lets the enabled gates execute a transition in a mutually exclusive manner.

The second process–type library is used for the reduced case analysis. In addition to all the state and selection variables of the previous case, a process type in the reduced case has two more selection variables for each fanin component it has to inform the corresponding fanin gate if it is a disabler and enabler input for this gate or not. It also has another selection variable called *ample* which is set to true only if the gate is an ample gate at the current state. We also modify AM in the reduced case to import the *ample* selection variables of the gates. If at a state, there exists a gate whose *ample* is true, then AM of the reduced case only allows this gate to execute. If none of the *ample* selection variables are true, then it again lets all the enabled gates execute mutually exclusively.

In both cases, COSPAN is run on the system and checks the property that no gate's hazard_flag is ever set.

5 Experiments and Discussion

Currently, there is no tool implemented to instantiate a system using the process type libraries. Therefore, we experimented with a hand-generated scalable family

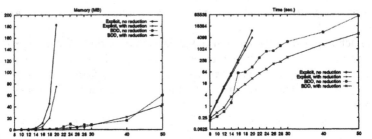

Fig. 1. Memory and time. The x–axis are the number of C–elements in the circuit.

of FIFO queue circuits implemented using Muller C-elements and inverting and non–inverting buffers.

Neither the original system nor the reduced system is reported to be hazardous. As the figure illustrated, the reduction ratio of the number of states (i.e. the number of states in the original system divided by the number of states in the reduced system) is exponential though with a smaller degree than the increase in the number of states. As a consequence of this reduction, the time required to analyze the reduced system is always less than that required to analyze the original system and the reduced case always uses less memory, but it is still exponential.

In the symbolic (BDD) case, we do not see any stable correlation between reduced and original cases in terms of memory usage. The reduced analysis, however, is faster after the scale goes above 16 C–elements despite the fact that the partial reduced system being verified is more complicated.

Although somewhat promising, more experiments are needed to better judge the merits of the proposed reduction approach.

References

1. J.R. Burch, E.M. Clarke, D.E. Long, K.L. McMillan and D.L. Dill, *Symbolic model checking for sequential circuit verification*, IEEE Transactions on CAD, Vol.13, No.4, April 1994, pp.401–424.

2. K.L. McMillan, *A technique of state space search based on unfolding*, Formal Methods in System Design, Vol.6, pp.45–65, 1995.

3. P.A. Beerel, J.R. Burch and T.H.-Y. Meng, *Checking Combinational Equivalence of Speed-Independent Circuits*, in Formal Methods on System Design, May 1998.

4. D.L. Dill, *Trace theory for automatic hierarchical verification of speed–independent circuits*, ACM Distinguished Dissertations, 1989.

5. R.P. Kurshan, *Computer-aided verification of coordinating processes*, Princeton University Press, Princeton, New Jersey, 1994.

6. D. Peled, *Combining partial order reductions with on the fly model checking*, 6th CAV, June 1994.

7. A. Valmari, *A stubborn attack on state space explosion*, 2nd CAV, 1990, pp.25–42.

8. R.P. Kurshan, V. Levin, M. Minea, D. Peled and H. Yenigun, *Static Partial Order Reduction*, TACAS, 1998.

Yet Another Look at LTL Model Checking*

Klaus Schneider

University of Karlsruhe, Department of Computer Science,
Institute for Computer Design and Fault Tolerance (Prof. D. Schmid),
P.O. Box 6980, 76128 Karlsruhe, Germany,
email: Klaus.Schneider@informatik.uni-karlsruhe.de,
http://goethe.ira.uka.de/~schneider

Abstract *A subset of* LTL *is presented that can be translated to ω-automata with only a linear number of states. The translation is completely based on closures under temporal and boolean operators. Moreover, it is shown how this enhancement can be combined with traditional translations so that all* LTL *formulas can be translated. Exponential savings are possible in terms of reachable states, as well as in terms of runtime and memory requirements for model checking.*

1 Introduction

Nearly all decision procedures for the linear-time temporal logic LTL are based on a translation to equivalent ω-automata. Given an LTL formula Φ, the states of the corresponding ω-automaton $\mathcal{A}_\Phi$ are usually given as the powerset of the elementary formulas, i.e., of the set of all subformulas of Φ that start with a temporal logic operator. The transition relation and the acceptance condition of the automaton is determined by the fixpoint characterization of the formulas in the corresponding state (cf. [6,1]).

Clarke, Grumberg and Hamaguchi [2] pointed out that for LTL model checking, there is no reason to construct the automaton $\mathcal{A}_\Phi$ explicitly. Instead, they directly abbreviated each elementary subformula φ of Φ by a new state variable ℓ_φ. The transition relation and the acceptance condition are then directly given in terms of these state variables. This yields in a translation procedure that runs in time $O(|\Phi|)$ and whose result can be directly used for symbolic model checking.

The number of possible states is however of order $O(2^{|\Phi|})$ since any elementary formula of Φ may double the set of reachable states. Although this exponential blow-up can not be circumvented in general, it can be avoided for many formulas, when the automaton is derived by means of closures (see [6] for an example): Given that we have already derived an automaton $\mathcal{A}_\varphi$ for φ, and we have a propositional formula ψ, the presented closure theorems tell us how to construct automata for $X\varphi$, $[\psi \; \underline{U} \; \varphi]$, and $[\varphi \; B \; \psi]$ by only introducing one additional state.

A forerunner of this approach is due to Jong [3] who already used closure theorems for X and F. We extend his idea to *all* temporal operators, i.e. also to the binary ones. The closures are used to keep the automaton as small as possible by avoiding the introduction of too much states and acceptance constraints. Our translation can still be performed within $O(|\Phi|)$ time and still computes the transition relation of the automaton $\mathcal{A}_\varphi$ in a symbolic manner similar to [2].

---

* This work has been financed by the DFG project 'Verification of embedded systems' in the priority program 'Design and design methodology of embedded systems'.

However, the mentioned closures can only be applied if ψ is propositional, so that there is an intimate relationship between the logic LeftCTL* [5,6] and the closure theorems. LeftCTL* is a branching-time temporal logic that is equal expressive as CTL, but has a much richer syntax, so that a lot of LTL formulas that are equivalent to some CTL formulas do even syntactically belong to LeftCTL*. Using the closure theorems, we will see that any formula of LTL ∩ CTL can be translated to a deterministic Büchi automaton. A similar result has been obtained by Kupferman and Vardi, who showed in [7] that each formula of the intersection of LTL with the alternation-free μ-calculus (a superset of CTL in some sense) can be translated to a deterministic Büchi automaton.

In [6], it has been shown how given CTL* model checking problems can be handled by the extraction of a LeftCTL* formula. It has already been remarked there that the procedure presented in [6] is a generalization of the usual translation of LTL to ω-automata (in the sense of [1]). Our overall approach is therefore as follows: We apply the 'extraction procedure' of [6] to a given LTL formula Φ. This essentially computes an ω-automaton $\mathcal{A}_\Phi$ for the parts of Φ that do not belong to LeftCTL* and some formula $\Psi \in$ LTL ∩ LeftCTL*. Ψ is then translated by means of the closure theorems to another ω-automaton $\mathcal{A}_\Psi$, so that the resulting automaton $\mathcal{A}_\Phi \times \mathcal{A}_\Psi$ is equivalent to Φ. The construction of $\mathcal{A}_\Psi$ does not introduce any fairness constraints at all, whereas the traditional translation generates $O(|\Phi|)$ fairness constraints that evidently increase the runtime requirements of the later model checking.

Due to lack of space, the paper is not self-contained. It refers to some definitions and algorithms given in [5,6,1] (available from http://goethe.ira.uka.de/~schneider).

2 Translating LeftCTL* ∩ LTL to ω-Automata by Closures

We describe ω-automata by formulas of the form $\mathcal{A}_\exists (Q, \Phi_\mathcal{I}, \Phi_\mathcal{R}, \Phi_\mathcal{F})$, where Q is the set of state variables, so that the states of the automaton are given as subsets of Q. $\Phi_\mathcal{I}$ is a propositional formula over Q and describes the initial states (all sets that 'satisfy' $\Phi_\mathcal{I}$). Similarly, $\Phi_\mathcal{R}$ describes the transition relation, and $\Phi_\mathcal{F}$ the acceptance condition. In particular, we consider the following types of acceptance conditions, where all formulas Φ, Φ_k, Ψ_k are propositional over Q, and hence denote sets of states ($\mathfrak{B}^+(\varphi_1, \ldots, \varphi_n)$ denotes the set of formulas that can be constructed with the boolean connectives $\wedge$ and $\vee$ from the formulas $\varphi_1, \ldots, \varphi_n$):

Büchi Acceptance:	$GF\Phi$
Generalized Büchi Acceptance:	$\mathfrak{B}^+(GF\Phi_1, \ldots, GF\Phi_m)$
Generalized Co-Büchi Acceptance:	$\mathfrak{B}^+(FG\Phi_1, \ldots, FG\Phi_m)$

It is well-known that generalized co-Büchi automata are less expressive than Büchi automata. The latter are closed under all boolean operations, but the former are not closed under complement. Moreover, generalized co-Büchi automata can be made deterministic (with an exponential blow-up) [8], while Büchi automata can not be made deterministic [4].

We next show how the LTL formulas that are also LeftCTL* [5] formulas can be translated to generalized nondeterministic co-Büchi automata with only $O(|\Phi|)$ states and an acceptance condition of length $O(|\Phi|)$. The grammar rules of LeftCTL* are as follows (see [5] for the semantics):

$$S ::= \text{ Variables } | \neg S | S \wedge S | S \vee S | \mathsf{E}P_E | \mathsf{A}P_A$$
$$P_E ::= S | \neg P_A | P_E \wedge P_E | P_E \vee P_E | \mathsf{X}P_E | \mathsf{G}S | \mathsf{F}P_E$$
$$| [P_E \ \mathsf{W} \ S] | [S \ \mathsf{U} \ P_E] | [P_E \ \mathsf{B} \ S] | [P_E \ \underline{\mathsf{W}} \ S] | [S \ \underline{\mathsf{U}} \ P_E] | [P_E \ \underline{\mathsf{B}} \ S]$$
$$P_A ::= S | \neg P_E | P_A \wedge P_A | P_A \vee P_A | \mathsf{X}P_A | \mathsf{G}P_A | \mathsf{F}S$$
$$| [P_A \ \mathsf{W} \ S] | [P_A \ \mathsf{U} \ S] | [S \ \mathsf{B} \ P_E] | [P_A \ \underline{\mathsf{W}} \ S] | [P_A \ \underline{\mathsf{U}} \ S] | [S \ \underline{\mathsf{B}} \ P_E]$$

We call a formula without path quantifiers A and E a P_E-formula if it can be derived from the nonterminal P_E with the above grammar rules. The translation of these P_E-formulas to ω-automata by closures is done by the following equations (any LeftCTL* formula can be reduced to one with only the operators X, $\underline{\mathsf{U}}$ and B):

Lemma 1 (Temporal Closure of Generalized Co-Büchi Automata). *Given a formula* Φ *of* $\mathfrak{B}^+(\mathsf{FG}\Phi_1, \ldots, \mathsf{FG}\Phi_n)$ *and a variable* p. *Define for any set* Q *of variables the formula* $\Phi_{\overline{Q}} := \bigwedge_{q \in Q} \neg p$. *Then, the following equations are valid:*

- *for propositional* φ, *we have* $\varphi = \mathcal{A}_\exists \begin{pmatrix} \{p, q\}, \neg p \wedge q, \\ (\mathsf{X}p) \wedge (\mathsf{X}q = q \wedge (p \vee \varphi)), \\ \mathsf{FG}q \end{pmatrix}$

- $\mathsf{X}[\mathcal{A}_\exists (Q, \Phi_\mathcal{I}, \Phi_\mathcal{R}, \Phi)] = \mathcal{A}_\exists \begin{pmatrix} Q \cup \{p\}, \neg p \wedge \Phi_{\overline{Q}}, \\ [\neg p \wedge \Phi_{\overline{Q}} \wedge \mathsf{X}p \wedge \mathsf{X}\Phi_\mathcal{I}] \vee [p \wedge \Phi_\mathcal{R} \wedge \mathsf{X}p], \\ \Phi \end{pmatrix}$

- $[\varphi \ \underline{\mathsf{U}} \ \mathcal{A}_\exists (Q, \Phi_\mathcal{I}, \Phi_\mathcal{R}, \Phi)] = \mathcal{A}_\exists \begin{pmatrix} Q \cup \{p\}, \neg p \wedge \Phi_{\overline{Q}}, \\ [\neg p \wedge \Phi_{\overline{Q}} \wedge \varphi \wedge \mathsf{X}(\neg p \wedge \Phi_{\overline{Q}})] \vee \\ [\neg p \wedge \Phi_{\overline{Q}} \wedge \varphi \wedge \mathsf{X}(p \wedge \Phi_\mathcal{I})] \vee \\ [p \wedge \Phi_\mathcal{R} \wedge \mathsf{X}p], \\ \Phi \end{pmatrix}$

- $[\mathcal{A}_\exists (Q, \Phi_\mathcal{I}, \Phi_\mathcal{R}, \Phi) \ \mathsf{B} \ \varphi] = \mathcal{A}_\exists \begin{pmatrix} Q \cup \{p\}, \neg p \wedge \Phi_{\overline{Q}}, \\ [\neg p \wedge \Phi_{\overline{Q}} \wedge \neg \varphi \wedge \mathsf{X}(\neg p \wedge \Phi_{\overline{Q}})] \vee \\ [\neg p \wedge \Phi_{\overline{Q}} \wedge \neg \varphi \wedge \mathsf{X}(p \wedge \Phi_\mathcal{I} \wedge \neg \varphi)] \vee \\ [p \wedge \Phi_\mathcal{R} \wedge \mathsf{X}p], \\ \mathsf{FG}(\neg p \wedge \Phi_{\overline{Q}}) \vee \Phi \end{pmatrix}$

Instead of giving a formal proof, we only explain the intuition that is behind the closure under $\underline{\mathsf{U}}$: a new initial state (encoded by $\neg p \wedge \Phi_{\overline{Q}}$) is added with a self-loop (encoded by $\neg p \wedge \Phi_{\overline{Q}} \wedge \varphi \wedge \mathsf{X}(\neg p \wedge \Phi_{\overline{Q}})$) that is enabled under φ. Also, there are transitions (encoded by $\neg p \wedge \Phi_{\overline{Q}} \wedge \varphi \wedge \mathsf{X}(p \wedge \Phi_\mathcal{I})$) to any initial state $\Phi_\mathcal{I}$ that are also enabled under φ. If p has become true, it will stay for all the future time true and therefore enables the transitions $\Phi_\mathcal{R}$ of the (sub)automaton $\mathcal{A}_\exists (Q, \Phi_\mathcal{I}, \Phi_\mathcal{R}, \Phi)$ (encoded by $p \wedge \Phi_\mathcal{R} \wedge \mathsf{X}p$). The acceptance condition Φ requires that we can not stay forever in the added state $\neg p \wedge \Phi_{\overline{Q}}$, which must however be allowed for the weak $\underline{\mathsf{U}}$ operator (and also for the B operator). The closures under X and B are constructed in a similar manner.

Note that only one further state is added by the above closures (we use a wasteful encoding that introduces a new state variable for the encoding of the new state). Similar closures hold also for generalized Büchi automata, and also for conjunction and disjunction. Assume that the functions $\Omega_\mathsf{X}, \Omega_{\underline{\mathsf{U}}}, \Omega_\mathsf{B}, \Omega_\wedge, \Omega_\vee$ implement the closures

```
function extract_P_E(φ)
  case φ of
    is_prop(φ)      :  return ({}, φ);
    φ₁ ∧ φ₂         :  (E₁, ψ₁) ≡ extract_P_E(φ₁); (E₂, ψ₂) ≡ extract_P_E(φ₂);
                       return (E₁ ∪ E₂, ψ₁ ∧ ψ₂);
    φ₁ ∨ φ₂         :  (E₁, ψ₁) ≡ extract_P_E(φ₁); (E₂, ψ₂) ≡ extract_P_E(φ₂);
                       return (E₁ ∪ E₂, ψ₁ ∨ ψ₂);
    Xφ₁             :  (E₁, ψ₁) ≡ extract_P_E(φ₁); return (E₁, Xψ₁);
    [φ₁ B φ₂]       :  (E₁, ψ₁) ≡ extract_P_E(φ₁); (E₂, ψ₂) ≡ tableau(φ₂);
                       return (E₁ ∪ E₂, [ψ₁ B ψ₂]);
    [φ₁ U φ₂]       :  (E₁, ψ₁) ≡ tableau(φ₁); (E₂, ψ₂) ≡ extract_P_E(φ₂);
                       return (E₁ ∪ E₂, [ψ₁ U ψ₂]);

function close(φ)
  case φ of q
    is_prop(φ)      :  return Π_prop(φ);
    φ₁ ∧ φ₂         :  return Ω_∧(close(φ₁), close(φ₂));
    φ₁ ∨ φ₂         :  return Ω_∨(close(φ₁), close(φ₂));
    Xφ₁             :  return Ω_X(close(φ₁));
    [φ₁ B b]        :  Ω_B(b, close(φ₁))
    [φ₁ U b]        :  Ω_U(b, close(φ₁))

function Closed_Tableau(Φ)
  ({ℓ₁ = φ₁, ..., ℓₙ = φₙ}, Ψ) := extract_P_E(NNF(Φ));
  Φ_R := ⋀ⁿᵢ₌₁ trans(ℓᵢ, φᵢ);
  Φ_F := ⋀ⁿᵢ₌₁ fair(ℓᵢ, φᵢ);
  A∃ (Q, Ψ_I, Ψ_R, Ψ_F) := close(Ψ);
  return A∃ (Q ∪ {ℓ₁, ..., ℓ_b}, Φ_I ∧ Ψ_I, Φ_R ∧ Ψ_R, Φ_F ∧ Ψ_F);
```

Figure 1. Algorithm for translating LTL to ω-automata by means of closures

under X, $\underline{U}$, B, $\wedge$, and $\vee$, respectively. Π_{prop} computes the ω-automaton for propositional formulas. Using these functions, the function call close(Φ) with a P_E-formula Φ for the algorithm given in figure 1 yields in a nondeterministic generalized co-Büchi automaton $\mathcal{A}_\exists (Q, \Phi_\mathcal{I}, \Phi_\mathcal{R}, \Phi_\mathcal{F})$ that is equivalent to Φ. This automaton has $O(|\Phi|)$ states and an acceptance condition of length $O(|\Phi|)$.

In [8], it is shown that generalized co-Büchi automata can be made deterministic, and that we can reduce the acceptance condition to the normal form $\bigvee_{i=0}^{f} \mathsf{FG}\Phi_i$ where even $f = 0$ will do. Hence, we can compute for any LeftCTL* formula a deterministic co-Büchi automaton. As P_A is dual to P_E, and deterministic Büchi automata are the complements of deterministic co-Büchi automata, this means that we can compute for any P_A formula a deterministic Büchi automaton only by application of the above closure theorems. Hence, we have the following result (similar to [7]).

Theorem 2. *The following facts are valid:*

1. *For all $\Phi \in P_E$ there is an equivalent nondeterministic generalized co-Büchi automaton with $O(|\Phi|)$ states and an acceptance condition of length $O(|\Phi|)$.*

2. For all $\Phi \in P_E$ there is an equivalent deterministic co-Büchi automaton with $2^{O(|\Phi|)}$ states.

3. For all $\Phi \in P_A$ there is an equivalent deterministic Büchi automaton with $2^{O(|\Phi|)}$ states.

3 Translating LTL to ω-Automata by Closures

To translate an arbitrary LTL formula Φ, we abbreviate each subformula φ of Φ that violates the grammar rules of P_E by a new variable ℓ_φ (see function extract_$P_E(\varphi)$ of figure 1). The resulting P_E-formula Ψ can then be translated by means of the closures to a nondeterministic generalized co-Büchi automaton $\mathcal{A}_\Psi$. Of course, it has to be assured that ℓ behaves always equivalent to φ. According to the product model checking approach [6], we therefore add transitions and fairness constraints according to [6] (see also [1] for the functions trans and fair). Hence, we obtain the following result:

Theorem 3. *For any structure $\mathcal{K}$, any state s of $\mathcal{K}$, and any quantifier-free formula Φ, the following is equivalent for $\mathfrak{A} := \mathcal{A}_\exists\, (Q, \Phi_\mathcal{I}, \Phi_\mathcal{R}, \Phi_\mathcal{F}) =$ Closed_Tableau(Φ) ($\mathcal{K}_\mathfrak{A}$ is the Kripke structure associated with $\mathfrak{A}$):*

- *there is a fair path π starting in s such that $(\mathcal{K}, \pi) \models \Phi$ holds*
- *there is an initial state s_0 of $\mathfrak{A}$ such that there is a path $\pi \times \pi_\mathfrak{A}$ through $\mathcal{K} \times \mathcal{K}_\mathfrak{A}$ such that $(\mathcal{K} \times \mathcal{K}_\mathfrak{A}, (s, s_0)) \models \Phi_\mathcal{F}$ holds.*

Moreover, $\Phi_\mathcal{F}$ is of the form $(\bigwedge_{i=1}^n \mathsf{GF}\Phi_i) \wedge \Psi$, where Ψ is built-up with conjunctions and disjunctions of $\mathsf{FG}\Psi_i$ formulas.

There is also the possibility to abbreviate more subformulas than necessary (for extracting a P_E formula), so that one can gradually choose between the traditional translation and the closure based ones for optimizations.

References

1. K. Schneider and D. Hoffmann. A HOL conversion for translating linear time temporal logic to ω-automata. In *Higher Order Logic Theorem Proving and its Applications*, LNCS, Nice, France, September 1999. Springer Verlag.
2. E.M. Clarke, O. Grumberg, and K. Hamaguchi. Another look at LTL model checking. In *Conference on Computer Aided Verification (CAV)*, LNCS 818, pp. 415–427, Standford, California, USA, June 1994. Springer-Verlag.
3. G.G de Jong. An automata theoretic approach to temporal logic. In *Computer Aided Verification (CAV)*, LNCS 575, pp. 477–487, Aalborg, July 1991. Springer-Verlag.
4. W. Thomas. Automata on infinite objects. In J. van Leeuwen, editor, *Handbook of Theoretical Computer Science*, volume B, pages 133–191, Amsterdam, 1990. Elsevier Science Publishers.
5. K. Schneider. CTL and equivalent sublanguages of CTL*. In C. Delgado Kloos, editor, *IFIP Conference on Computer Hardware Description Languages and their Applications (CHDL)*, pp. 40–59, Toledo, Spain, April 1997. IFIP, Chapman and Hall.
6. K. Schneider. Model checking on product structures. In *Formal Methods in Computer-Aided Design*, LNCS 1522, pp. 483–500, Palo Alto, CA, November 1998. Springer Verlag.
7. O. Kupferman and M.Y. Vardi. Freedom, weakness, and determinism: From linear-time to branching-time. In *IEEE Symposium on Logic in Computer Science*, 1998.
8. K. Wagner. On ω regular sets. *Information and control*, 43:123–177, 1979.

Verification of Finite-State-Machine Refinements Using a Symbolic Methodology

Stefan Hendricx and Luc Claesen

Imec vzw/Katholieke Universiteit Leuven
kapeldreef 75, B-3001 Heverlee (Belgium)
Stefan.Hendricx@imec.be

Abstract. The top-down design of VLSI-systems typically features a step-wise refinement of intermediate solutions. Even though these refinements do usually not preserve time-scales, current formal verification approaches are largely based on the assumption that both specification and implementation utilize the same scales of time. In this paper, a symbolic methodology is presented to verify the step-wise refinement of finite state machines, allowing for possible differences in timing-granularity.

1 Introduction

The top-down design of electronic VLSI-systems can, in general, be characterized in terms of well-defined (and mostly well-automated) refinement steps. Resource allocation, partitioning, control- and datapath synthesis and logic optimization are probably the best-known refinement steps encountered. With the aid of these refinements, high-level behavioural specifications of VLSI-systems — e.g. microprocessors — are progressively 'compiled' into physical implementations.

Despite the well-understood nature of these refinements, the history of electronic hardware design clearly illustrates that both the designer and the design-process may still be subject to a wide variety of shortcomings. Designers may suffer the occasional lapse of attention, design specifications may be ambiguous, incomplete or misinterpreted and there may even be bugs in the design-software itself. As a consequence, the possibility of refinement errors must always be kept in mind. Yet even with well-considered and far-reaching design, simulation and testing procedures, incidents such as the Pentium affair demonstrate that errors can still go undetected until the final release of the product [2].

The ever-growing complexity of modern VLSI-designs is partly to be blamed for limiting the effectiveness of traditional simulation-based verification and testing. Complex designs can no longer be verified exhaustively within realistic design-times and with non-exhaustiveness, there always exists the possibility that certain design-flaws remain undiscovered. Because of its crucial — and self-evident — role in producing qualitative designs, the ability to guarantee hardware correctness must therefore be procured through other alternatives. The most promising alternative is offered by formal verification [1]. Among other advantages, formal approaches allow exhaustive verification within a more-or-less acceptable time-frame. Not surprisingly, formal verification is currently one the focal points in electronic hardware-design and several applications — commercial as well as academic — have already been reported on.

Motivation The majority of formal techniques proposed so far, however, usually start from the assumption that both specification and implementation utilize the same scales of time. Whereas such an assumption is indeed valid for applications such as equivalence checking and finite-state-machine comparison, it is clearly not always applicable. In general, refinement does not preserve time-scales and timing-granularity tends to increase during the design-process. At the algorithmic level, for instance, executing a microprocessor-instruction may be specified in terms of a single (algorithmic) clock-cycle. At the physical level, however, executing the same instruction may span multiple physical clock-cycli.

In an effort to overcome this shortcoming, we recently proposed a symbolic approach dedicated to verifying the step-wise refinement of finite state machines, allowing for possible differences in timing [3].

Outline This research-paper presents a brief overview of our symbolic methodology. The next section discusses the underlying principles of our approach. In the poster accompanying this paper, the practical aspects of our approach are illustrated at the hand of a simple *paper-and-pen* example. The poster also demonstrates the step-wise refinement and verification of a simple microprocessor.

2 Symbolic Verification Methodology

The intuitive notion underlying our verification approach is very simple; when a finite state machine M_r is supposed to be a refinement of a state machine M, it is expected that a close relationship exists between a (non-empty) subset of M_r's states and the state-set of M. Based on this idea, the concept of a mapping-function is introduced.

2.1 User-defined Mapping-functions

Let S and S_r denote the high-level and low-level state-spaces respectively. In addition, let $S_M \subseteq S_r$ be the (non-empty) subset of all low-level states that can be related to the high-level states. Specifically, the *mapping* $\Psi : (S_M \subseteq S_r) \to S$ then denotes a function that maps — or relates — states in S_M to states in S.

In case of a refinement-relationship, Ψ exhibits the following properties:

- Ψ maps each state in S_M to exactly *one* state in S.
- Ψ may map multiple states in S_M to the same state in S.
- Ψ preserves the high-level outputs. That is, if $s = \Psi(s_r)$, the outputs relevant to the high-level state s should have the same values in both s and s_r.

In general, it may be possible to determine more than one viable candidate for Ψ between any given pair of finite state machines. In this paper, however, we assume that the mapping of interest is supplied by the user.

2.2 State-set-bounded Trajectories

Together with the mapping-information provided by the user, the concept of *State-Set-Bounded* trajectories plays a key-role in our methodology.

Definition 1. *A trajectory* $\Delta_{s_0 \to s_n}$ *of a finite state machine M is a sequence of state-transitions* $s_0 \xrightarrow{\tau_0} s_1 \xrightarrow{\tau_1} \ldots \xrightarrow{\tau_{n-1}} s_n$ ($n \geq 1$) *of M.*

Definition 2. *A trajectory* $\Delta_{s_0 \to s_n}$ *of M is* bounded *by states* s_0 *and* s_n, *if* s_0 *and* s_n *appear only at the beginning or ending of the trajectory.*

Definition 3. *Assume S is a subset of states of M. A* state-set-bounded *trajectory or SSB-trajectory* $\Delta_{s_0 \to s_n}^{S}$ *of M is a trajectory* $\Delta_{s_0 \to s_n}$ *bounded by states* $s_0, s_n \in S$ *and containing no other states of S.*

In short, to verify that a finite state machine M_r is a refinement of a state machine M, we will examine if each low-level *SSB*-trajectory associated with the user-defined set S_M implies the existence of a high-level transition in M.

2.3 Finite-state-machine Refinements

Let M and M_r be finite state machines with state-spaces S and S_r respectively. In addition, let Ψ be a (user-defined) function, which maps states in $S_M \subseteq S_r$ to states in S, such that for all low-level states $s_{r1}, s_{r2} \in S_M$, there exist high-level states $s_1, s_2 \in S$, for which $\Psi(s_{r1}) = s_1$ and $\Psi(s_{r2}) = s_2$.

M_r is called a refinement of M, if for each SSB-trajectory $\Delta_{s_{r1} \to s_{r2}}^{S_M}$ between states s_{r1} and s_{r2}, a high-level transition T exists between s_1 and s_2, such that

$$Eval(\Delta_{s_{r1} \to s_{r2}}^{S_M}) \Rightarrow T_{s_1 \to s_2} \tag{1}$$

The function *Eval( )* denotes the symbolic evaluation of the trajectories involved, with respect to the *de facto* stability conditions implied by the high-level state machine — i.e. during a transition, input signals are assumed to be stable.

2.4 Symbolic Evaluation of SSB-trajectories

Symbolic manipulation techniques and fixed-point calculations enable us to derive Boolean expressions for the symbolic evaluation of the SSB-trajectories associated with the mapping Ψ. As explained in [3], we simply need to consider a fixed-point for the generalized (n-steps) transition-relation $\delta^n(\mathbf{x}, \mathbf{y})$.

Informally, $\delta^n(\mathbf{x}, \mathbf{y})$ denotes the Boolean condition to reach an arbitrary state $\mathbf{y}$ in a sequence of n or *less* state-transitions, starting from state $\mathbf{x}$ (assuming stability of inputs during this sequence). For $n = 1$, $\delta(\mathbf{x}, \mathbf{y})$ simply corresponds to the conventional transition-relation of a finite state machine. For $n > 1$, we have the following recursive definition:

$$\delta^n(\mathbf{x}, \mathbf{y}) = \delta^{n-1}(\mathbf{x}, \mathbf{y}) \vee \left(\sum_{\forall \mathbf{q}} \delta^{n-1}(\mathbf{x}, \mathbf{q}) \wedge \delta^*(\mathbf{q}, \mathbf{y}) \right) \tag{2}$$

Above, $\delta^*(\mathbf{x}, \mathbf{y})$ denotes the (1-step) transition-relation of a modified state machine M^*. [1] For a detailed discussion on M^*, the reader is again refered to [3].

For a finite state machine M_r and a user-defined set S_M, we can show that there exists a finite value n_f, such that for all $m \geq n_f$,

$$\delta^{n_f}(\mathbf{x}, \mathbf{y}) \equiv \delta^m(\mathbf{x}, \mathbf{y}) \tag{3}$$

Using this fixed-point, a practical expression [2] can be derived for the symbolic evaluation of *all* SSB-trajectories associated with S_M:

$$Eval(\Delta^{S_M}_{\mathbf{x} \to \mathbf{y}}) \equiv \delta^{n_f}_{M_r}(\mathbf{x}, \mathbf{y}) \wedge \chi_{S_M}(\mathbf{x}) \wedge \chi_{S_M}(\mathbf{y}) \tag{4}$$

Finally, expression 4 enables us to re-formulate the verification problem at hand (see section 2.3). To verify that M_r is a refinement of M — under the user-defined mapping $\Psi : (S_M \subseteq S_r) \to S$ — we need to prove that:

$$\delta^{n_f}_{M_r}(\mathbf{x}, \mathbf{y}) \wedge \chi_{S_M}(\mathbf{x}) \wedge \chi_{S_M}(\mathbf{y}) \Rightarrow \delta_M(\Psi(\mathbf{x}), \Psi(\mathbf{y})) \tag{5}$$

So far, we have successfully applied expression 5 to verify — among others — the step-wise refinement of a simple microprocessor. That application example is illustrated in the poster accompanying this paper.

3 Conclusion

In this paper, a symbolic methodology to verify the step-wise refinement of finite state machines was presented. Key-elements to our approach are user-defined mapping-functions, that keep track of time-scale differences between the state-machines under consideration.

4 Acknowledgment

The research presented in this paper was supported by a scholarship from the Flemish Institute for the promotion of Scientific-Technological Research in Industry (IWT).

References

[1] P. Camurati and P. Prinetto. Formal Verification of Hardware Correctness: Introduction and Survey of Current Research. *IEEE Computer*, 21(7):8–19, July 1988.

[2] T. Coe, Mathisen T., C. Moler, and V. Pratt. Computational Aspects of the Pentium Affair. *IEEE Computational Science & Engineering*, pages 18–31, spring 1995.

[3] S. Hendricx and L. Claesen. Symbolic Multi-Level Verification of Refinement. In *Nineth Great Lakes Symposium on VLSI*, Ann Arbor, MI, 4-6 March 1999. IEEE Computer Society Press.

---

[1] Each transition in M starting in a state $s \in S_M$ is replaced in M^* by a default transition to s itself. All other transitions (and states) of M are preserved in M^*.

[2] In expressions 4 and 5, $\chi_{S_M}(\;)$ represents the characteristic function of the set S_M.

Refinement and Property Checking in High-Level Synthesis Using Attribute Grammars

George Economakos and George Papakonstantinou

National Technical University of Athens
Dept. of Electrical and Computer Engineering
Zographou Campus, GR-15773 Athens, Greece
george@cslab.ece.ntua.gr

Abstract. Recent advances in fabrication technology have pushed the digital designers' perspective towards higher levels of abstraction. Previous work has shown that attribute grammars, used in traditional compiler construction, can also be effectively adopted to describe in a formal and uniform way high-level hardware compilation heuristics, their main advantages being modularity and declarative notation. In this paper, a more abstract form of attribute grammars, relational attribute grammars, are further applied as a framework over which formal hardware verification is performed along with synthesis. The overall hardware design methodology proposed is a novel idea that supports provable correct designs.

1 Introduction

Over the last twenty years, advances in circuit fabrication technology have increased device densities and as a consequence, they have increased design complexity. To manage continuously emerging tasks, designers have moved towards higher levels of abstraction, which are closer to the way they conceive their work. However, each design must be described, eventually, at the lowest level (e.g. layout masks), in order to be fabricated. The transformation from one level of abstraction to the next is performed by various synthesis processes. All such processes need some kind of validation for their results. This validation can be performed by *formal verification* [11], mainly to prove that a transformation from one state to another is correct (*refinement checking*), that two states, initial and transformed, are equivalent (*equivalence checking*) or that the transformed state satisfies certain conditions (*property checking*).

High-level synthesis [7, 9, 10, 13], is defined as the transformation of behavioral circuit descriptions into *register-transfer level* (RTL) structural descriptions that implement the given behavior while satisfying user defined constraints. Even though it has been introduced over twenty years ago, it has recently gained acceptance because the lower level tools have matured enough to support it. However, a lot of problems are still open.

Attribute grammars (AGs), were devised by Knuth [8] as a tool for the formal specification of programming languages. However, in the general case, an AG can be seen as a mapping from the language described by a context free grammar (CFG) into a user defined domain. The main advantage of AGs over other formal specification methods is that they can also be used as an executable method, for the automatic construction of programs to implement the specified mapping [12].

Attempting to overcome the inefficiencies of conventional high-level synthesis and propose a unifying formal framework, an AG formalism describing scheduling heuristics was proposed in [4], which operates by decorating the parse tree of a behavioral circuit description with appropriate attributes. Recently [3, 6, 5], this methodology was realized into the AGENDA integrated design environment that supports top-down implementation of behavioral descriptions using the VHDL hardware description language [2]. Overall, the main advantages of AGs as a formal specification and implementation high-level synthesis formalism are modularity and the declarative notation used for implementation.

In this paper, an extended grammar based methodology is given, which can support two kinds of formal verification, refinement and property checking. It is based on the definition of a more abstract form of AGs, the *Relational Attribute Grammars* (RAGs) [1]. This methodology is a great improvement since it supports provable correct high-level synthesis transformations using a simple, formal and uniform specification and implementation formalism.

2 Attribute Grammars in Synthesis

High-level synthesis transformations can be performed during semantic analysis using AGs. For example, scheduling is performed by decorating the nonterminal symbols of the parse subtree corresponding to primitive operations, with an attribute that is evaluated as the control step at which each operation will be performed. By altering the semantics, the evaluation rules are altered and thus, different heuristics are implemented. For example, consider the ASAP scheduling algorithm. Using AGs, ASAP scheduling is performed by attaching special attributes to all primitive operator parsing syntactic rules, like the following:

$$operation \rightarrow operand_1 \ operator \ operand_2 \qquad (1)$$

ASAP scheduling requires that each output must be scheduled in the next control step after all its inputs have been scheduled. This can be accomplished by using an attribute to pass scheduling information (the control step at which the operator is scheduled) from inputs to outputs, with the following semantic rule attached to (1):

$$operation.ASAPcs = MAX(operand_1.ASAPcs, operand_2.ASAPcs) + 1$$

The scheduling information, along with all other information about each primitive operation, is inserted at each such rule, into a special, list type attribute, that passes information from all leaves of the parse tree to the root. So evaluation of the whole AG, results in a root attribute that accumulates the scheduled CDFG of the behavioral description.

3 Attribute Grammars in Refinement Checking

After evaluating the synthesis AG of the previous section, a special attribute of the root of the parse tree contains the scheduled CDFG of the given behavioral description. However, the resulting architectural implementation may or may not be correct. To prove its correctness, a *specification* [1] must be constructed, which can verify certain conditions of the synthesis or refinement process. Generally speaking, a specification is a set of formulas for each nonterminal symbol of the underlying grammar, where all free variables are attributes of that symbol. Each formula may be true or false. A specification is said to be *inductive* if, for any production rule $p = X_0 \rightarrow X_1 \ldots X_n$, when the specifications of all $X_i, i = 1 \ldots n$ are true and all attributes of the rule have been evaluated, the specification of X_0 can be proven to be true. When a specification is inductive, the AG is correct with respect to it.

For the circuit implementation of a behavioral description to produce correct outputs, one condition must hold. For each input of operator o_i found in the scheduled CDFG and assigned a value in a previous statement in the behavioral description, if this assignment has been scheduled at some control step s_i, o_i must be scheduled later than s_i. In other words, variable dependencies of the original description have not been violated. This can be proven to hold by proving that the synthesis AG is correct with respect to a corresponding specification.

4 Attribute Grammars in Property Checking

After evaluating the synthesis AG of the first section, the scheduled CDFG, which can be seen as a rough architectural implementation (assuming a greedy allocation), is contained in the special attribute of the root of the parse tree. Even though the refinement process may be correct, checked with inductive specifications as described in the previous section, the implementation may not satisfy some design constraints, because inappropriate synthesis algorithms have been used. In that case, design constraints can be tested using *semantic conditions*. Semantic conditions are relations that the attributes of some production must satisfy for attribute evaluation to be valid. Semantic conditions differ from inductive specifications because they check only local (within a single production) conditions. Since, attributes contain implementation details, attribute relations directly reflect implementation properties. If a relation is true, the corresponding property holds and so, semantic conditions can be used for property checking.

5 Experimental Results and Conclusion

Experiments have been conducted, synthesizing provable correct hardware modules. The traditional way to validate the results are through simulation, behavioral or presynthesis as well as postsynthesis. For each example at least a few hours were needed to produce output waveforms for different inputs and validate them. For large circuit implementations, with large number of gates, simulation must be very thorough in order to find errors. On the contrary, using AGENDA all examples were synthesized with a single iteration through the design process, requiring milliseconds on a Sun Ultra SPARC 140Mhz. Moreover, the final implementations were proven correct following the proof method presented in this paper. For each different AG, the proof is needed only once, while, when simulation is used, each example must be validated. The number of AG code to be proven correct is much smaller than the number of gates in the implemenation, so the problems seem to be less complicated, and most of them are trivial (no need for any proof). All these advantages can dramatically increase the designer's productivity.

References

1. P. Deransart and J. Maluszynski. *A Grammatical View of Logic Programming.* MIT Press, 1993.
2. G. Economakos and G. Papakonstantinou. Exploiting the use of VHDL specifications in the AGENDA high-level synthesis environment. In *24th EUROMICRO Conference, Workshop on Digital System Design*, pages 91–98. EUROMICRO, 1998.
3. G. Economakos, G. Papakonstantinou, K. Pekmestzi, and P. Tsanakas. Hardware compilation using attribute grammars. In *Advanced Research Working Conference on Correct Hardware Design and Verification Methods*, pages 273–290. IFIP WG 10.5, 1997.
4. G. Economakos, G. Papakonstantinou, and P. Tsanakas. An attribute grammar approach to high-level automated hardware synthesis. *Information and Software Technology*, 37(9):493–502, 1995.
5. G. Economakos, G. Papakonstantinou, and P. Tsanakas. AGENDA: An attribute grammar driven environment for the design automation of digital systems. In *Design Automation and Test in Europe Conference and Exhibition*, pages 933–934. ACM/IEEE, 1998.
6. G. Economakos, G. Papakonstantinou, and P. Tsanakas. Incorporating multi-pass attribute grammars for the high-level synthesis of ASICs. In *Symposium on Applied Computing*, pages 45–49. ACM, 1998.
7. D. Gajski, N. Dutt, A. Wu, and S. Lin. *High-Level Synthesis.* Kluwer Academic Publishers, 1992.
8. D. E. Knuth. Semantics of context-free languages. *Mathematical Systems Theory*, 2(2):127–145, 1968.
9. Y-L. Lin. Recent development in high level synthesis. *ACM Transactions on Design Automation of Electronic Systems*, 2(1):2–21, 1997.
10. M. C. McFarland, A. C. Parker, and R. Camposano. The high-level synthesis of digital systems. *Proceedings of the IEEE*, 78(2):301–318, 1990.
11. K. L. McMillan. Fitting formal methods into the design cycle. In *31st Design Automation Conference*, pages 314–319. ACM/IEEE, 1994.
12. J. Paaki. Attribute grammar paradigms - a high-level methodology in language implementation. *ACM Computing Surveys*, 27(2):196–255, 1995.
13. R. A. Walker and S. Chaudhuri. High-level synthesis: Introduction to the scheduling problem. *IEEE Design & Test of Computers*, 12(2):60–69, 1995.

A Systematic Incrementalization Technique and Its Application to Hardware Design

Steven D. Johnson*, Yanhong A. Liu**, and Yuchen Zhang

Indiana University Computer Science Department
sjohnson@cs.indiana.edu

Abstract. A transformation method based on *incrementalization* and value *caching*, generalizes a broad family of loop refinement techniques. This method and *CACHET*, an interactive tool supporting it, are presented. Though highly structured and automatable, better results are obtained with intelligent interaction, which provides insight and proofs involving term equality. Significant performance improvements are obtained in many representative program classes, including iterative schemes that characterize Today's hardware specifications. Incrementalization is illustrated by the derivation of a hardware-efficient nonrestoring square-root algorithm.

KEYWORDS AND PHRASES: Formal methods, hardware verification, design derivation, formal synthesis, transformational programming, floating point operations.

1 Introduction

Incrementalization [3, 5, 4] is a generalization of program refinement techniques such as *strength reduction*, in which partial results are introduced to optimize looping computations. *CACHET* [2, 6] is a prototype refinement tool developed to explore incrementalization strategies. In this paper [1], we look at its application to a representative problem in hardware specification. A *nonrestoring integer square root* algorithm was previously used by O'Leary, Leeser, Hickey, and Aagaard [7] to illustrate the use of a theorem prover the step-wise refinement of a hardware implementation. We applied CACHET to the same problem in order to compare how the critical insights needed to justify an implementation are discovered and applied under deductive and derivational modes of formal reasoning. In either case, the implementation proof depends on just a few algebraic identities which must be provided by the (presumably human) external tool user. One of the problems inherent to formalized reasoning is the often overwhelming logical context in which relatively simple key facts must be applied,

---

* Supported, in part, by the National Science Foundation under grant MIP-9601358.
** Supported, in part, by the National Science Foundation under grant CCR-9711253

which includes not only the complex formal proof and design objects, but also the strategy being followed to achieve the verification goal. In 1993, Windley, Leeser, and Aagard pointed out that numerous hardware verification case studies have been found to follow a common proof plan [8]. Incrementalization might also be seen as a "super duper" derivation tactic, but it is one that is applicable a class of generally recursive specification patterns, of which hardware is a special subclass.

2 Systematic Incrementalization

The incrementalization method is an interplay between two kinds of function extension (Figure 1). $F: W \rightarrow V$ plays the role of the specification being transformed; $\oplus: Y \times W \rightarrow W$ is some *state mutator*, a combination of elementary operations applied to F's argument The *incrementalization of F with respect to* $\oplus$ is a function F' that computes $F(w \oplus y)$ given the value of $F(w)$. The idea is this: given a specification for F by which computing $F(w \oplus y)$ involves a recursive call to $F(w)$, we want to specify how $F(w)$ is used in calculating the final result. *Caching* extends a function to return auxiliary results. $F: W \rightarrow V$ is extended to $\overline{F}: W \rightarrow V^k$, so that $\overline{F}(w) = \langle F(w), v_2, \dots, v_k \rangle$. What we are really after is $\overline{F}'$, the incrementalization of the caching extension of F, in which cached values are exploited to optimize across recursive calls.

3 Application to *sqrt* [7]

Incrementalization applied to a singly tail-recursive function (i.e., *while-loop*) is known as *strength reduction*. Take $\oplus$ to be the "body" of the loop, so that, unless F terminates, incrementalizing F with respect to $\oplus$ is yields $F'(\oplus(x), F(x)) = F(\oplus(\oplus(x)))$. Thus, incrementalization is tantamount to loop unrolling, and caching accumulates partial values for use across iterations.

$\oplus: Y \times W \rightarrow W$	*original*	*incrementalized*
original	$F: W \rightarrow V$	$F': W \times Y \times V \rightarrow V$ $F'(w, y, F(w)) = F(w \oplus y)$
caching	$\overline{F}: W \rightarrow V^n$ $\overline{F}(w) = v_1$ **where** $\langle v_1, v_2, \dots \rangle = \overline{F}(w)$	$\overline{F}': W \times Y \times V^n \rightarrow V^n$ $\overline{F}'(w, y, \overline{F}(w)) = \overline{F}(w \oplus y)$

Fig. 1. Components of incrementalization and identities relating cached, incrementalized, and cached-incrementalized variants of F.

We applied CACHET to the specification of *sqrt* used by O'Leary, et. al, to obtain the same implementation. The source and target expression are shown in statement form in Figure 2, left and right respectively. The *sqrt* algorithm is expressed in the form $F(n, m, i) = \oplus(n, m, i) = \langle n, M(n, m, i), i - 1 \rangle$, where M is the state mutator incrementalized in Figure 3. At five points in this CACHET derivation, judgment was exercised that we would regard as requiring insight. These judgments were of two forms, the application of an algebraic identity (' $\overset{!?}{=}$ ') or the invocation of an invariant assertion (' $\overset{!?}{\Longleftrightarrow}$ '). Facts (d) and (e) are used in *after* incrementalization as the result is incorporated and the surrounding algorithm is simplified.

$$(a) \quad n - (m \pm u)^2 \overset{!?}{=} n - m^2 \mp 2mu - u^2$$

$$(b) \quad w' = (u')^2 = \left(\tfrac{1}{2}u\right)^2 \overset{!?}{=} \tfrac{1}{4}w$$

$$(c) \quad 2m'n' = 2(m+u)(\tfrac{1}{2}u) \overset{!?}{=} \tfrac{2}{2}mu + \tfrac{2}{2}u^2 = \tfrac{1}{2}v + w$$

$$(d) \quad i' \geq 0 \iff i \geq 1 \overset{!?}{\iff} u \geq 2 \iff u^2 \geq 4 \iff w \geq 4$$

$$(e) \quad i' = -1 \iff i = 1 \overset{!?}{\iff} u = 1$$

```
n, i, m := input, (l - 2), 2^{l-1};          p, v, w := input, 0, 2^{2(l-1)};
while i ≥ 0 do                                while (w ≥ 1) do
    p := n - m²;                                  if p > 0 then
    if p > 0 then                                     p, v, w := p - v - w; v/2 + w, w/4
        m := m + 2^i                              else if p < 0 then
    else if p < 0 then                                p, v, w := p + v - w, v/2 - w, w/4
        m := m - 2^i;                            else
    i := i - 1;                                       v, w := v/2, w/4;
    output := m                                output := v
```

Fig. 2. Specification and implementation of nonrestoring *sqrt*

References

1. Steven D. Johnson, Yanhong A. Liu, and Yuchen Zhang. A systematic incrementalization technique and its application to hardware design. Computer Science Department Technical Report 524, Indiana University, June 1999.
2. Yanhong A. Liu. CACHET: An interactive, incremental-attribution-based program transformation system for deriving incremental programs. In *Proceedings of the 10th Knowledge-Based Software Engineering Conference*, pages 19–26, Boston, Massachusetts, November 1995. IEEE CS Press, Los Alamitos, Calif.
3. Yanhong A. Liu. Principled strength reduction. In Richard Bird and Lambert Meertens, editors, *Algorithmic Languages and Calculi*, pages 357–381. Chapman & Hall, London, U.K., 1997.

4. Yanhong A. Liu, Scott D. Stoller, and Tim Teitelbaum. Static caching for incremental computation. *ACM Trans. Program. Lang. and Syst.*, 20(2):1–40, March 1998.
5. Yanhong A. Liu and Tim Teitelbaum. Systematic derivation of incremental programs. *Sci. Comput. Program.*, 24(1):1–39, February 1995.
6. Yanhong Annie Liu. *Incremental Computation: A Semantics-Based Systematic Transformational Approach.* PhD thesis, Department of Computer Science, Cornell University, Ithaca, New York, January 1996.
7. John O'Leary, Miriam Leeser, Jason Hickey, and Mark Aagaard. Non-restoring integer square root: A case study in design by principled optimization. In Ramayya Kumar and Thomas Kropf, editors, *Proceedings of the 2nd International Conference on Theorem Provers in Circuit Design: Theory, Practice, and Experience*, volume 901 of *Lecture Notes in Computer Science*, pages 52–71, Bad Herrenalb (Black Forest), Germany, September 1994. Springer-Verlag, Berlin.
8. Phillip Windley, Mark Aagard, and Miriam Leeser. Towards a super duper hardware tactic. In Jeffery J. Joyce and Carl Seger, editors, *Higher-Order Logic Theorem Proving and its Applications*, volume 780 of *Lecture Notes in Computer Science*. Springer-Verlag, August 1993.

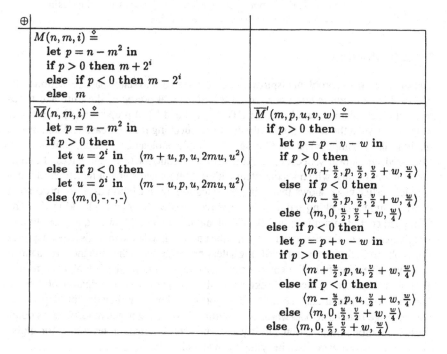

Fig. 3. Incrementalization of *sqrt* [7]

Bisimulation and Model Checking*

Kathi Fisler and Moshe Y. Vardi**

Department of Computer Science, Rice University
Houston, TX 77005-1892
{kfisler, vardi}@cs.rice.edu

Abstract. State space minimization techniques are crucial for combating state explosion. A variety of verification tools use bisimulation minimization to check equivalence between systems, to minimize components before composition, or to reduce a state space prior to model checking. This paper explores the third use in the context of verifying invariant properties. We consider three bisimulation minimization algorithms. From each, we produce an on-the-fly model checker for invariant properties and compare this model checker to a conventional one based on backwards reachability. Our comparisons, both theoretical and experimental, lead us to conclude that bisimulation minimization does not appear to be viable in the context of invariance verification because performing the minimization requires as many, if not more, computational resources as model checking the unminimized system through backwards reachability.

1 Introduction

The state-explosion problem inspires extensive research into state-space reduction techniques. *Bisimulation minimization* [6], a technique that preserves the truth and falsehood of all μ-calculus (and hence all CTL*, CTL, and LTL) properties [4], is particularly attractive in the context of symbolic model checking for two reasons. First, bisimulation can be computed as the fixpoint of a simple boolean expression, so it is easily expressed symbolically. Second, unlike many other reduction techniques, it can be computed automatically, which is consistent with the automated spirit of model checking.

Our earlier work shows that using bisimulation minimization as a pre-processing phase to model-checking reduces the resources needed for model checking [2]. The combined cost of minimization followed by model checking, however, appears to outweigh the costs of simply model checking the unminimized system. This paper explores this problem in the context of verifying safety properties, which is the most fundamental model checking task. We convert three bisimulation minimization algorithms into BDD-based, on-the-fly model checkers for safety properties. A combination of theoretical and experimental analyses on these algorithms show that they do not improve resource usage over basic backwards reachability. This strong negative result casts doubt on bisimulation minimization's utility as a state-space reduction technique for global finite-state transition systems in symbolic model checking.

---

* The full version of this paper appears as a technical report [3].

** Supported in part by NSF grants CCR-9628400 and CCR-9700061, and by a grant from the Intel Corporation. Part of this work was done when the second author was a Varon Visiting Professor at the Weizmann Institute of Science.

2 Three Bisimulation Minimization Algorithms

Bisimulation minimization algorithms partition a state space into equivalence classes such that states in the same class agree on whether the invariance holds and on their next-state transitions to other classes. These algorithms follow a common outline. First, they partition the states into two blocks: those which satisfy the invariance and those that do not. Next, they repeatedly split existing blocks into new ones until all states in a block agree on their next-state transitions to other blocks. If some states in a block B_1 reach states in B_2 and some do not, B_2 is called a *splitter* of B_1. The minimized system contains one state from each block (now an equivalence class) in the final partition.

The naïve bisimulation minimization algorithm has two shortcomings in the context of symbolic model checking. First, it computes the *relation*, rather than the individual equivalence classes. The BDD for the relation requires twice as many variables as the BDDs for the classes. Second, the naïve algorithm fails to distinguish between reachable and unreachable blocks. Several bisimulation algorithms address one or both of these problems. Our work considers the algorithms by Paige and Tarjan (henceforth PT) [7], Bouajjani, Fernandez, and Halbwachs (henceforth BFH) [1], and Lee and Yannakakis (henceforth LY) [5]. We chose these algorithms for the following reasons:

- **PT:** Has the best provable worst-case running time of bisimulation minimization algorithms that do not distinguish between reachable and unreachable blocks.
- **BFH:** Improves on PT by choosing only reachable blocks to split on each iteration; however, it may split an unreachable block that was split off from the reachable block being split in the current iteration.
- **LY:** Improves on BFH by never splitting an unreachable block.

By splitting few, if any, unreachable blocks, the BFH and LY algorithms are tailored to verification contexts. The PT algorithm, although not so tailored, is interesting because it chooses splitters instead of blocks to split (LY and BFH do the latter). None of the algorithms is fully symbolic. LY and BFH operate on a symbolically-represented transition system and produce an explicit-state minimized transition system. PT is originally expressed for explicit state systems; we converted it to the same hybrid symbolic/explicit style as LY and BFH.

Converting these algorithms to on-the-fly model checkers for safety properties involved adding an extra flag to each block to aid in detecting failed properties. We also restricted the PT-based algorithm to only split reachable blocks. The full paper [3] proves that our new algorithms behave as the originals when the tested properties hold, and always discover failures when they do not.

3 Theoretical and Experimental Analysis

We want to compare our new minimizing model checkers to backwards reachability (henceforth BR). BR has a similar flavor to bisimulation minimization. Starting from a partition into good and bad states, BR repeatedly identifies good states that reach the bad states and puts them into a new "frontier" set. Our results formally prove that the contents of the successive frontier sets bear close, and often exact, resemblance to the

	State	BR			LY			BFH			PT		
	Vars	Iter	Time	Mem	Iter	Time	Mem	Iter	Time	Mem	Iter	Time	Mem
gigamax	16	6	1.8	5.59	5	2.3	5.60	6	2.1	5.59	6	2.0	5.58
eisenberg*	17	19	1.1	3.80	18	3.3	3.99	19	9.3	4.48	270	9.3	4.28
abp	19	11	0.9	3.81	10	2.3	3.85	11	2.8	3.82	19	2.0	3.86
bakery*	20	58	1.3	3.70	57	6.5	3.87	58	126.9	9.67	212	7.8	4.55
treearb4	23	24	3.5	4.28	23	16.9	5.18	24	99.0	6.14	232	118.3	6.06
elev23	32	1	3.9	8.45	1	4.2	8.54	1	4.4	8.51	1	4.0	8.43
coherence1	37	5	3.6	6.28	4	85.5	22.0	5	33.0	20.0	23	29.5	8.55
coherence2	37	14	9.3	7.81	13	279.3	31.0	14	174.8	21.0	166	567.4	18
coherence3*	37	5	6.5	7.96	4	84.2	20.0	5	24.4	11.0	9	7.9	7.89
coherence4*	37	5	7.3	8.58	4	78.2	18.0	5	34.4	11.0	685	13.8H	68
elev33	45	1	7.0	11.0	1	444.5	17.0	1	443.8	17.0	1	7.2	11
elev43	56	1	11.9	15.0	1	1590.1	42.0	1	1661.0	39.0	1	12.2	15
tcp*	80	1	3.1	7.83	1	3.6	8.06	1	3.0	8.08	1	3.2	7.83

Table 1. Experimental comparison of the algorithms. A $*$ after an experiment name indicates that the tested invariant property fails. The Iter columns indicates how many iterations an algorithm took before locating an error or reaching a fixpoint. The units for the Time and Mem columns are seconds and megabytes, respectively. An H after the time indicates hours, rather than seconds.

---

contents of the new blocks computed in each iteration of bisimulation minimization. However, each algorithm computes these similar sets in different ways. We therefore need a way to compare and possibly predict the algorithms' relative performance.

Lower bounds on the numbers of operations that each algorithms performs provide a means of comparison. We have derived the bounds below in terms of several variables: n, the number of iterations BR needs to terminate; M, image operations; I, intersection operations; U, union operations; D, set-difference operations; and E, equality tests.

BR: $n * (M + U + D + 2E + I)$ BFH: $(M + I + 2E) * \frac{n^2 + 3n}{2} + n * D$
LY: $(n - 1) * (5M + 4I + 3D + 4E)$ PT: $n * (2M + D + I + E)$

We have proven that we can bound the number of iterations that each algorithm takes from the number of BR iterations: LY needs one less iteration than BR, BFH uses the same number of iterations as BR, and PT requires at least as many iterations as BR.

In predicting the behavior of the four algorithms for symbolic model checking with BDDs, the number of image computations (variable M) should be most useful. The bounds indicate that BFH requires fewer image computations than LY on runs requiring fewer than five iterations. However, BFH's performance should degrade as the number of iterations gets larger. Predicting PT's performance from the bounds is more difficult because there is no upper bound on the number of iterations it performs relative to BR. Based on variable M, we expect BR to have the best performance overall.

Table 1 presents time and memory statistics for running these algorithms on a suite of invariant properties, some of which hold and some of which do not. Our experimental framework uses VIS (version 1.2) [8] as a front-end to obtain the BDDs for the transition relation, initial state, and bad states from Verilog designs and CTL invariant properties.

Our code for each algorithm uses VIS's routines for performing image computations (using partitioned transition relations). In order to have precise control over the BR experiments, we used our own implementation of BR. All runs were performed on an UltraSparc 140 with 128 megabytes of memory running Solaris 5.5.1.

These results show that BR has better time and memory usage than the three adapted minimization algorithms in all cases except elev23, in which PT achieved a negligible savings in memory over BR. As predicted, BFH does perform worse than LY on the designs that require the largest numbers of iterations (eisenberg, bakery, and treearb4). The difference between BFH and LY is most pronounced for bakery, which required the most iterations. BFH generally requires less memory and time than LY on the smaller examples. The one example that falls outside of our predictions is coherence2: on this example, BFH appears to compute images of much smaller sets than LY, which avoided blowup in the intermediate BDDs. Despite taking many more iterations, PT does surprisingly well in comparison to BFH and LY. With the exception of example coherence4, PT has comparable or better memory performance. Its time performance is comparable or better on all examples but treearb4, coherence2, and coherence4. Thus, under our optimization of not splitting unreachable blocks, choosing unreachable blocks as splitters does not seem to hurt PT's performance.

Combining the theoretical and experimental evidence, using bisimulation minimization either as a part of, or as a pre-processor to, model checking invariant properties of transition systems does not appear to be a viable approach. This does not imply that bisimulation has no role in verification contexts. Minimization in a compositional verification framework may make certain verification problems tractable that would not be so without minimization. Similarly, minimization can be used to collapse infinite-state systems into finite ones for purposes of exhaustive analyses. It remains to be seen what other implications our results have for the use of bisimulation in verification. We plan to continue our investigation in the context of verification of liveness properties.

References

1. Bouajjani, A., J.-C. Fernandez and N. Halbwachs. Minimal model generation. In *Proc. Intl. Conference on Computer-Aided Verification (CAV)*, pages 197–203. Springer-Verlag, 1990.
2. Fisler, K. and M. Y. Vardi. Bisimulation minimization in an automata-theoretic verification framework. In *Proc. Intl. Conference on Formal Methods in Computer-Aided Design (FMCAD)*, pages 115–132. Springer-Verlag, 1998.
3. Fisler, K. and M. Y. Vardi. Bisimulation and model checking (extended version). Technical report TR99-339, Rice University, Department of Computer Science, 1999. Available at http://www.cs.rice.edu/CS/Verification/.
4. Hennessy, M. and R. Milner. Algebraic laws for nondeterminism and concurrency. *Journal of ACM*, 32:137–161, 1985.
5. Lee, D. and M. Yannakakis. Online minimization of transition systems. In *Proc. 24th ACM Symposium on Theory of Computing*, pages 264–274, Victoria, May 1992.
6. Milner, R. *A Calculus of Communicating Systems*. Springer Verlag, Berlin, 1980.
7. Paige, R. and R. Tarjan. Three partition refinement algorithms. *SIAM Journal on Computing*, 16:973–989, 1987.
8. The VIS Group. VIS: A system for verification and synthesis. In *Proc. of the 8th Intl. Conference on Computer Aided Verification (CAV)*, pages 428–432. Springer Verlag, July 1996.

Circular Compositional Reasoning about Liveness

K. L. McMillan

Cadence Berkeley Labs

Abstract. Compositional proofs about systems of many components often involve apparently circular arguments. That is, correctness of component A must be assumed when verifying component B, and *vice versa*. The apparent circularity of such arguments can be resolved by induction over time. However, previous methods for such circular compositional proofs apply only to safety properties. This paper presents a method of circular compositional reasoning that applies to liveness properties as well. It is based on a new circular compositional rule implemented in the SMV proof assistant. The method is illustrated using Tomasulo's algorithm for out-of-order instruction execution. An implementation is proved live for arbitrary resources using compositional model checking.

1 Introduction

Compositional model checking methods [7, 4] reduce the verification of large systems to smaller, localized verification problems. This is necessary because model checking is limited by the "state explosion problem". When reasoning compositionally about two processes A and B, it is often necessary to assume correctness of A to verify B and *vice versa*. The apparent circularity of such arguments can be resolved by induction over time. However, existing methods for such circular compositional proofs [2, 3, 7] apply only to safety properties. Nonetheless, mutual dependence of liveness properties does occur in real systems. Consider, for example, the problem of multiple execution units in an instruction set processor. At some times, the instruction in unit A may depend on the result of the instruction in unit B, and at other times the inverse relation may hold. Thus, in order to prove that unit A is live (always eventually produces a result), we must assume the B is live, and *vice versa*. Here, we introduce a compositional technique that allows this kind of circular compositional reasoning. In essence, it makes explicit the induction over time implied in the above approach, by assuming property P *only* up to time $t - 1$ when proving Q at time t, and *vice versa*. This condition (Q up to $t - 1$ implies P up to t) is expressible in temporal logic. Thus, the proof obligations incurred using this method can be discharged by model checking. This proof method has been integrated with others in a proof assistant based on the SMV model checking system. The integration of the circular proof rule with various reduction techniques, including *symmetry reduction, temporal case splitting* and *data type reduction* makes it possible to verify liveness of systems with unbounded arrays, as we illustrate, using Tomasulo's algorithm as an example. The approach is only sketched here – a more extensive treatment can be found in [1].

2 Circular compositional proofs

Using the standard linear temporal logic (LTL) as our framework, we will formalize an inference rule for circular compositional reasoning. Note that there is no notion of process or process composition in this system. As in, for example, TLA [5], a process is simply viewed as a temporal proposition.

Now, suppose that we have a collection of formulas P, and we would like to prove Gp for all $p \in P$. We first fix a well founded order $\prec$ on the formulas in P. Intuitively, if $p \prec p'$ then we may assume p up to time t when proving p' at time t, otherwise we may assume p only up to time $t - 1$. For any proposition $p \in P$, we denote by Δ_p the set of propositions assumed up to time t when proving p, and by Θ_p the set assumed at time t. Every element of Θ_p must be less than p, according to $\prec$. However, any $p' \in P$ (including p itself) may be an element of Δ_p. This is what allows us to construct (apparently) circular arguments. The notion that "p' up to time $t - 1$ implies p at time t" can be expressed in as a formula in LTL, as in the following theorem:

Theorem 1. *Given sets Γ, P of formulas, a well founded order $\prec$ on P, and, for all $p \in P$, sets $\Delta_p \subseteq \Theta_p \subseteq P$, such that $q \in \Delta_p$ implies $q \prec p$, if*

$$\models \Gamma \Rightarrow \neg(\Delta_p \ U \ (\Theta_p \wedge \neg p))$$

for all $p \in P$, then $\models Gp$ for all $p \in P$.

That is, to prove Gp for all p, it suffices to prove $\Gamma \Rightarrow \neg(\Delta_p \ U \ (\Theta_p \wedge \neg p))$ for all p. Note that the latter are linear temporal formulas whose validity we can verify by model checking methods.

Proof graphs The SMV system applies the above theorem in the context of a *proof graph*. This supports both assumption/guarantee style reasoning, where we assume A at all time to prove B at all time, and circular compositional reasoning, where we assume A only up to time t or $t - 1$. A proof graph is a directed graph (V, E), where the vertices V are the propositions to be proved, and an edge $(p, p') \in E$ indicates that proposition p is to be assumed when proving p'. A subset $E^+ \subseteq E$ of the edges are identified as *unit delay edges*. If $(p, p') \in E^+$, then p is assumed only up to time $t - 1$ when proving p' at time t. Although the proof graph may be cyclic, the graph $(V, E \setminus E^+)$ must have no infinite backward paths. If V is finite, this means that every cycle must contain at least one unit delay edge. Each strongly connected component of (V, E) corresponds to an application of theorem 1. Thus, every formula on a cycle must be of the form Gp, so that we may apply the theorem. Formulas not on a cycle may be of any form, however.

Proof obligations are constructed as follows. Suppose that some proposition $Gp \in V$ is on a cycle. Let C be the strongly connected component containing Gp. Let Δ_p be the set of all propositions $Gp' \in C$ such that $(Gp', Gp) \in E$. Let Θ_p be the subset of these such that $(Gp', Gp) \notin E^+$. Let Γ_p be the set of propositions q, such that $(q, Gp) \in E$ and $q \notin C$. On the other hand, if a proposition q is not on a cycle, then let Γ_q be the set of q' such that $(q', q) \in E$.

Theorem 2. *Let C be the union of the strongly connected components of (V, E). If*
  - *for all $Gp \in C$: $\sigma \models \Gamma_p \Rightarrow \neg(\Delta_p \ U \ (\Theta_p \wedge \neg p))$*
  - *for all $q \notin C$: $\sigma \models \Gamma_q \Rightarrow q$*

then for all $p \in V$: $\sigma \models p$.

In order to apply circular compositional reasoning in SMV, we have only to supply the set of properties to be proved and the proof graph. From these, the above theorem can be used to construct a sufficient set of proof obligations in the form of LTL formulas.

3 Verifying a version of Tomasulo's algorithm

As an example, we now consider how the circular compositional approach can be used to prove liveness of an implementation of Tomasulo's algorithm. This design, and its functional verification are described elsewhere in this volume [6]. Here, we assume familiarity with that material, and consider only the verification of liveness. In fact, the liveness proof follows the structure of the functional proof almost exactly.

We prove that an instruction in any reservation station eventually terminates. As in the functional proof, the first step is to break the problem into two lemmas. The first lemma states that the operands required by an instruction in a reservation station eventually arrive. The second states that the result of an instruction in a reservation station eventually returns on the result bus (that is, the reservation station is eventually cleared). We use operand liveness to prove result liveness and *vice versa*. Here is the SMV specification of the operand liveness lemma (for the opra operand):

```
forall (i in TAG)
   live1a[i] : assert G (st[i].valid -> F st[i].opra.valid);
```

That is, for all reservation stations i, if station i is valid (contains an instruction) then its opra operand will eventually be valid (hold a value and not a tag). A similar lemma is stated for the oprb operand. The result liveness lemma is just as simply stated:

```
forall (i in TAG)
   live2[i] : assert G (st[i].valid -> F ~st[i].valid);
```

That is, for all reservation stations i, if station i is valid it is eventually cleared. Note that the reservation station is cleared when its result returns on the result bus.

Operand liveness We use the same case splits for the liveness proof as for the functional proof. That is, to prove liveness of operands arriving at consumer reservation station k, we consider a particular producer reservation station i and a particular intermediate register j. To prove a given case, we need to use only reservation stations i and k, and register j. This case split is specified in SMV as follows (for the opra operand):

```
forall(i,k in TAG; j in REG)
   subcase live1a[k][i][j] of live1a[k]
   for st[k].opra.tag = i & aux[i].srca = j;
```

To prove that operands of consumer reservation station k eventually arrive, we have to assume that the producer reservation station i eventually produces a result. On the other hand, we also have to assume that the operands of an instruction eventually arrive in order to prove that it eventually produces a result. This is where the circular compositional rule comes into play. Note that the producer instruction always enters the machine at least one time unit before the consumer instruction. Thus, to prove that the consumer operand eventually arrives for instructions arriving at time t, it is sufficient to assume that results eventually arrive for producer instructions arriving up to time $t-1$. Thus, we add a unit arc to the proof graph, as follows (for the opra operand):

```
forall (i,k in TAG; j in REG) using (live2[i]) prove live1a[k][i][j];
```

That is, when proving live1a[k][i][j] at time t, we assume live2[i] up to time $t-1$ (parentheses indicate the unit delay). As in the functional verification, the default data type reductions automatically eliminate all but reservation stations i, k and register j, and also reduce the types of register indices and tags to two and three values respectively. Symmetry automatically reduces the n^3 cases we need to verify to just 2 representative cases (for $i = k$ and $i \neq k$).

Result liveness For the result liveness lemma, we again consider the possible paths of a data item from producer to consumer. In this case, operands are sent from reservation station i to some execution unit j. The result then returns on the result bus tagged for reservation station i, which in turn clears the reservation station. We would therefore like to split into cases based on the execution unit. This presents a problem, since at the time the instruction enters the reservation station, the execution unit is not yet determined. Nonetheless, it is possible to split cases on a future value of variable, using the following declaration:

```
forall(i in TAG; j in EU)
    subcase live2[i][j] of live2[i]
        for (aux[i].eu = j) when st[i].issued;
```

That is, for each execution unit j, a result must eventually arrive if the instruction *will* be in execution unit j at the next time the reservation station is found to be in the issued state. Note that when simply is a derived temporal operator. SMV recognizes that at any time v = i when q must be true for at least one value of i, and thus that the given set of cases is complete.

To prove that the instruction in a given execution unit eventually terminates, we must assume that its operands eventually arrive. Thus, we add the following arcs to the proof graph:

```
forall (i in TAG; j in EU) using live1a[i], live1b[i] prove live2[i][j];
```

Note that here, we do not use a unit delay arc. This is allowable, since all the cycles in the proof graph are broken by a unit delay arc. Note, we must also apply appropriate fairness constraints to the arbiters in the design to prove liveness. This is discussed in detail in [1].

Verification The result of applying the above described proof decomposition is a set of proof subgoals that can be solved by model checking. Our implementation is shown to satisfy the given lemmas for an arbitrary number of registers, reservation stations, and execution units. All told, there are 5 model checking subgoals, with a maximum of 20 state variables. The overall verification time (including the generation of proof goals and model checking) is 3.2 CPU seconds (on 233MHz Pentium II processor).

4 Conclusions

Mutual dependence of liveness properties does occur in practice, for example in processors with multiple execution units. The standard assumption/guarantee approach to compositional verification cannot be followed in such a situation. However, we can use an appropriate circular compositional rule, combined with model checking, to prove such mutually dependent liveness properties by induction over time. Such a rule was obtained by extending the technique for safety verification in [7].

References

1. http://www-cad.eecs.berkeley.edu/~kenmcmil/papers/1999-02.ps.gz, Feb. 1999.
2. M. Abadi and L. Lamport. Composing specifications. *ACM Trans. on Prog. Lang. and Syst.*, 15(1):73–132, Jan. 1993.
3. R. Alur and T. A. Henzinger. Reactive modules. In *11th annual IEEE symp. Logic in Computer Science (LICS '96)*, 1996.
4. R. Alur, T. A. Henzinger, F. Mang, S. Qadeer, S. K. Rajamani, and S. Tasiran. Mocha: Modularity in model checking. In *CAV '98*, number 1427 in LNCS, pages 521–25. Springer-Verlag.
5. L. Lamport. The temporal logic of actions. Research report 79, Digital Equipment Corporation, Systems Research Center, Dec. 1991.
6. K. L. McMillan. Verification of infinite state systems by compositional model checking. this volume.
7. K. L. McMillan. Verification of an implementation of Tomasulo's algorithm by compositional model checking. In *CAV '98*, number 1427 in LNCS, pages 100–21. Springer-Verlag, 1998.

Symbolic Simulation of Microprocessor Models Using Type Classes in Haskell

Nancy A. Day, Jeffrey R. Lewis, and Byron Cook

Oregon Graduate Institute, Portland, OR, USA
{nday, jlewis, byron}@cse.ogi.edu

Abstract. We present a technique for doing symbolic simulation of microprocessor models in the functional programming language Haskell. We use polymorphism and the type class system, a unique feature of Haskell, to write models that work over both concrete and symbolic data. We offer this approach as an alternative to using uninterpreted constants. When the full generality of rewriting is not needed, the performance of symbolic simulation by evaluation is much faster than previously reported symbolic simulation efforts in theorem provers.

Symbolic simulation of microprocessor models written in the Haskell programming language [13] is possible without extending the language or its compilers and interpreters. Compared to rewriting in a theorem prover, symbolically simulating via evaluation of a program is generally much faster. Haskell's type class system allows a symbolic domain to be substituted for a concrete one without changing the model or explicitly passing the operations on the domain as parameters. Algebraic manipulations of values in the symbolic domain carry out simplifications similar to what is accomplished by rewriting in theorem provers to reduce the size of terms in the output.

Symbolic simulation in Haskell involves constructing a symbolic domain to represent the operations of the model. The values of this domain are syntactic representations of the machine's behavior. For example, using a recursive data type, an appropriate symbolic domain for numeric computations is:

```
data Symbo =
      Const Int
    | Var String
    | Plus Symbo Symbo
    | Minus Symbo Symbo
    | Times Symbo Symbo
```

Haskell's type class system allows us to make the operations that manipulate data be overloaded on both concrete and symbolic data. A type class groups a set of operations by the common type they operate over. For example, part of the Num class definition is:

```
class Num a where
    (+) :: a -> a -> a
    (-) :: a -> a -> a
    (*) :: a -> a -> a
    fromInt :: Int -> a
```

Parentheses indicate that the operation is infix. The parameter a after the name of the class is a placeholder for types belonging in this class. The function fromInt turns integers into values of type a. This capability is very useful when moving to the symbolic domain because it means existing uses of constant integers do not have to be converted by hand into their representation in the symbolic domain – fromInt is automatically applied to them by the parser.

In Haskell, the type Int is declared to be an instance of the Num class. We also declare our symbolic data as an instance of Num using:

```
instance Num Symbo where
    x + y           = x 'Plus' y
    x - y           = x 'Minus' y
    x * y           = x 'Times' y
    fromInt x       = Const x
```

Now without changing the microprocessor model, we can execute it for either concrete data or symbolic data.

The symbolic domain must behave consistently with the concrete domain. For the case of numbers, there are algebraic laws that hold for the concrete domain that can be used to simplify the output of symbolic simulation. These rules can be implemented for the symbolic domain by augmenting the instance declaration for Symbo with cases that describe the algebraic rules. For example, instead of just having the rule x + y = x 'Plus' y, we have:

```
Var x + Var y = if (x == y) then Const 2 * Var x
                    else Var x 'Plus' Var y
Const x + Const y = Const (x + y)
Const 0 + y       = y
x + Const 0       = x
x + y             = x 'Plus' y
```

When control values in a program are symbolic, the output of symbolic simulation captures the multiple execution paths that the program could have followed. To deal with symbolic control values, we extend the idea of a state to include branches representing multiple execution paths. This leads us to have a symbolic representation of Boolean terms that are used to decide the branches. We introduce an abstraction of the "if-then-else" operation because the first argument to the if operator may be symbolic. A multi-parameter type class captures the behavior of our new ifc. A multi-parameter type class constrains multiple types in a single class instantiation. In the case of ifc, we parameterize the type of the first argument (the deciding value) separately from the type of the other arguments. The result of the function has the same type as the second and third arguments.

```
class Conditional a b where
    ifc :: a -> b -> b -> b
```

The normal "if-then-else" operator is an instance of this class with the type parameter a being Bool (concrete Booleans).

Further details and examples can be found in Day, Lewis, and Cook [7]. One of these examples walks through symbolic simulation in Haskell of the simple, non-pipelined, state-based processor model found in Moore [12].

We have also worked on symbolic simulation of a superscalar, out-of-order with exceptions, pipelined microprocessor model in the Haskell-based hardware description language Hawk [3, 11]. We are now able to simulate symbolic data flow for programs running on the model. We are currently extending the Hawk library to handle symbolic control paths as well. Because it is stream-based, the model does not have explicit access to its state. Hawk models usually process transactions, which capture the state of an instruction as it progresses through the pipeline. The key to having symbolic control flow is to have trees of transactions flowing along the wires rather than just simple transactions. Instead of simply having a top-level branching of state, the branching must be threaded through the entire model, just as transactions are. This means that most components will need to understand how to handle trees of transactions. We are exploring how to best use a transaction type class to define easily a new instance of transactions that are trees.

Symbolic simulation can be carried out with uninterpreted constants using rewriting in a theorem prover (e.g., [8, 9]) or using more specialized techniques such as symbolic functional evaluation [5]. Rewriting requires searching a database of rewrite rules and potentially following unused simplifications [12]. Constructors in our symbolic domains play the same role as uninterpreted constants in a logical model. Because our approach simply involves executing a functional program, we do not suffer a performance penalty for symbolic simulation compared with concrete simulation. Running on a platform roughly two and half times faster than Moore [12], we achieved performance of 58 300 instructions per second compared to ACL2's performance of 235 instructions per second (with hints) for the same non-pipelined, state-based processor model.

Type classes avoid the need to pass the operations to all the components of the model as in Joyce [10]. The type classes keep track of the higher-order function parameters that Joyce grouped in "representation variables".

The approach described in this paper is closely related to work on Lava [1], another Haskell-based hardware description language. Lava has explored using Haskell features such as monads to provide alternative interpretations of circuit descriptions for simulation, verification, and generation of code from the same model. Our emphasis has been more on building symbolic simulation on top of the simulation provided by the execution of a model as a functional program.

Graph structures such as Binary decision diagrams (BDDs) [2] and Multiway decision diagrams (MDGs) [4] are canonical representations of symbolic formulae. In more recent work, we have investigated linking symbolic simulation in Haskell directly with decision procedures for verification to take advantage of the reduced size of representations in these packages [6].

The infrastructure required for using symbolic values and maintaining a symbolic state set is reusable for simulation of different models. We believe the approach presented in this paper may be applied in other languages with user-defined data types, polymorphism, and overloading. However, a key requirement is that overloading work over polymorphic types. Few programming languages support this, although a different approach using parameterized modules, as in SML, might also work well. Haskell's elegant integration of overloading with type inference makes symbolic simulation easy.

For their contributions to this research, we thank Mark Aagaard of Intel; Dick Kieburtz, John Launchbury, and John Matthews of OGI; and Tim Leonard, and Abdel Mokkedem of Compaq. The authors are supported by Intel, U.S. Air Force Material Command (F19628-93-C-0069), NSF (EIA-98005542) and the Natural Science and Engineering Research Council of Canada (NSERC).

References

1. P. Bjesse, K. Claessen, M. Sheeran, and S. Singh. Lava: Hardware design in Haskell. In *ACM Int. Conf. on Functional Programming*, 1998.
2. R. E. Bryant. Graph-based algorithms for Boolean function manipulation. *IEEE Transactions on Computers*, C-35(8):677-691, August 1986.
3. B. Cook, J. Launchbury, and J. Matthews. Specifying superscalar microprocessors in Hawk. In *Workshop on Formal Techniques for Hardware*, 1998.
4. F. Corella, Z. Zhou, X. Song, M. Langevin, and E. Cerny. Multiway decision graphs for automated hardware verification. Technical Report RC19676, IBM, 1994. Also *Formal Methods in Systems Design*, 10(1), pages 7-46, 1997.
5. N. A. Day and J. J. Joyce. Symbolic functional evaluation. To appear in TPHOLs'99.
6. N. A. Day, J. Launchbury, and J. Lewis. Logical abstractions in Haskell. Submitted for publication.
7. N. A. Day, J. R. Lewis, and B. Cook. Symbolic simulation of microprocessor models using type classes in Haskell. Technical Report CSE-99-005, Oregon Graduate Institute, 1999.
8. D. A. Greve. Symbolic simulation of the JEM1 microprocessor. In *FMCAD*, volume 1522 of *LNCS*, pages 321-333. Springer, 1998.
9. J. Joyce. *Multi-Level Verification of Microprocessor Based Systems*. PhD thesis, Cambridge Comp. Lab, 1989. Technical Report 195.
10. J. J. Joyce. Generic specification of digital hardware. In *Designing Correct Circuits*, pages 68-91. Springer-Verlag, 1990.
11. J. Matthews, B. Cook, and J. Launchbury. Microprocessor specification in Hawk. In *International Conference on Computer Languages*, 1998.
12. J. Moore. Symbolic simulation: An ACL2 approach. In *FMCAD*, volume 1522 of *LNCS*, pages 334-350. Springer, 1998.
13. J. Peterson and K. Hammond, editors. *Report on the Programming Language Haskell*. Yale University, Department of Computer Science, RR-1106, 1997.

Exploiting Retiming in a Guided Simulation Based Validation Methodology

Aarti Gupta[1] and Pranav Ashar[1] and Sharad Malik[2]

[1] CCRL, NEC USA, Princeton, NJ
[2] Princeton University, Princeton, NJ

1 Introduction

There has been much interest recently in combining the strengths of formal verification techniques and simulation for functional validation of large designs [6]. Typically, a formal *test model* is first obtained from the design. Then, test sequences which satisfy certain coverage criteria are generated from the test model, which are simulated on the design for functional validation. In this paper, we focus on automatic abstractions for obtaining the test model from the design for simulation vector generation under the transition tour coverage model. Since most efforts using guided simulation have concentrated only on state/transition coverage, without relating these to error coverage of the original design, there is hardly any notion of preserving correctness, which has made it hard to use abstraction effectively.

We introduce one possible notion of abstraction correctness in this context – *transition tour error coverage completeness* of the abstract test model with respect to the original design. Intuitively, the abstraction is correct if it preserves coverage of those errors that can be captured by a transition tour. In other words, the test sequences generated from a transition tour on the abstract test model should cover the complete set of those design errors that are covered by test sequences generated from any transition tour on the original design. We have focused on the transition tour methodology in this paper, since it seems to be the most prevalent mode of generating test sequences. The notion can be potentially extended to other modes as well.

In particular, we propose the use of *Maximal Peripheral Retiming (MPR)* [8], where the number of internal (non-peripheral) latches is minimized via retiming. Subsequently all latches that can be retimed to the periphery are automatically abstracted in the test model, thereby reducing its complexity. We prove the correctness of such an abstraction under certain conditions, where correctness is regarded as preserving design errors that can be detected by using a transition tour on the original design.

2 Retiming for Abstraction

An *input peripheral latch* p_i is a latch whose fanin q is either a primary input or the output of another input peripheral latch, such that q fans out only to p_i. An *output peripheral latch* p_o is a latch which fans out only to either primary outputs or other output peripheral latches. An *internal latch* l_{int} is a latch which is neither an input peripheral latch, nor an output peripheral latch. The application of primary inputs is merely delayed by input latches – they do not affect reachability of rest of the state space. Similarly, output latches merely delay the availability of primary outputs. Internal latches determine the reachability of a circuit's state space.

A design implementation is considered to have an *error* with respect to its specification if the implementation produces an incorrect output for some input sequence. In general, a transition tour cannot capture all errors, since it covers all single transitions only, and not all transition sequences [3]. This is a basic limitation of using transition tours – an error may get detected several transitions after it is excited, and only along a specific path in the state transition graph. If this path is not selected in the transition tour, the error will not be covered. Furthermore, depending on what particular paths are selected, different transition tours may cover different sets of errors.

In order to not tie our analysis to a particular choice of a transition tour, we focus on those errors that can be covered by *any* transition tour of a given design implementation. Such an error can be detected as a difference in the observed output on a transition from a state, regardless of how that state was reached. We also do not restrict our analysis to systems with special properties that allow *all* errors to be covered by a transition tour [3]. We propose the following definitions as the criteria for correctness of an abstraction in this context:

Definition 1: Transition tour error coverage completeness: A model T has transition tour error coverage completeness with respect to another model D, if all errors in D which can be covered by any transition tour on D, are also errors in T and can be covered by some transition tour on T.

Definition 2: An abstraction $\mathcal{A}$ is correct if the abstract test model $T = \mathcal{A}(D)$ has transition tour error capture completeness with respect to the original design implementation D.

In particular, we propose the use of *Maximal Peripheral Retiming (MPR)* as an abstraction which satisfies this correctness property under certain conditions. In general terms, an MPR is a retiming where as many state elements as possible are moved to the periphery of the circuit. Consequently, there are as few internal state elements as possible. In the rest of this paper use the term "latches" to refer to all forms of state elements (latches/registers/flipflops). Once MPR is done on the design, consider the following two abstractions:

- $\mathcal{A}_{eq}$: Removal of all p_o, and removal of an *equal* number of p_i across all inputs.
- $\mathcal{A}_{neq}$: Removal of all p_o, and removal of *all* p_i across all inputs.

First note that the I/O-preserving nature of a retiming transformation itself guarantees that an error observed on a transition from a reachable state s in the original circuit will be preserved as a corresponding error from an equivalent state s' in the retimed circuit. It follows that the removal of a peripheral latch does not affect the presence of such an error, provided the latch was *correctly positioned* to start with. The position of a peripheral latch implies that it can not affect the value of an output (only its timing). It follows that a transition tour on the abstract circuit will eventually reach s' (given the initial conditions) and cover the erroneous transition. We say that a peripheral latch is *correctly positioned*, if it has been separately validated that this position of the latch at the circuit periphery is consistent with intended behavior. In a sense, we are decomposing the burden of checking design correctness into (i) detecting errors in the peripheral latches, and (ii) detecting errors in rest of the circuit. It is our belief that the former task can be handled separately, and in some cases more efficiently, than the latter. In the typical case, a comparison with a higher level description of the design, or designer input may be called for to establish the correcness of the peripheral latches.

2.1 Proofs of Correctness

Theorem 1. *Removal of correctly positioned output peripheral latches preserves error detection for the circuit.*

Proof. The only purpose served by correctly positioned output peripheral latches is to buffer the primary outputs. It is clear that removal of these latches has no impact on either the state space visited during a transition tour, or on its input sequences, thereby preserving detection of errors. ∎

Theorem 2. *Removal of an equal number of correctly positioned input peripheral latches across all inputs preserves error detection for the circuit.*

Proof. For the abstract test model, let n_{pi} be the number of input peripheral latches removed from all inputs. Given m inputs, let $l_{i,j}, 1 \leq i \leq m, 1 \leq j \leq n_{pi}$ denote the initial value on the jth input peripheral latch for the ith input. Let state s be reachable by an input sequence Σ in the original design. Then, there exists an equivalent state s' reachable in the abstract model by the input sequence $\Sigma' = \sigma_1, \sigma_2, \ldots, \sigma_{n_p i}, \Sigma$, where σ_j is the input vector $l_{1,j} l_{2,j} \ldots l_{m,j}$. Since s and s' are equivalent, if there is an error from state s on input a, there will be an error from state s' on input a. Furthermore, since all reachable states and all transitions from those states are covered by a transition tour, this error will be detected by any transition tour with $\sigma_1, \sigma_2, \ldots, \sigma_{n_p i}$ as the initial prefix. Finally, during simulation on the original design, Σ' should be padded by $n_p i$ dummy inputs at the end, in order to account for the delay due to the original input peripheral latches with respect to the observed outputs. ∎

Theorem 3. *Removal of all correctly positioned input peripheral latches preserves error detection for the circuit.*

Proof. For the abstract test model, let n_i be the number of input peripheral latches removed from the ith input. Given m inputs, let $l_{i,j}, 1 \leq i \leq m, 1 \leq j \leq n_i$ denote the initial value on the jth input peripheral latch for the ith input. Now consider a state s reachable by an input sequence $\Sigma = \sigma_1, \sigma_2, \ldots, \sigma_r$ in the original design, where each $\sigma_k = \sigma_{1,k} \sigma_{2,k} \ldots \sigma_{m,k}$ denotes the vector of inputs 1 through m. Due to the equivalence of blocks C and C', there exists an equivalent state s' reachable in the abstract model by the input sequence $\Sigma' = \sigma'_1, \sigma'_2, \ldots, \sigma'_{r+max(n_i)}$, where $\sigma'_k = \sigma'_{1,k} \sigma'_{2,k} \ldots \sigma'_{m,k}$ is the input vector constructed in such a way that:

$$
\begin{aligned}
\sigma'_{i,k} &= l_{i,k} && \text{if } k \leq n_i \\
&= \sigma_{i,k-n_i} && \text{if } n_i < k \leq r + n_i \\
&= - \text{ (don't care)} && \text{if } k > r + n_i
\end{aligned}
$$

Since s and s' are equivalent, same reasoning as for Theorem 2 follows. ∎

Note that in practice, Σ is not known *a priori* since we want to avoid performing a transition tour on the entire design implementation. Instead, during the generation of the transition tour Σ' on the abstract test model, the initial prefix captures the constraints imposed by the initial values of those input peripheral latches that are removed by the abstraction. Also note that though our analysis has been presented in terms of abstraction from the original design implementation to a test model, it also holds for abstraction from any concrete test model to an abstract one.

3 Algorithms for Maximal Peripheral Retiming (MPR)

Given a circuit C, we would like to derive a circuit C' by a retiming such that the number of latches at the periphery of C' is maximized. We refer to this problem as *maximal peripheral retiming* (MPR). We will consider two flavors of this, MPR_{eq} and MPR_{neq} corresponding to the abstractions $\mathcal{A}_{eq}$ and $\mathcal{A}_{neq}$ in Section 2. In MPR_{eq}, only an equal number of latches at each primary input will eventually be abstracted, so the remainder must be counted as internal latches. In MPR_{neq} there is no such restriction. In the worst case, if none of the latches move to the periphery, then the final circuit is the same as the initial circuit. Similar ideas are used in automatic test pattern generation for sequential circuits [1, 4].

Let us consider the algorithm for MPR_{neq} first. The conventional retiming algorithm [7] for latch minimization with the cost of peripheral latches assigned to zero will suffice to minimize the number of internal latches by attempting to push as many as possible to the periphery.

The algorithm for MPR_{eq} is handled with a minor modification to the circuit. A dummy node d_1 is added to the circuit with fanout to each of the primary inputs i_n.

Another dummy node d_0 is added with no fanin and a single fanout to d_1. The cost of the edges is as follows: (1) for the (d_0, d_1) edge it is 0 (2) for the (d_1, i_n) edges it is infinity (3) for the output periphery edges it is 0 (4) for all other edges, it is the same as in the original circuit. This will force a minimization of internal latches in the circuit by trying to push as many latches to the output periphery and on the (d_0, d_1) edge. We can also maximize the number of latches at the primary outputs, by biasing the peripheral retiming in that direction by picking appropriate bus widths, e.g. 0 for output peripheral edges, 1 for input peripheral edges, and some large number (greater than the maximum fanin of any gate) for internal edges.

Handling Enables and Multiple Clocks/Phases: To maximize the ability to move latches to the periphery, paths with unbalanced phases are balanced by adding dummy latches in appropriate locations. We can do this since the dummy latches do not affect the logical behavior. Only the clock period requirement is affected – which is immaterial to our application.

4 Case Study: The DLX Processor

We used a Verilog RTL implementation (without floating-point and exception-handling instructions) [2] of the popular DLX processor [5]. It uses a standard 5-stage pipeline consisting of the fetch, decode, execute, memory and write-back stages. The model after manual abstraction consisted of 92 latches. Upon applying MPR, we could abstract 32 latches that were already at the input periphery, 31 latches that were already at the output periphery, and 8 latches that could be retimed to the periphery. It was also interesting to examine the 21 remaining latches which could not be peripherally retimed. There are two main structures that prevent such retiming – (i) self-loops (ii) reconvergent paths with different number of latches. Typical examples of these for the DLX design are the self-loop for dumping the fetched instruction in case of a taken branch, and the reconvergent structure used to select the ALU sources. To summarize, of the 92 original latches, 71 were abstracted out by use of MPR, resulting in a final test model of 21 latches, 25 primary inputs and 31 primary outputs. The Verilog code was converted to an FSM description, and the implicit transition relation representation of the final model was obtained in about 10 seconds on an Ultrasparc (166 MHz.) workstation with 64Mb. main memory.

References

1. A. Balakrishnan and S. T. Chakradhar. Software transformations for sequential test generation. In *Fourth Asian Test Symposium*, 1995.

2. P. Franzon. Digital computer technology and design: Fall 1994 project. Private Communication, 1996.

3. A. Gupta, S. Malik, and P. Ashar. Toward formalizing a validation methodology using simulation coverage. In *Proc. 34th Design Automation Conf.*, pages 740–745, June 1997.

4. R. Gupta, R. Gupta, and M. A. Breuer. BALLAST: A methodology for partial scan design. In *Proceedings of the International Symposium on Fault Tolerant Computing*, pages 118–125, June 1989.

5. J. L. Hennessy and D. A. Patterson. *Computer Architecture: A Quantitative Approach*. Morgan Kaufmann, 1990.

6. R. C. Ho, C. H. Yang, M. A. Horowitz, and D. L. Dill. Architecture validation for processors. In *Proc. 22nd Annual International Symposium on Computer Architecture*, June 1995.

7. Charles E. Leiserson and James B. Saxe. Retiming Synchronous Circuitry. *Algorithmica*, 6(1):5–36, 1991.

8. S. Malik, E. Sentovich, R. K. Brayton, and A. Sangiovanni-Vincentelli. Retiming and resynthesis: Optimizing sequential networks with combinational techniques. *IEEE Tran. on CAD of Integrated Circ. and Sys.*, 10(1):74–84, Jan. 1991.

Fault Models for Embedded Systems*
(Extended Abstract)

Jens Chr. Godskesen**

The IT-University in Copenhagen
DK–2400 NV, Denmark
jcg@itu.dk

Abstract. In this paper we present the notion of an *input fault model* for embedded systems and outline how to obtain *minimal complete* test suites for a fault model. The system software is expected to be *embedded* and specified as a finite state machine.

1 Introduction

Testing a system correct with respect to a specification is often referred to as *conformance testing*. The goal is to derive, hopefully automatically a set of tests that may help to ensure the conformance of the system against the specification. The tests should at least enjoy the property that if the system does not pass all the tests then it is not in conformance with its specification.

For finite state machines (FSM's) it has been shown ([1]) that conformance between a specification S and its implementation I can be tested by a test suite that is polynomial in the size of the number of states n and input symbols in S provided I is a FSM with n states. Unfortunately, if I has more than n states the size of the test suite will be exponential. For industrial sized examples it may often be that S contains millions of states and hundreds of input symbols so even if S and I are of equal size the conformance testing problem is intractable.

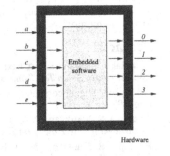

Fig. 1. A conceptualization of embedded systems.

In this paper we put forward an alternative theory for testing *embedded systems*. Embedded systems are dedicated systems like mobile phones, hi-fi equipment,

---

* The full version of this paper is obtainable from http://www.itu.dk/~jcg

** Supported by a grant from the Danish Technical Research Council.

remote controls etc., where the software is embedded. Conceptually an embedded system may be regarded as the *composition* of the two system components: the embedded software and the hardware as depicted in figure 1. The inputs from the system environment to the software have to pass through the hardware towards the software via *connections* (the unlabelled arrow arrows) and likewise the outputs generated by the software have to pass via connections through the hardware. Connections are considered to be purely abstract notions, they may have no physical counterpart. Each input and output has its own connection.

Exploiting the ability to automatically generate code from specifications and assuming compilers to be correct it would be reasonable to expect the generated software to be correct with respect to its specification. The hardware we shall regard as a black box interfacing the embedded software through the connections. As a consequence we shall not be able to refer directly to hardware errors. Given these two assumptions a fault may indirectly be modeled as erroneous connections. Therefore, what has to be tested in order to establish the system correctness is *the non-existence of faults manifested as errors in the connections*. Hence the focus is on testing the composition of the two system components instead of testing the behaviour of the whole system.

In figure 1 a fault could be that the connection for input b is missing. For a mobile phone this may correspond to some button being disconnected such that the software will never receive the input, making the pressing of the button cause no effect. In order to be able to handle faults we introduce the notion of a *fault model* and we outline a framework for testing embedded systems by means of such models. Fault models in a testing framework has appeared elsewhere in the literature, see e.g. [2]. However, although similar their notion is in the setting of conformance testing.

2 Finite state machines

A deterministic *finite state machine* is a five tuple $M = (\Sigma, \mathcal{E}, \Omega, \tau, s_M)$ where Σ is a finite set of states, $\mathcal{E}$ a finite set of inputs, Ω a finite set of outputs, and τ a transition function, $\tau : \Sigma \times \mathcal{E} \to \Sigma \times \mathcal{P}(\Omega)$. $s_M \in \Sigma$ is the *initial state*. We let s range over Σ, α over $\mathcal{E}$, ω over Ω, ι over $\mathcal{P}(\mathcal{E})$, o over $\mathcal{P}(\Omega)$, and σ over $\mathcal{E}^*$. [1] If σ' is a prefix of σ we write $\sigma' \preceq \sigma$. $|\sigma|$ denotes the length of σ. ϵ is the empty sequence. $\sigma \downarrow \iota$ denotes an abstraction of σ where all occurrences of $\alpha \in \iota$ are removed, that is $\sigma \downarrow \iota$ may be inductively defined by $\sigma \downarrow \iota = \epsilon$ if $\sigma = \epsilon$, $\sigma \downarrow \iota = \sigma' \downarrow \iota$ if $\sigma = \sigma' \alpha$ and $\alpha \in \iota$, and $\sigma \downarrow \iota = (\sigma' \downarrow \iota)\alpha$ if $\sigma = \sigma' \alpha$ and $\alpha \notin \iota$.

We write $s \xrightarrow{\alpha/o} s'$ and $s \xrightarrow{\alpha} s'$ if $\tau(s, \alpha) = (s', o)$. If $s_i \xrightarrow{\alpha_i} s_{i+1}$ for $i = 1, \dots, n$ we write $s_1 \xrightarrow{\sigma} s_{n+1}$ where $\sigma = \alpha_1 \dots \alpha_n$. We shall regard s as a function

---

[1] $\mathcal{P}$ is the power set constructor.

$s : \mathcal{E}^* \to \mathcal{P}(\Omega)$ defined by $s(\epsilon) = \emptyset$ and $s(\sigma\alpha) = o$ if $s \xrightarrow{\sigma} s'$ and $s' \xrightarrow{\alpha/o} s''$. We shall write $M(\sigma)$ for $s_M(\sigma)$. If for all σ, $s(\sigma) = s'(\sigma)$ we write $s \approx s'$. If $s_M \approx s_{M'}$ we write $M \approx M'$. If not $s \approx s'$ ($M \approx M'$) we write $s \not\approx s'$ ($M \not\approx M'$).

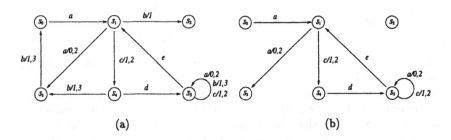

(a)　　　　　　　　　　　　(b)

Fig. 2. (a) A finite state machine. (b) The FSM in (a) with input b removed. [2]

The machine $M[\iota]$ denotes M where any transition $s \xrightarrow{\alpha/o} s'$ in M is replaced by $s \xrightarrow{\alpha/\emptyset} s$ whenever $\alpha \in \iota$. Letting M denote the machine in figure 2(a), $M[\{b\}]$ is the machine in figure 2(b). For any set of inputs $\{\alpha_1, \ldots, \alpha_n\}$ we shall adopt the convention of writing $M[\alpha_1, \ldots, \alpha_n]$ instead of $M[\{\alpha_1, \ldots, \alpha_n\}]$.

M is α-*active* if $M(\sigma\alpha) \neq \emptyset$ for some σ. M is ι-active if it is α-active for some $\alpha \in \iota$. The machine in figure 2(b) is not b-active whereas the one in figure 2(a) is. M is α-*distinguishable* if $M(\sigma_1\alpha\sigma_2) \neq M(\sigma_1\sigma_2)$ for some σ_1 and σ_2. M is ι-distinguishable if it is α-distinguishable for some $\alpha \in \iota$. M is ι-*sensitive* if it is ι-active or ι-distinguishable. The FSM in figure 2 are e-distinguishable because $M(acdeaa) \neq M(acdaa)$, hence they are e-sensitive although not e-active.

Lemma 1. $M \not\approx M[\iota]$ *if and only if* M *is ι-sensitive.*

3　Fault models

The purpose of a fault model is to capture the faults that may occur among the connections in an embedded system. For instance, we would like that a fault corresponding to the disconnection of the b-input in figure 1 can be handled by a fault model. In general we let any M' where $M \not\approx M'$ be a fault relative to M, and we define a *fault model* for M as a set of machines $\mathcal{M}^M$ such that for all $M' \in \mathcal{M}^M$, $M \not\approx M'$.

---

[2] Empty outputs and transitions to the same state with empty output are left out.

The fault that a set of inputs ι is not connected can be modeled by $M[\iota]$. Let $\mathcal{I} \subseteq \mathcal{P}(\mathcal{P}(\mathcal{E})/\emptyset)$. Then, if M is ι-sensitive for any $\iota \in \mathcal{I}$, the fault model for unconnected input for M is defined by $\mathcal{M}_{\mathcal{I}}^M = \{M[\iota] \mid \iota \in \mathcal{I}\}$. Intuitively, $\mathcal{M}_{\mathcal{I}}^M$ represents any one implementation that for some $\iota \in \mathcal{I}$ contains the disrupted input connections in ι, but otherwise behaves as expected by M. Letting M denote the FSM in figure 2(a) then the fault model below is an input fault model for M. It contains beyond the multi-fault $M[b, d]$ the single-faults $M[c]$, $M[d]$, and $M[e]$.

$$\mathcal{M}_{\mathcal{I}}^M = \mathcal{M}_{\{\{c\},\{d\},\{e\},\{b,d\}\}}^M \tag{1}$$

4 Tests

A *test* is a finite sequence of input events $\sigma \in \mathcal{E}^*$. A *test suite* $T \subseteq \mathcal{E}^*$ is a finite set of tests. The application of a test and a test suite respectively to M' with respect to some reference M is defined by

$$apply_M(M', \sigma) = \begin{cases} pass & \text{if } \forall \sigma' \preceq \sigma.\ M(\sigma') = M'(\sigma') \\ fail & \text{otherwise} \end{cases}$$

$$apply_M(M', T) = \begin{cases} pass & \text{if } \forall \sigma \in T.\ apply_M(M', \sigma) = pass \\ fail & \text{otherwise} \end{cases}$$

We say that a test suite T is *sound* for a fault model $\mathcal{M}^M$ if for any $\sigma \in T$ there exists some $M' \in \mathcal{M}^M$ such that $apply_M(M', \sigma) = fail$. A test suite T is *exhaustive* for $\mathcal{M}^M$ if for all $M' \in \mathcal{M}^M$, $apply_M(M', T) = fail$. Finally, we say that a test suite T is *complete* for $\mathcal{M}^M$ if it is sound and exhaustive for $\mathcal{M}^M$.

Given a set of tests T we let $max(T) = \{\sigma \in T \mid \forall \sigma' \in T.\ |\sigma| \geq |\sigma'|\}$ and $min(T) = \{\sigma \in T \mid \forall \sigma' \in T.\ |\sigma| \leq |\sigma'|\}$. For a family of set of tests $\langle T_i \rangle_{i \in I}$ for some index set I we write $T \in \langle T_i \rangle_{i \in I}$ whenever $T = \{\sigma_i \mid i \in I\}$ where $\sigma_i \in T_i$ for all $i \in I$. When it is clear from the context we let $\langle T_i \rangle_{i \in I}$ denote the set of set of tests $\{T \mid T \in \langle T_i \rangle_{i \in I}\}$.

In order to obtain complete test suites for a fault model we want to obtain a set of tests for each fault in the model that characterizes the fault. A fault $M[\iota]$ can be characterized by $T_\iota^M = \bigcup_{\alpha \in \iota} \{\sigma\alpha \mid M(\sigma\alpha) \neq \emptyset\} \cup \bigcup_{\alpha \notin \iota} \{\sigma\alpha \mid M(\sigma\alpha) \neq M(\sigma{\downarrow}\iota)\}$. Letting M be the FSM in figure 2(a) it is obvious that $ac \in T_{\{c\}}^M$, $acda \in T_{\{d\}}^M$, $acdeaa \in T_{\{e\}}^M$, and $ab \in T_{\{b,d\}}^M$.

Lemma 2. $apply_M(M[\iota], \sigma) = fail$ *if and only if* $\exists \sigma' \in T_\iota^M.\ \sigma' \preceq \sigma$.

Given a family of characterizing sets of tests, one for each fault in a fault model $\mathcal{M}_{\mathcal{I}}^M$, a complete test suite can be defined.

Theorem 1. *Let $\mathcal{M}_\mathcal{I}^M$ be a fault model. Then any $T \in \langle T_\iota^M \rangle_{\iota \in \mathcal{I}}$ is complete for $\mathcal{M}_\mathcal{I}^M$.*

No complete test suite for $\mathcal{M}_\mathcal{I}^M$ needs to contain more tests than the number of member in the family $\langle T_\iota^M \rangle_{\iota \in \mathcal{I}}$. Letting $\mathcal{M}_\mathcal{I}^M$ be the input fault model defined by equation 1, the test suite $\{ac, acda, acdeaa, ab\}$ is complete for $\mathcal{M}_\mathcal{I}^M$ because $ac \in T_{\{c\}}^M$, $acda \in T_{\{d\}}^M$, $acdeaa \in T_{\{e\}}^M$, and $ab \in T_{\{b,d\}}^M$. However, since $acda \in T_{\{b,d\}}^M \cap T_{\{d\}}^M$ it turns out that also $\{ac, acda, acdeaa\}$ is complete for $\mathcal{M}_\mathcal{I}^M$.

5 Minimal complete test suites

A test suite T is *minimal complete* for $\mathcal{M}_\mathcal{I}^M$ if $T \in \langle min(T_\iota) \rangle_{\iota \in \mathcal{I}'}$ for some $\mathcal{I}' \subseteq \mathcal{I}$, if T is complete for $\mathcal{M}_\mathcal{I}^M$, and if no $T' \subset T$ is complete for $\mathcal{M}_\mathcal{I}^M$.

For any family of set of tests $\langle T_i \rangle_{i \in I}$ we write $T \in \mu \langle T_i \rangle_{i \in I}$ if $T \in \langle T_i \rangle_{i \in I}$ and

$$\forall \sigma \in T.\ \exists i \in I.\ T \cap min(T_i) = \{\sigma\} \tag{2}$$

When it is clear from the context we let $\mu \langle T_i \rangle_{i \in I}$ denote the set of tests $\{T \mid T \in \mu \langle T_i \rangle_{i \in I}\}$. For a family of set of tests $\langle T_i \rangle_{i \in I}$ and a set of tests $T \in \langle min(T_i) \rangle_{i \in I}$ we let $\mu_{\langle T_i \rangle_{i \in I}}(T)$ denote a subset of T such that $\mu_{\langle T_i \rangle_{i \in I}}(T) \in \mu \langle T_i \rangle_{i \in I}$.[3] Letting M be the FSM in figure 2(a) it can be shown that $\{ac, acda, acdeaa, ab\}$ belongs to $\mu \langle T_\iota^M \rangle_{\iota \in \mathcal{I}}$ where $\mathcal{I} = \{\{c\}, \{d\}, \{e\}, \{b,d\}\}$.

Define for any $\langle T_i \rangle_{i \in I}$ the function $\Phi_{\langle T_i \rangle_{i \in I}} : \langle min(T_i) \rangle_{i \in I} \to \mathcal{P}(\mathcal{E}^*)$ by

$$\Phi_{\langle T_i \rangle_{i \in I}}(T) = \begin{cases} \emptyset & \text{if } T = \emptyset \\ T' \cup \Phi_{\langle T_i \rangle_{i \in I''}}(T'') & \text{otherwise} \end{cases}$$

where $T' = \mu_{\langle T_i \rangle_{i \in I'}}(max(T))$ for $I' = \{i \in I \mid \exists \sigma \in max(T).\ \sigma \in T_i\}$, and where $T'' = \{\sigma_i \in T \mid i \in I''\}$ for $I'' = \{i \in I \mid \forall \sigma \in T'.\ \forall \sigma' \preceq \sigma.\ \sigma' \notin T_i\}$, whenever $T = \{\sigma_i \mid i \in I\}$.

The functions may as results give minimal complete test suites.

Theorem 2. *Whenever $\mathcal{M}_\mathcal{I}^M$ is a fault model and $T \in \langle min(T_\iota^M) \rangle_{\iota \in \mathcal{I}}$. Then $\Phi_{\langle T_\iota^M \rangle_{\iota \in \mathcal{I}}}(T)$ is minimal complete for $\mathcal{M}_\mathcal{I}^M$.*

Intuitively, a minimal complete test suite for a fault model is constructed by first selecting a set of tests T containing a smallest test for each fault in the fault

---

[3] $\mu_{\langle T_i \rangle_{i \in I}}$ can be considered a function $\mu_{\langle T_i \rangle_{i \in I}} : \langle min(T_i) \rangle_{i \in I} \to \mu \langle T_i \rangle_{i \in I}$ because the resulting set of tests can be defined by keeping all tests in the argument that satisfy equation 2 and thereafter successively removing tests from the argument as needed due to some lexicographical ordering of the elements in $\mathcal{E}^*$ until the definition of $\mu \langle T_i \rangle_{i \in I}$ is satisfied. This also shows the existence of tests in $\mu \langle T_i \rangle_{i \in I}$ if $I \neq \emptyset$.

model. The test for a fault is selected from the set of tests that characterizes the fault. Secondly, all the longest tests in T are collected and among those some tests may be eliminated such that equation 2 is satisfied. This gives as a result T' which will be contained in the resulting test suite. All faults in the model that may be detected by at least one of the remaining largest tests are disregarded and not considered further in the computation. The process is repeated for the faults still to be considered until no fault in the model needs consideration. As an example, letting M be as defined in figure 2, it can be shown that $\Phi_{(T_i^M)_{i \in \mathcal{I}}}(\{ac, acda, acdeaa, ab\}) = \{acdeaa\}$ where $\mathcal{I} = \{\{c\}, \{d\}, \{e\}, \{b, d\}\}$.

In this paper we have put forward an approach for test generation of embedded systems with software that is modeled as a deterministic FSM. The approach differs from the traditional conformance testing approaches in that the focus is on testing the composition of the embedded software and the hardware. That is, it is the interface (as defined in section 1) between the hardware and the software that is to be tested.

Currently we are involved in ongoing work where algorithms for generating minimal complete test suites based on fault models are developed.

References

1. T.S Chow. Testing software desing models by finite-state machines. *IEEE Transactions on Software Engeneering*, 4(3):178–187, 1978.
2. A. Petrenko, N. Yevtushenko, and G. v. Buchmann. Fault models for testing in context. In *Proceedings of the First Joined International Conference on Formal Description Techniques for Distributed Systems and Communication Protocols and Protocol Specification, Testing and Verification (FORTE/PSTV'96)*, pages 163–178, University of Kaiserslautern, Department of Informatics, October 1996. Chapman & Hall.

Validation of Object-Oriented Concurrent Designs by Model Checking*

Klaus Schneider, Michaela Huhn, and George Logothetis

University of Karlsruhe, Department of Computer Science
Institute for Computer Design and Fault Tolerance (Prof. Dr.-Ing. D. Schmid)
P.O. Box 6980, 76128 Karlsruhe, Germany
email: {schneide|huhn|logo}@informatik.uni-karlsruhe.de
http://goethe.ira.uka.de/

1 Introduction

Reusability and evolutivity are important advantages to introduce object-oriented modeling and design also for embedded systems [1,2]. For this domain, one of the most important issues is to validate the interactions of a set of objects with concurrent methods. We apply model checking (see [3] for a survey) for the *systematic debugging of concurrent designs* to detect errors in the behavior and interactions of the object community. As we assume a fixed finite maximal number of objects and also finite data types, we can only show the correctness for finite instances and detect only errors that appear in such a finite setting. Nevertheless, the approach is useful for embedded systems, where the system's size is limited by strong hardware constraints. Moreover, we claim that most errors in the concurrent behavior already occur with a small number of components. To handle larger designs, we emphasize that it is often obvious that several attributes of an object do not affect the property of interest. Thus, there is no need to model the objects completely. This obvious abstraction leads to significantly better results because the resulting models are smaller. More sophisticated abstractions can be found e.g. in [4,5]. In the next section, we briefly explain how to derive in general a finite state system from an object-oriented concurrent design. Then, we illustrate the method by a case study taken from [6].

2 From Concurrent Objects to Finite State Machines

We assume a class C with attributes $a_1, \ldots, a_n$ of types $\alpha_1, \ldots, \alpha_n$, respectively, and methods $\tau_0, \ldots, \tau_m$ to manipulate the attributes (there may be constructor and destructor methods). For an object O of class C, we denote the method τ_i invoked for O by $O.\tau_i$ and the value of the attribute a_i of O by $O.a_i$.

We are not interested in inheritance or typing issues, but in the concurrent behavior: the methods $O.\tau_i$ are implemented as threads, i.e., they may be invoked in parallel. Their execution may be interleaved or truely concurrent. The methods work on the same

---

* This work has been financed by the DFG priority program 'Design and Design Methodology of Embedded Systems'.

memory (namely the attributes of $\mathcal{O}$). The methods $\mathcal{O}.\tau_i$ are not necessarily atomic, instead they consist of a sequence of atomic operations. Hence, a method $\mathcal{O}.\tau_i$ may be suspended, aborted or interrupted by another method $\mathcal{O}.\tau_j$. A major concern is that the concurrent execution does not lead to inconsistencies or runtime faults. This problem is nontrivial since concurrent threads may modify the same attributes $\mathcal{O}.a_i$ or interrupt such modifications before they are completed. For our finite state machine abstraction, three different kinds of abstractions have to be applied systematically:

- *Infinite data types* are abstracted to a finite data domain. E.g., integers are mapped to bitvectors of a certain length n.
- While *objects may come and go dynamically*, the finite state machine model requires to fix a maximal finite number of objects that invariantly exist from the beginning. Hence, we allocate for each class a maximal finite number of objects in advance, and model construction and destruction such that at each point of time at most the maximal number of these objects are in use.
- All methods $\mathcal{O}_j.\tau_i$ are modeled as finite state machines $\mathcal{A}_{j,i}$ that may interact with each other. As the entire system must be finite-state, we have to *restrict the maximal number of methods that may run in parallel at a point of time*. Similar to the construction and destruction of objects, we can however model that threads are dynamically started, suspended, aborted, or that they terminate.

Due to the description level and the properties of interest, a granularity of atomicity has to be chosen. For instance, for a system design given in Java, it is reasonable to assume Java statements as atomic. Also, the treatment of write-clashes has to be modeled: One could either nondeterministically select a value or determine the value by a resolution function as common in many concurrent languages. Our model is able to cope with any of these solutions.

In contrast to other formal approaches to object-oriented designs, we consider methods as non-atomic and allow the concurrent methods running on the same object. Hence, our model is closer to practical implementations in C++ or Java and even takes the thread management of the operating system into account. In particular, we are interested in whether the design is robust wrt. the suspension and activation of threads. We emphasize that the construction of the finite-state model can be done automatically, when the maximal number of concurrent threads and concurrently existing objects are fixed, and the mapping from infinite data types to finite ones is given.

3 Interrupt-Transparent Lists

We now present a case study taken from an embedded operating system kernel [6]. The objects are single-linked lists. The methods to modify the list may be interrupted at any time. Therefore, naive sequential list operations can not be used, since the list could become inconsistent (in particular, some items get lost). Due to the lack of space, we only consider the enqueue method. C++ implementations are given in figure 1.

Cargo-objects consist of two pointers: next points to the first list element, and tail points to the last one. The Cargo constructor assigns next to 0, and tail to the

```
class Chain                          void Cargo::enqueue(Chain* item)
  (public:Chain* next);                (Chain *last;
                                        Chain *curr;
class Cargo:public Chain             s1 : item->next = 0;
  (public:                           s2 : last = tail;
    Chain* tail;                     s3 : tail = item;
    Cargo();                         s4 : if (last->next)
    void enqueue(Chain* item);       s5 :  { curr = last;
  );                                 s6 :     while (curr->next)
                                     s7 :        curr = curr->next;
Cargo::Cargo()                       s8 :     last = curr;
  (next = 0;                              }
   tail = (Chain*) this;)            s9 : last->next = item;
                                        }
```

Figure1. Implementation of Interrupt-Transparent Lists

current object. To ensure robustness against interrupts, we need a sophisticated implementation of the enqueue method due to [6]. Given a Chain object item, enqueue resets the next-pointer of the argument (s1), and stores the current end of the list (s2). In line s3, it assigns the tail-pointer to the argument. If the thread is not interrupted last->next must be 0 at s3. Otherwise an interrupting enqueue thread may have inserted some elements. Then, the current end of the list has to be found (loop s5, ..., s8). Finally, the argument is added to the list (s9).

Modeling Lists of Bounded Length. To derive a finite-state model, we consider a maximal number of n enqueue threads E_1, ..., E_n manipulating in an interleaving execution one Cargo object. Analogous to the C++ code, we abstract from the items and simply enumerate them: E_i inserts item i. The tail pointer of the Cargo object is represented by the variable $tail$, the next pointer by $pt[0]$. The C++ expressions a->next and a->next->next correspond in our model with the terms $pt[a]$ and $pt[pt[i]]$, respectively. We start with the initial values $pt[0] = 0$ and $tail = 0$.

$$\text{type enum } Item: 1, \dots, n; \qquad \text{var } pt : \text{array } Address \text{ of } Address;$$
$$\text{type enum } Address: 0, \dots, n; \qquad \text{var } tail : Address;$$

Modeling the Enqueue Threads. Each E_i is modeled as a finite state machine that manipulates two local variables $last_i$ and $curr_i$ of type Address. The states directly correspond C++ code of figure 1. States s_0 and s_{10} are added to model that E_i has not yet started or has already terminated.

s_0 :

$s_1 : pt[i] := 0$

$s_2 : last_i := tail$

$s_3 : tail := i$

$s_4 : \text{if } pt[last_i] = 0 \text{ goto } s_9 \text{ else goto } s_5$

$s_5 : curr_i := last_i$

$s_6 : \text{if } pt[curr_i] = 0 \text{ goto } s_8 \text{ else goto } s_7$

$s_7 : curr_i := pt[curr_i]; \text{ goto } s_6$

$s_8 : last_i := curr_i$

$s_8 : last_i := curr_i$

$s_9 : pt[last_i] = i$

$s_{10} :$

As the threads $E_1, \ldots E_n$ run in an interleaved manner, at most one thread manipulates the Cargo object at each point of time. Without loss of generality, we assume that E_i starts before E_j iff $i < j$. If E_j starts when E_i is still running, E_i will be interrupted, since E_j has a higher priority due to the interruption. The interruption is modeled by the following signals: $run_i := \bigvee_{k=1}^{9} E_i.s_k$, $end_i := E_i.s_{10}$, $ac_i := \bigwedge_{k=i+1}^{n} \neg run_k$, and $perm_i := \bigwedge_{k=1}^{i-1} start_k$. run_i holds iff thread E_i is currently running. end_i holds iff thread E_i has terminated. ac_i indicates that E_i is activated, i.e., no thread E_k with higher priority ($i < k$) is currently running. Finally, $perm_i$ implements the priorities, i.e. E_i may only start if all E_j with $j < i$ have started before. Using these control signals, we obtain the state transition diagram as given in figure 2.

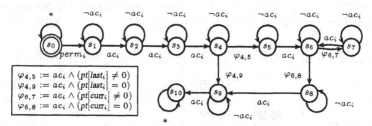

Figure2. The enqueue thread as finite state process

Properties to be Checked The properties we checked using McMillan's SMV system[1] are as follows:

S_1: All items are enqueued: $\mathsf{G}\left[\left(\bigwedge_{i=1}^{n} end_i\right) \rightarrow \left(\bigwedge_{i=1}^{n} \bigvee_{j=0}^{n} pt[j] = i\right)\right]$

S_2: All items are enqueued at most once:

$$\mathsf{G}\left[\left(\bigwedge_{i=1}^{n} end_i\right) \rightarrow \left(\bigwedge_{i=1}^{n} \bigvee_{j=0}^{n} pt[j] = i \rightarrow \bigwedge_{k=0}^{n} pt[k] = i \rightarrow k = j\right)\right]$$

S_3: Absence of deadlocks: $\mathsf{G}\left[\bigwedge_{i=1}^{n} start_i \wedge perm_i \rightarrow \mathsf{F}end_i\right]$

S_4: The threads are started in order $E_1, E_2, \ldots E_n$: $\mathsf{G}\bigwedge_{i=1}^{n-1} E_i.s_0 \rightarrow \bigwedge_{j=i+1}^{n} E_j.s_0$

S_5: If $E_1, \ldots E_k$ have terminated *before* any of the threads $E_{k+1}, \ldots E_n$ has started, the first k elements have been enqueued:

$$\mathsf{G}\bigwedge_{k=1}^{n-1}\left[\left(\bigwedge_{i=1}^{k} end_i\right) \wedge \left(\bigwedge_{i=k+1}^{n} E_i.s_0\right) \rightarrow \bigwedge_{j=0}^{k} \bigvee_{i=0}^{k} pt[i] = j\right]$$

$\bigwedge_{j=0}^{k} \bigvee_{i=0}^{k} pt[i] = j$ means $pt[0], \ldots, pt[k]$ is a permutation of $0, \ldots, k$, which means that each element of $0, \ldots, k$ is contained in the list.

---

[1] Available from http://www-cad.eecs.berkeley.edu/~kenmcmil/

S_6: If $E_1, \ldots E_k$ have terminated *before* any of the threads $E_{k+1}, \ldots E_n$ has started, then the elements $1, \ldots, k$ will occur before the elements $k+1, \ldots, n$ in the final list. Moreover, the k-prefix of the list is stable from the moment where all $E_1, \ldots E_k$ have terminated:

$$\mathsf{G} \bigwedge_{k=1}^{n-1} \left[\left(\bigwedge_{i=1}^{k} end_i \right) \wedge \left(\bigwedge_{i=k+1}^{n} E_i.s_0 \right) \to \mathsf{G} \bigwedge_{j=1}^{k} \bigvee_{i=0}^{k} pt[i] = j \right]$$

The experiments were made on a Intel Pentium with 450 MHz and 512 MByte main memory under Linux. For the results marked with *, we had to use the dynamic sifting option of SMV which increases the runtime due to optimization of the BDD sizes. All properties were checked with moderate resources, besides the deadlock detection.

Property/threads	Runtime [seconds]	BDD Nodes	Property/threads	Runtime [seconds]	BDD Nodes
$S_1/3$	2.02	244974	$S_4/3$	1.58	147067
$S_1/4$	7.68	584141	$S_4/4$	21.20	1976314
$S_1/5$	95.29	7682664	$S_4/5^*$	554.55	823411
$S_2/3$	2.02	244993	$S_5/3$	1.04	134491
$S_2/4$	7.64	584580	$S_5/4$	8.19	599318
$S_2/5$	95.97	7683911	$S_5/5$	105.75	9592257
$S_3/3$	82.73	661349	$S_6/3$	2.67	239445
$S_3/4$	3 40.10	1897469	$S_6/4$	17.69	1216240
$S_3/5^*$	20383.78	30593433	$S_6/5$	205.25	10724191

Acknowledgment. We thank Prof. Schröder-Preikschat, Ute and Olaf Spinczyk from Magdeburg University, and Friedrich Schön from GMD-FIRST for the example and helpful discussions.

References

1. W. Nebel and G. Schumacher. Object-oriented hardware modelling: Where to apply and what are the objects. In *Euro-Dac '96 with Euro-VHDL '96*, 1996.
2. J.S. Young, J. MacDonald, M. Shilman, P.H. Tabbara, and A.R. Newton. Design and specification of embedded systems in Java using successive, formal refinement. In *Design Automation Conference (DAC'98)*, 1998.
3. E.M. Clarke, O. Grumberg, and D.E. Long. Model checking. volume 152 of *Nato ASI Series F*. Springer-Verlag, 1996.
4. E. Clarke, O. Grumberg, and D. Long. Model checking and abstraction. *ACM Transactions on Programming Languages and systems*, 16(5):1512–1542, September 1994.
5. C. Loiseaux, S. Graf, J. Sifakis, A. Bouajjani, and S. Bensalem. Property preserving abstractions for the verification of concurrent systems. *Formal Methods in System Design*, 6:1–35, February 1995.
6. F. Schön and W. Schröder-Preikschat. On the interrupt-transparent synchronization of interrupt-driven code. Arbeitspapiere der GMD, GMD Forschungszentrum Informatik, 1999.

Author Index

Lecture Notes in Computer Science

For information about Vols. 1–1622
please contact your bookseller or Springer-Verlag

Vol. 1667: J. Hlavička, E. Maehle, A. Pataricza (Eds.), Dependable Computing – EDCC-3. Proceedings, 1999. XVIII, 455 pages. 1999.

Vol. 1668: J.S. Vitter, C.D. Zaroliagis (Eds.), Algorithm Engineering. Proceedings, 1999. VIII, 361 pages. 1999.

Vol. 1670: N.A. Streitz, J. Siegel, V. Hartkopf, S. Konomi (Eds.), Cooperative Buildings. Proceedings, 1999. X, 229 pages. 1999.

Vol. 1671: D. Hochbaum, K. Jansen, J.D.P. Rolim, A. Sinclair (Eds.), Randomization, Approximation, and Combinatorial Optimization. Proceedings, 1999. IX, 289 pages. 1999.

Vol. 1672: M. Kutylowski, L. Pacholski, T. Wierzbicki (Eds.), Mathematical Foundations of Computer Science 1999. Proceedings, 1999. XII, 455 pages. 1999.

Vol. 1673: P. Lysaght, J. Irvine, R. Hartenstein (Eds.), Field Programmable Logic and Applications. Proceedings, 1999. XI, 541 pages. 1999.

Vol. 1674: D. Floreano, J.-D. Nicoud, F. Mondada (Eds.), Advances in Artificial Life. Proceedings, 1999. XVI, 737 pages. 1999. (Subseries LNAI).

Vol. 1675: J. Estublier (Ed.), System Configuration Management. Proceedings, 1999. VIII, 255 pages. 1999.

Vol. 1976: M. Mohania, A M. Tjoa (Eds.), Data Warehousing and Knowledge Discovery. Proceedings, 1999. XII, 400 pages. 1999.

Vol. 1677: T. Bench-Capon, G. Soda, A M. Tjoa (Eds.), Database and Expert Systems Applications. Proceedings, 1999. XVIII, 1105 pages. 1999.

Vol. 1678: M.H. Böhlen, C.S. Jensen, M.O. Scholl (Eds.), Spatio-Temporal Database Management. Proceedings, 1999. X, 243 pages. 1999.

Vol. 1679: C. Taylor, A. Colchester (Eds.), Medical Image Computing and Computer-Assisted Intervention – MICCAI'99. Proceedings, 1999. XXI, 1240 pages. 1999.

Vol. 1680: D. Dams, R. Gerth, S. Leue, M. Massink (Eds.), Theoretical and Practical Aspects of SPIN Model Checking. Proceedings, 1999. X, 277 pages. 1999.

Vol. 1682: M. Nielsen, P. Johansen, O.F. Olsen, J. Weickert (Eds.), Scale-Space Theories in Computer Vision. Proceedings, 1999. XII, 532 pages. 1999.

Vol. 1683: J. Flum, M. Rodríguez-Artalejo (Eds.), Computer Science Logic. Proceedings, 1999. XI, 580 pages. 1999.

Vol. 1684: G. Ciobanu, G. Păun (Eds.), Fundamentals of Computation Theory. Proceedings, 1999. XI, 570 pages. 1999.

Vol. 1685: P. Amestoy, P. Berger, M. Daydé, I. Duff, V. Frayssé, L. Giraud, D. Ruiz (Eds.), Euro-Par'99. Parallel Processing. Proceedings, 1999. XXXII, 1503 pages. 1999.

Vol. 1687: O. Nierstrasz, M. Lemoine (Eds.), Software Engineering – ESEC/FSE '99. Proceedings, 1999. XII, 529 pages. 1999.

Vol. 1688: P. Bouquet, L. Serafini, P. Brézillon, M. Benerecetti, F. Castellani (Eds.), Modeling and Using Context. Proceedings, 1999. XII, 528 pages. 1999. (Subseries LNAI).

Vol. 1689: F. Solina, A. Leonardis (Eds.), Computer Analysis of Images and Patterns. Proceedings, 1999. XIV, 650 pages. 1999.

Vol. 1690: Y. Bertot, G. Dowek, A. Hirschowitz, C. Paulin, L. Théry (Eds.), Theorem Proving in Higher Order Logics. Proceedings, 1999. VIII, 359 pages. 1999.

Vol. 1691: J. Eder, I. Rozman, T. Welzer (Eds.), Advances in Databases and Information Systems. Proceedings, 1999. XIII, 383 pages. 1999.

Vol. 1692: V. Matoušek, P. Mautner, J. Ocelíková, P. Sojka (Eds.), Text, Speech and Dialogue. Proceedings, 1999. XI, 396 pages. 1999. (Subseries LNAI).

Vol. 1693: P. Jayanti (Ed.), Distributed Computing. Proceedings, 1999. X, 357 pages. 1999.

Vol. 1694: A. Cortesi, G. Filé (Eds.), Static Analysis. Proceedings, 1999. VIII, 357 pages. 1999.

Vol. 1695: P. Barahona, J.J. Alferes (Eds.), Progress in Artificial Intelligence. Proceedings, 1999. XI, 385 pages. 1999. (Subseries LNAI).

Vol. 1696: S. Abiteboul, A.-M. Vercoustre (Eds.), Research and Advanced Technology for Digital Libraries. Proceedings, 1999. XII, 497 pages. 1999.

Vol. 1697: J. Dongarra, E. Luque, T. Margalef (Eds.), Recent Advances in Parallel Virtual Machine and Message Passing Interface. Proceedings, 1999. XVII, 551 pages. 1999.

Vol. 1698: M. Felici, K. Kanoun, A. Pasquini (Eds.), Computer Safety, Reliability and Security. Proceedings, 1999. XVIII, 482 pages. 1999.

Vol. 1699: S. Albayrak (Ed.), Intelligent Agents for Telecommunication Applications. Proceedings, 1999. IX, 191 pages. 1999. (Subseries LNAI).

Vol. 1700: R. Stadler, B. Stiller (Eds.), Active Technologies for Network and Service Management. Proceedings, 1999. XII, 299 pages. 1999.

Vol. 1701: W. Burgard, T. Christaller, A.B. Cremers (Eds.), KI-99: Advances in Artificial Intelligence. Proceedings, 1999. XI, 311 pages. 1999. (Subseries LNAI).

Vol. 1702: G. Nadathur (Ed.), Principles and Practice of Declarative Programming. Proceedings, 1999. X, 434 pages. 1999.

Vol. 1703: L. Pierre, T. Kropf (Eds.), Correct Hardware Design and Verification Methods. Proceedings, 1999. XI, 366 pages. 1999.

Vol. 1704: Jan M. Żytkow, J. Rauch (Eds.), Principles of Data Mining and Knowledge Discovery. Proceedings, 1999. XIV, 593 pages. 1999. (Subseries LNAI).

Vol. 1705: H. Ganzinger, D. McAllester, A. Voronkov (Eds.), Logic for Programming and Automated Reasoning. Proceedings, 1999. XII, 397 pages. 1999. (Subseries LNAI).

Vol. 1707: H.-W. Gellersen (Ed.), Handheld and Ubiquitous Computing. Proceedings, 1999. XII, 390 pages. 1999.

Vol. 1708: J.M. Wing, J. Woodcock, J. Davies (Eds.), FM'99 – Formal Methods. Proceedings Vol. I, 1999. XVIII, 937 pages. 1999.

Vol. 1709: J.M. Wing, J. Woodcock, J. Davies (Eds.), FM'99 – Formal Methods. Proceedings Vol. II, 1999. XVIII, 937 pages. 1999.

Vol. 1710: E.-R. Olderog, B. Steffen (Eds.), Correct System Design. XIV, 417 pages. 1999.

Vol. 1718: M. Diaz, P. Owezarski, P. Sénac (Eds.), Interactive Distributed Multimedia Systems and Telecommunication Services.. Proceedings, 1999. XI, 386 pages.